A Special Issue of
The Quarterly Journal of
Experimental Psychology: Section B
Comparative and Physiological Psychology

The Role of the Medial Temporal Lobe in Memory and Perception: Evidence from Rats, Nonhuman Primates and Humans

Edited by

Kim S. Graham
MRC Cognition and Brain Sciences Unit, Cambridge, UK
and
David Gaffan
Department of Experimental Psychology,
Oxford University, Oxford, UK

Routledge
Taylor & Francis Group
LONDON AND NEW YORK

First published 2005 by Psychology Press

Published 2018 by Routledge
2 Park Square, Milton Park, Abingdon, Oxon OX14 4RN
52 Vanderbilt Avenue, New York, NY 10017

First issued in paperback 2018

Routledge is an imprint of the Taylor & Francis Group, an informa business

British Library Cataloguing in Publication Data
A catalogue record for this book is available from the British Library

This book is also a special issue of the journal *Quarterly Journal of Experimental Psychology Section B: Comparative and Physiological Psychology* and forms Issues 3 and 4 of Volume 58B (2005).

Cover by Hybert Design, White Waltham, Berkshire, UK
Typeset by Techset Composition Limited, Salisbury, Wiltshire, UK

ISSN   0272–4995
ISBN 13: 978–1–138–87328–5 (pbk)
ISBN 13: 978–1–84169–998–1 (hbk)

Contents

THE QUARTERLY JOURNAL OF EXPERIMENTAL PSYCHOLOGY
2005, 58B (3/4), 193–201

The role of the medial temporal lobe in memory and perception: Evidence from rats, nonhuman primates and humans

Kim S. Graham and David Gaffan

*MRC Cognition and Brain Sciences Unit, Cambridge, UK, and
Department of Experimental Psychology, Oxford, UK*

The articles reported in this Special Issue are based on talks given at the Experimental Psychology Society (EPS) meeting held in Oxford, March 31st–April 2nd 2004. Six of the authors (Aggleton, Buckley, Bussey, Hampton, Holdstock, and Lee) presented as part of a symposium on the role of the perirhinal cortex in memory and perception, organized to coincide with the delivery of the 2004 Bartlett Lecture by Dr Karalyn Patterson. Three of the other contributions stem from related talks given at the same EPS meeting (Bright, E. A. Gaffan, and Rolls), with the final paper (Henson) summarizing relevant findings from functional neuroimaging.

The main aim of our EPS symposium was "to bring together a number of researchers working on perirhinal cortex (in rats, nonhuman primates, and humans) in order to debate the controversial issues surrounding the role of medial temporal lobe (MTL) structures in memory and perception". This aim remains the central thread of this Special Issue with a broadening of the topic to include all MTL regions that might be critically involved in aspects of memory. Consistent with our goal to facilitate an awareness of cross-disciplinary research in this area, especially in the UK, the papers reported here include lesion studies (Eacott & E. A. Gaffan) and early gene imaging in rats (Aggleton & Brown), electrophysiological (Rolls, Franco, & Stringer) and lesion studies in nonhuman primates (Buckley; Bussey, Saksida, & Murray; Hampton), and lesion studies (Holdstock; Lee, Barense, & Graham) and functional neuroimaging in human participants (Bright, Moss, Stamatakis, & Tyler; Henson), as well as touching on computational modelling approaches (Bussey et al.; Rolls et al.). One obvious issue that remains, of course, is how to make sense of these methodologically diverse contributions, and Murray (who acted as discussant for our EPS symposium) has put together a thought-provoking final chapter that highlights the main consistencies and discrepancies from these reviews, as well as providing some thoughts about how we can progress in the future (Murray, Graham, & Gaffan).

Correspondence should be addressed to Dr Kim S Graham, MRC Cognition and Brain Sciences Unit, 15 Chaucer Road, Cambridge CB2 2EF, UK. Email: kim.graham@mrc-cbu.cam.ac.uk

© 2005 The Experimental Psychology Society
DOI:10.1080/02724990544000059

So what theoretical issues are considered in the Special Issue? One theme that is evident in all papers—whether explicitly stated or not—is whether all the regions within the MTL function as a unitary system with a single function. The general conclusion drawn by most of the researchers here is that the different brain areas in the MTL play unique, and dissociable, roles in memory. Although the papers presented here seem quite consistent in their approach to this issue, the unitary view was widely accepted until at least 1994 (D. Gaffan, 1994) and is still held by some authorities (Squire, Stark, & Clark, 2004).

Considering human memory initially, there is general acceptance that the MTL is critical for the acquisition of new episodic (event-based) and semantic (factual) memories. Support for this view stems from virtually 50 years of neuropsychological studies, including a number that involve the most famous amnesic case, HM (Scoville & Milner, 1957). HM underwent bilateral temporal lobe resection, including removal of the hippocampal complex, entorhinal cortex, medial temporal polar cortex, and ventral perirhinal cortex (Corkin, Amaral, Gonzalez, Johnson, & Hyman, 1997) for intractable epilepsy. He subsequently became densely amnesic and he described his phenomenological experience as if he was constantly waking from a dream.

Many other cases, with a variety of aetiologies, have been reported and investigated in detail (see Lee et al., this issue, for details of some of these patients), and the overall conclusion from these experiments would be that the different regions within the MTL seem to function as a unitary memory system for consciously available events and facts (Squire, 1992). The role played by the MTL memory system in memory retrieval is proposed to be temporary: Over time, via repeated reinstatement of previous experiences, links between elements comprising the new experiences are generated in neocortical brain regions, and the retrieval of the memory eventually becomes independent of the MTL (a process termed memory consolidation). Critically, this view predicts no specialization within the MTL or, at the very least, proponents argue that none of our current psychological fractionations of memory—for example, episodic versus semantic, or recollection (retrieval of information associated with an event) versus familiarity (a feeling of knowing that something was seen before)—adequately maps onto the division of labour in the MTL (Manns, Hopkins, Reed, Kitchener, & Squire, 2003; Squire et al., 2004; Wixted & Squire, 2004).

Recent studies have, however, challenged this unitary assumption (Aggleton & Shaw, 1996; Holdstock et al. 2002; Vargha-Khadem et al., 1997; Yonelinas et al., 2002). Vargha-Khadem et al. (1997) reported three cases with developmental amnesia that showed damage to the hippocampus bilaterally after early trauma. These developmental cases revealed a striking pattern on tests of episodic and semantic memory: While episodic memory was profoundly impaired, semantic memory was disproportionately better. These patients also showed a striking discrepancy between performance on tests of recall (poor retrieval of a previously learned prose passage) and recognition memory (good memory for which of two faces was studied previously), a pattern that has also been reported in some, but not all, adult cases (Aggleton & Shaw, 1996; Holdstock et al., 2002).

Holdstock (this issue) addresses this topic in more detail describing a series of studies in a single case, YR, which show that recognition memory or, more specifically, familiarity for previously studied episodes, may be critically dependent upon perirhinal cortex and the dorso-medial thalamic nucleus, but not upon the hippocampus. For example, Holdstock reports that, over 43 separate tests of item recognition, YR performed on average only 0.5

standard deviations below the control mean, while on 34 tests of recall, her mean performance was over 3.5 standard deviations below the control average (Mayes, Holdstock, Isaac,
Hunkin, & Roberts, 2002). Notably, YR's recognition memory for pairs of words and faces
was also unimpaired (mean performance for four tests, 0.7 *SD*s below the control mean),
while her performance on 18 tests requiring memory for associations between information
(objects–locations, faces–voices, etc.) was poor (2.9 *SD*s below control mean). Holdstock
concludes that the hippocampus is critically involved in recall and in recognition for associations between different types of information, but not for item or within-type associative
recognition memory.

Henson (this issue) considers the neural basis of recognition memory from the stand-
point of functional neuroimaging (fMRI) in healthy volunteers. Henson's review
concentrates on the neural substrates involved in encoding (the difference in pattern of
activation associated with studied items later remembered versus those that have been subsequently forgotten) and retrieval (the difference in pattern of activation between test items
recognized as previously studied, hits, versus not studied, new). Using the anatomical
models available from the animal literature (Aggleton & Brown, 1999), he considers two types
of possible MTL activity: that restricted to the hippocampus and that seen in surrounding
medial temporal cortex (which encompasses the parahippocampal and rhinal cortices).
Inconsistent with Holdstock's (this issue) single case study (see also Aggleton & Brown,
1999; Vargha-Khadem et al., 1997), Henson concludes that, while structures within the
MTL are clearly involved in memory-related differences (both at encoding and at retrieval),
there is little indication, at present, that the hippocampus and other MTL structures might
be differentially involved in recognition memory. Nor were there any clear differences when
considering the activation patterns in terms of anterior versus posterior or left versus right
MTL regions. This lack of anatomical differentiation in haemodynamic response would
be consistent with a model that assumes no functional division of labour in the MTL
(Squire et al., 2004).

Intriguingly though, Henson (this issue) notes some trends in the data, including (a) that
scenes are particularly good at eliciting memory-related MTL activations (compared to
words), (b) that the hippocampus and posterior medial temporal cortex (parahippocampal
cortex) seem to be involved in encoding and retrieving source information and associations
between distinct items, and (c) that anterior medial temporal cortical regions, possibly
perirhinal cortex, often show decreased responses for old relative to new items during tests
of recognition memory. A good example of how MTL regions might play different roles in
mnemonic processing comes from the recent paper by Davachi, Mitchell, and Wagner
(2003) that investigated the role of MTL regions in subsequent memory for items (knowing
that a word had been studied previously) versus source (knowing that a studied word had
been imagined with a scene versus imagining the sound of the word if read backwards). An
activation in a left anterior middle temporal cortical region, possibly perirhinal cortex,
predicted subsequent item memory, but did not differ according to which source condition
the word had been imagined at study. By contrast, bilateral hippocampus and left posterior
medial temporal cortical regions were more active for items in which the source task was
successfully identified than for stimuli correctly recognized with no source information
present (see also Ranganath et al., 2004). These studies highlight how functional neuroimaging
can be used to tease apart the role of different MTL brain regions in memory, although as

noted by Henson there are still significant methodological issues that remain to be addressed in order to maximize the potential of this methodology.

Within the animal literature (see Buckley, this issue), a similar change of heart seems to have taken place, although notably some researchers (D. Gaffan, 2001; Horel, 1978) never fully embraced the MTL memory system account. Early studies, such as by Mishkin (1978), found that combined amygdala and hippocampal lesions in nonhuman primates impaired recognition memory (delayed–match–to–sample) at long delay lengths, a result that is compatible with the human literature, as patients, like HM, also had significant recognition memory impairments. As noted by Buckley, however, the combined amgydala/ hippocampal lesions that resulted in delayed–match–to–sample impairments were confounded by damage to other regions in the medial temporal lobe, notably the perirhinal cortex. If the amygdala and hippocampus were left intact, and the perirhinal and entorhinal cortex lesioned (Meunier, Bachevalier, Mishkin, & Murray, 1993), delayed–match–to–sample deficits almost as severe as those identified in the earlier studies were evident. In particular, the perirhinal cortex was shown to be critical for object recognition memory, with little evidence of impairment after hippocampal lesions (Murray & Mishkin, 1998). Buckley also reviews his own and D. Gaffan's (1994) work, which highlights dissociations in function depending upon where there is damage to the monkey temporal lobe: impairments in colour discrimination, but not object recognition memory, after lesions to the middle temporal gyrus, deficits in object and scene recognition memory but not colour discrimination or spatial discrimination learning after perirhinal cortex lesions, and impairments in simple spatial discrimination after fornix transection.

Some of the most convincing evidence for functional specialization in the MTL comes from Aggleton and Brown's (this issue) and Eacott and E. A. Gaffan's (this issue) contributions to the Special Issue. Both of these papers clearly highlight the dissociability of different MTL regions, while acknowledging the close anatomical links that exist between these brain areas. By measuring levels of Fos, the protein product of the c-fos gene (which is expressed throughout the temporal lobe), Aggleton and Brown sought to identify how closely perirhinal cortex and hippocampus worked together to support spatial and object recognition memory. Consistent with their lesion work (Aggleton & Brown, 1999; Aggleton, Kyd, & Bilkey, 2004), c-fos activity in some hippocampal subfields, but not perirhinal cortex, was increased following tests requiring spatial learning, while increases in c-fos activity in perirhinal cortex, but not the hippocampus, was evident when rats were presented with novel but not familiar visual objects. Spatial rearrangement of familiar objects also resulted in expression of c-fos in the hippocampus but not in the perirhinal cortex. These studies confirm that the double dissociations evident in rat lesion studies can be replicated using a very different methodological technique.

Eacott and E. A. Gaffan's article (this issue) focuses on lesion studies in the rat, but in contrast to Aggleton and Brown, compares the function of rat perirhinal and postrhinal cortex. Postrhinal cortex is thought to play a critical role in spatial processing, and Eacott and E. A. Gaffan ask how this type of processing is different from that carried out in the hippocampus and perirhinal cortex. They conclude that the perirhinal cortex is involved in learning about features that represent objects, while postrhinal cortex plays a role in learning about within-scene positions and contexts. The hippocampus, by contrast, seems to be critical for associations between objects, their positions and contexts, a memory that could be considered similar to human episodic memory. For example, in an ingenious experiment rats were first exposed to two different objects in one presentation phase (context) followed by the same two objects (with their locations swapped) in a different context. After a delay, the rats were placed back in one of the contexts that contained two copies of one of the objects. One of these objects had never been seen in the location in the presented context: There was a novel configuration of object, position and context. While rats with perirhinal and postrhinal lesions were able to demonstrate memory for these novel configurations, fornix-lesioned rats did not perform above chance even at short (2-s) delays, though they could perform object–place memory tasks (albeit at short delays, 5 s).

The two rat papers conclusively demonstrate, therefore, across a series of different studies that MTL regions serve independent functions in memory and, together with the lesion work in nonhuman primates (Buckley, this issue; Bussey et al., this issue), send a clear signal to human researchers about what components of memory might be differentially dependent upon distinct MTL regions.

Many of the articles in this Special Issue also address the different types of memory that seem to be dependent upon MTL regions, in particular focusing on the role of the perirhinal cortex in recognition memory (Henson, this issue; Holdstock, this issue), paired associate learning (Eacott & E. A. Gaffan, this issue), and concurrent discrimination (Buckley, this issue; Bussey et al., this issue; Lee et al., this issue). Rolls and colleagues (Rolls et al., this issue) focus on a longer term form of familiarity memory, in which the responses of some perirhinal neurons gradually increase over hundreds of repeated presentations of novel stimuli. The degree of response to these novel items eventually becomes as large as that seen to familiar stimuli. The authors discuss how a neural network model based on Hebbian associative learning can explain these findings and propose that this process allows the perirhinal cortex to form unique representations of complex stimuli (see also Bussey et al., this issue). Rolls and colleagues touch upon other types of neuronal response seen in the perirhinal cortex (Brown & Aggleton, 2001): in particular, findings that perirhinal neurons seem capable of short-term responses to stimuli that reflect recency (less response to repeated stimuli regardless of whether these are only slightly or highly familiar), familiarity (decreased response to familiar compared to novel stimuli on both a first and a second presentation), and novelty (a decrease in activity after a novel stimulus is repeated for the second time). It is not immediately clear how these different types of response map onto each other or whether these response decrements and increases can be considered the physiological basis for the types of psychological process that we believe contribute to episodic and semantic memory for example, familiarity-based memory (Bright et al., this issue; Henson, this issue; Holdstock, this issue; Rolls et al., this issue).

Although approached from very different methodological standpoints the articles by Bright et al. (this issue), Buckley (this issue), Bussey et al. (this issue), and Lee et al. (this issue) also seem to implicate the perirhinal cortex in the representation of complex object stimuli, expanding the function of this region beyond memory processing. Reviewing a series of their published studies, Bussey et al. (this issue) convincingly demonstrate that feature ambiguity—a property of visual discriminanda whereby a feature is part of both a rewarded and nonrewarded stimuli—can significantly influence performance in animals with perirhinal lesions. More specifically, animals with perirhinal lesions performed normally when there was no feature ambiguity between presented stimuli, but were significantly impaired when feature ambiguity was maximally present. While not manipulating feature ambiguity per se, Buckley reports similarly compelling findings in which monkeys with perirhinal lesions performed poorly on concurrent discriminations in which objects were presented in one of three different views (Buckley & D. Gaffan, 1998) and oddity tasks in which they had to indicate which object or face, presented as part of a set of different views of another object or face, was different. These findings highlight a contributory role for the perirhinal cortex in perceptual processing, as well as memory, although Hampton (this issue) challenges these conclusions with some clear arguments.

Hampton (this issue) argues that the set size effects reported by Buckley and D. Gaffan (1997) reflect a simple scaling effect—that is, rather than there being an increasing ratio of errors to problems as the number of discrimination problems is increased, instead there is evidence of a similar ratio across conditions with different discrimination sizes, a pattern that is most likely attributable to learning, rather than to any deficit in perception. Hampton also raises some questions about the interpretation of Buckley and D. Gaffan's (1998) data in which monkeys with perirhinal lesions were shown to be impaired in their ability to relearn discriminations requiring generalization across different views. Hampton proposes that if perceptual difficulties underlie the problems demonstrated by the monkeys, then one might expect greater deficits in relearning tests than seen in the initial learning phase. Although the monkeys made more errors in the relearning than in the initial transfer condition, Hampton is not convinced that there was a significant difference between the control and lesioned animals or that the lesioned animals did not have a problem with learning new associations between visual features and reward (which may explain the greater number of errors made by this group on the original transfer task).

Hampton (this issue) also stresses the possible confound of perception and memory in Buckley, Booth, Rolls, and D. Gaffan's (2001; Buckley, this issue) oddity experiments, in which perirhinally lesioned monkeys were impaired at indicating which object was the odd one out (i.e., a single object was presented with five different views of another object). Lee et al. (this issue) provide a partial answer to this debate. Lee et al. aim to address discrepancies between the nonhuman primate and human literature by applying similar tests in humans to those reported in animal lesion studies (e.g., concurrent discrimination utilizing feature ambiguity, Bussey, Saksida, & Murray, 2002, and oddity paradigms, Buckley et al., 2001). Strikingly, Lee finds virtually identical patterns in human subjects with perirhinal involvement as reported in monkeys with lesions to this region (i.e., impairments on concurrent discrimination for objects with high but not low feature ambiguity and clear deficits on oddity judgements for objects and faces). Lee et al. extend these results by revealing that bilateral damage to the hippocampus does not impair object discrimination learning or object oddity, but does result in significant difficulties with scene discrimination (even in conditions where there is minimal memory demand, Lee et al., 2005). Critically, in some of Lee et al.'s experiments trial-unique stimuli were adopted reducing the possibility that learning across trials could explain the deficits seen in the patients.

Holdstock (this issue) also asks whether perception is impaired in her amnesic patients, reviewing an experiment using delayed-matching-to-sample that failed to find the perceptual deficits that would be predicted by Lee et al.'s (2005) recent work. Holdstock proposes that these discrepant findings may be due to stimuli differences—as noted by Bussey et al. (this issue) and highlighted in Lee et al.'s concurrent discrimination experiment, it is important that perception (and memory) are tested using stimuli with feature ambiguity, or more specifically where there is a need to process conjunctions of object features. It remains possible that some of the controversy in the literature can be attributable to this issue, and further studies are clearly required.

In Bright et al.'s (this issue) contribution to this Special Issue, the authors address the issue of how complexity of stimuli (by which they mean the number and type of properties shared by stimuli, and the correlations between these properties) influences activation patterns within the anterior temporal lobe, including perirhinal cortex, using functional neuroimaging. The authors report that anterior temporal regions seem critically involved in fine-grained discriminations among objects, with greater activity evident for basic (dog, lion, table) versus domain (living, manmade) level naming of coloured photographs (Tyler et al., 2004) and within basic naming, greater activity for living over nonliving categories (Moss, Rodd, Stamatakis, Bright, & Tyler, 2005). Furthermore, patients with damage to the structures implicated in the functional neuroimaging comparisons show the predicted category-specific deficits (i.e., living items proving more difficult to discriminate than nonliving given their inherent featural overlap). Although Bright and colleagues approach their research question from a long-standing interest in the neural basis of semantic memory, it is exciting to see that the models being developed from the animal literature (Bussey et al., this issue) have applicability at the human level.

As mentioned earlier, the aim of this Special Issue was to highlight how cross-disciplinary approaches in the field of episodic and semantic memory are contributing to our understanding about the role of MTL structures in memory and perception. We believe that one of the most striking aspects of this Special Issue is the obvious convergence across studies, despite the major species differences that exist between rats, nonhuman primates, and humans, let alone the difficulties inherent in making cross-methodological comparisons. While there are issues of data interpretation, localization of brain damage or activation and differences in the theoretical framework adopted by researchers (e.g., perceptual-mnenomic/feature conjunction vs. conceptual structure account), a clear message, except from the human functional neuroimaging work on recognition memory, is that the MTL is not functioning as a unitary system, and that different areas within this brain region make unique contributions to memory, and possibly perception. The challenge for the future is to tease apart these contributions and to further understand how such closely related brain regions work together to support complex cognitive processes such as recollection, familiarity, semantic memory, and perception of complex objects and scenes.

REFERENCES

Aggleton, J. P., & Brown, M. W. (1999). Episodic memory, amnesia, and the hippocampal-anterior thalamic axis. *Behavioral Brain Sciences, 22,* 425–489.

Aggleton, J. P., & Brown, M. W. (this issue). Contrasting hippocampal and perirhinal cortex function using immediate early gene imaging. *Quarterly Journal of Experimental Psychology, 58B,* 218–233.

Aggleton, J. P., Kyd, R., & Bilkey, D. (2004). When is the perirhinal cortex necessary for the performance of spatial memory tasks? *Neuroscience and Biobehavioural Reviews, 28,* 611–624.

Aggleton, J. P., & Shaw, C. (1996). Amnesia and recognition memory: A re-analysis of psychometric data. *Neuropsychologia, 34,* 51–62.

Bright, P., Moss, H. E., Stamatakis, E. A., & Tyler, L. K. (this issue). The anatomy of object processing: The role of anteromedial temporal cortex. *Quarterly Journal of Experimental Psychology*, *58B*, 361–377.

Brown, M. W., & Aggleton, J. P. (2001). Recognition memory: What are the roles of the perirhinal cortex and hippocampus? *Nature Reviews Neuroscience*, *2*, 51–61.

Buckley, M. J. (this issue). The role of the perirhinal cortex and hippocampus in learning, memory, and perception. *Quarterly Journal of Experimental Psychology*, *58B*, 246–268.

Buckley, M. J., Booth, M. C. A., Rolls, E. T., & Gaffan, D. (2001). Selective perceptual impairments after perirhinal cortex ablation. *Journal of Neuroscience*, *21*, 9824–9836.

Buckley, M. J., & Gaffan, D. (1997). Impairment of visual object-discrimination learning after perirhinal cortex ablation. *Behavioral Neuroscience*, *111*, 467–475.

Buckley, M. J., & Gaffan, D. (1998). Learning and transfer of object–reward associations and the role of the perirhinal cortex. *Behavioral Neuroscience*, *112*, 15–23.

Bussey, T. J., Saksida, L. M., & Murray, E. A. (2002). Perirhinal cortex resolves feature ambiguity in complex visual discriminations. *European Journal of Neuroscience*, *15*, 365–374.

Bussey, T. J., Saksida, L. M., & Murray, E. A. (this issue). The perceptual-mnemonic/feature conjunction model of perirhinal cortex function. *Quarterly Journal of Experimental Psychology*, *58B*, 269–282.

Corkin, S., Amaral, D. G., Gonzalez, R. G., Johnson, K. A., & Hyman, B. T. (1997). HM's medial temporal lobe lesion: Findings from magnetic resonance imaging. *Journal of Neuroscience*, *17*, 3964–3979.

Davachi, L., Mitchell, J. P., & Wagner, A. D. (2003). Multiple routes to memory: Distinct medial temporal lobe processes build item and source memories. *Proceedings of the National Academy of Sciences, USA*, *100*, 2157–2162.

Eacott, M. J., & Gaffan, E. A. (this issue). The roles of the perirhinal cortex, postrhinal cortex, and the fornix in memory for objects, contexts, and events in the rat. *Quarterly Journal of Experimental Psychology*, *58B*, 202–217.

Gaffan, D. (1994). Scene-specific memory for objects: A model of episodic memory impairment in monkeys with fornix transection. *Journal of Cognitive Neuroscience*, *6*, 305–320.

Gaffan, D. (2001). Against memory systems. *Philosophical Transactions of the Royal Society*, *357*, 1111–1121.

Hampton, R. R. (this issue). Monkey perirhinal cortex is critical for visual memory, but not for visual perception: Re-examination of the behavioural evidence from monkeys. *Quarterly Journal of Experimental Psychology*, *58B*, 283–299.

Henson, R. (this issue). Amini-reviews of fMRI studies of human medial temporal lobe activity associated with recognition memory. *Quarterly Journal of Experimental Psychology*, *58B*, 340–360.

Holdstock, J. S. (this issue). The role of the human medial lobe in object recognition and object discrimination. *Quarterly Journal of Experimental Psychology*, *58B*, 326–339.

Holdstock, J. S., Mayes, A. R., Roberts, N., Cezayirli, E., Isaac, C. L., O'Reilly, R. C., et al. (2002). Under what conditions is recognition spared relative to recall after selective hippocampal damage in humans? *Hippocampus*, *12*, 341–351.

Horel, J. A. (1978). The neuroanatomy of amnesia. A critique of the hippocampal memory hypothesis. *Brain*, *101*, 403–445.

Lee, A. C. H., Barense, M. D., & Graham, K. S. (this issue). The contribution of the human medial temporal lobe to perception: Bridging the gap between animal and human studies. *Quarterly Journal of Experimental Psychology*, *58B*, 300–325.

Lee, A. C. H., Bussey, T. J., Murray, E. A., Saksida, L. M., Epstein, R. A., Kapur, N., et al. (2005). Perceptual deficits in amnesia: Challenging the medial temporal lobe 'mnemonic' view. *Neuropsychologia*, *43*, 1–11.

Manns, J. R., Hopkins, R. O., Reed, J. M., Kitchener, E. G., & Squire, L. R. (2003). Recognition memory and the human hippocampus. *Neuron*, *37*, 171–180.

Mayes, A. R., Holdstock, J. S., Isaac, C. L., Hunkin, N. M., & Roberts, N. (2002). Relative sparing of item recognition memory in a patient with adult-onset damage limited to the hippocampus. *Hippocampus*, *12*, 325–340.

Meunier, M., Bachevalier, J., Mishkin, M., & Murray, E.A. (1993). Effects on visual recognition of combined and separate ablations of the entorhinal and perirhinal cortex in rhesus monkeys. *Journal of Neuroscience*, *13*, 5418–5432.

Mishkin, M. (1978). Memory in monkeys is severely impaired by combined but not by separate removal of amygdala and hippocampus. *Nature*, *273*, 297–298.

Moss, H. E., Rodd, J. M., Stamatakis, E. A., Bright, P., & Tyler, L. K. (2005). Anteromedial temporal cortex supports fine-grained differentiation among objects. *Cerebral Cortex*, *15*, 616–627.

Murray, E. A., Graham, K. S., & Gaffan, D. (this issue). Perirhinal cortex and its neighbours in the medial temporal lobe: Contributions to memory and perception. *Quarterly Journal of Experimental Psychology, 58B*, 378–396.

Murray, E. A., & Mishkin, M. (1998). Object recognition and location memory in monkeys with excitotoxic lesions of the amygdala and hippocampus. *Journal of Neuroscience, 18*, 6568–6582.

Ranganath, C., Yonelinas, A. P., Cohen, M. X., Dy, C. J., Tom, S. M., & D'Esposito, M. (2004). Dissociable correlates of recollection and familiarity within the medial temporal lobes. *Neuropsychologia, 42*, 2–13.

Rolls, E. T., Franco, L., & Stringer, S. M. (this issue). The perirhinal cortex and long-term familiarity memory. *Quarterly Journal of Experimental Psychology, 58B*, 234–245.

Scoville, W. B., & Milner, B. (1957). Loss of recent memory after bilateral hippocampal lesions. *Journal of Neurology, Neurosurgery and Psychiatry, 20*, 11–21.

Squire, L. R. (1992). Memory and the hippocampus—a synthesis from findings with rats, monkeys and humans. *Psychological Review, 99*, 195–231.

Squire, L. R., Stark, C. E., & Clark, R. E. (2004). The medial temporal lobe. *Annual Review of Neuroscience, 27*, 279–306.

Tyler, L. K., Stamatakis, E. A., Bright, P., Acres, K., Abdallah, S., Rodd, J. M., et al. (2004). Processing objects at different levels of specificity. *Journal of Cognitive Neuroscience, 16*, 351–362.

Vargha-Khadem, F., Gadian, D. G., Watkins, K. E., Connelly, A., Van Paesschen, W., & Mishkin, M. (1997). Differential effects of early hippocampal pathology on episodic and semantic memory. *Science, 277*, 376–380.

Wixted, J. T., & Squire, L. R. (2004). Recall and recognition are equally impaired in patients with hippocampal damage. *Cognitive, Affective and Behavioural Neuroscience, 4*, 58–66.

Yonelinas, A. P., Kroll, N. E. A., Quamme, J. R., Lazzara, M. M., Sauve, M.-J., Widaman, K. F., et al. (2002). Effects of extensive temporal lobe damage or mild hypoxia on recollection and familiarity. *Nature Neuroscience, 5*, 1236–1241.

THE QUARTERLY JOURNAL OF EXPERIMENTAL PSYCHOLOGY
2005, 58B (3/4), 202–217

The roles of perirhinal cortex, postrhinal cortex, and the fornix in memory for objects, contexts, and events in the rat

M. J. Eacott

University of Durham, UK

E. A. Gaffan

University of Reading, UK

Investigation of the anatomical substructure of the medial temporal lobe has revealed a number of highly interconnected areas, which has led some to propose that the region operates as a unitary memory system. However, here we outline the results of a number of studies from our laboratories, which investigate the contributions of the rat's perirhinal cortex and postrhinal cortex to memory, concentrating particularly on their respective roles in memory for objects. By contrasting patterns of impairment and spared abilities on a number of related tasks, we suggest that perirhinal cortex and postrhinal cortex make distinctive contributions to learning and memory: for example, that postrhinal cortex is important in learning about within-scene position and context. We also provide evidence that despite the strong connectivity between these cortical regions and the hippocampus, the hippocampus, as evidenced by lesions of the fornix, has a distinct function of its own—combining information about objects, positions, and contexts.

Both the perirhinal and the postrhinal cortex of the rat receive multimodal inputs from areas earlier in the sensory processing streams and provide major parallel inputs into the hippocampal system. In addition, the perirhinal cortex and postrhinal cortex are reciprocally interconnected (Burwell & Amaral, 1998b; Burwell, Witter, & Amaral, 1995). However, there are important differences between the two areas: The perirhinal cortex receives its input predominantly from the object-based ventral processing stream (Burwell et al., 1995), while inputs to the postrhinal cortex originate in the parietal lobe and appear to carry spatial information (Burwell & Amaral, 1998a). The rat's postrhinal cortex is considered to be the homologue of the primate parahippocampal cortex (Burwell et al., 1995). However, with a few exceptions—for example, that the primate PeRh receives a greater proportion of its projections from visual areas than does the perirhinal cortex of the rat

Correspondence should be addressed to Dr M. J. Eacott, Department of Psychology, Science Laboratories, South Road, Durham. DH1 3LE, UK. Email: M.J.Eacott@durham.ac.uk

© 2005 The Experimental Psychology Society
DOI:10.1080/02724990444000203

(Suzuki & Amaral, 1994)—the general pattern of connectivity between the areas is broadly comparable in the rat and monkey (Burwell et al., 1995). Thus on the basis of connectivity alone there is evidence to suggest that the perirhinal cortex may play a role in object memory while the postrhinal cortex may be involved with spatial aspects of the environment. Evidence from the neuronal activity of these two areas also suggests a functional distinction. For example, neurons in perirhinal cortex of the rat respond selectively to objects (Zhu, Brown, & Aggleton, 1995a) and convey information about the familiarity of the objects (Zhu, Brown, McCabe, & Aggleton, 1995b), again suggesting a role in memory for objects. In contrast, the responsivity of neurons in postrhinal cortex suggests a different and possibly spatial role (Burwell & Hafeman, 2003; Vann, Brown, Erichsen, & Aggleton, 2000; Wan, Aggleton, & Brown, 1999). There is now wide consensus that the perirhinal cortex plays a crucial role in memory for objects (Aggleton & Brown, this issue; Buckley; this issue; Holdstock, this issue) and much support for the view that the cortex is also important for the perception of objects (Bussey & Saksida 2002; Bussey, Saksida, & Murray, 2003; Bussey, Saksida, & Murray, this issue; Eacott, Gaffan, & Murray, 1994; Eacott & Heywood, 1995; Murray, Bussey, Hampton, & Saksida, 2000). There is an emerging view that postrhinal cortex is involved in spatial processing, though in a different manner from the hippocampus itself (Burwell, Saddoris, Bucci, & Wiig, 2004). However, the precise roles that these two areas play and the relationship between these roles are less clear. Here we discuss a series of studies from our laboratories over recent years, which have investigated the functions of the peri and postrhinal cortex in the rat and have also, where relevant, contrasted these with hippocampal function.

Despite the consensus that perirhinal cortex may play a crucial role in the perception and memory for objects, specifying the exact nature of this role has been less straightforward. The role has variously been described as representation of objects (Bussey & Saksida, 2002), providing the precise specification of objects by associating together the various visual features inherent in particular visual objects (D. Gaffan, 1994) or a gestalt representation of a complete stimulus (Murray & Bussey, 1999). These descriptions attempt to characterize the pattern of impairments seen after perirhinal cortex lesions in that both rats and monkeys may perform poorly on tasks requiring whole–object information (Eacott, Machin, & Gaffan, 2001). Without such whole–object information, objects may be represented less precisely, resulting in a degraded memory for the stimulus. One source of support for this view came from evidence that animals could be impaired on tasks that involved large number of stimuli, but perform normally when there are few stimuli (Buckley & Gaffan, 1998b; Eacott et al., 1994). It can be argued that the more potentially confusable stimuli there are within memory, the more precisely each has to be represented in order to accurately discriminate between stimuli. With a relatively small pool of potential stimuli, a relatively impoverished representation of a stimulus may be sufficient to discriminate it in memory from other stimuli. As the pool of stimuli increases in size, so the precision of the specification has to rise, in order to avoid interference between similar stimuli (although see Hampton, this issue).

Eacott et al. (2001) examined the generality of the idea that perirhinal cortex is required whenever a stimulus must be precisely represented. Rats learned a very simple two-choice visual discrimination between a square and a rectangle in a computer-controlled Y-maze (E. A. Gaffan & Eacott, 1995). The discriminanda appeared on VDU screens situated at the end of the arms of the Y-maze. Food rewards were delivered for approaching the correct stimulus. Perirhinal-lesioned rats, as well as intact rats and sham-operated rats, were able to perform this simple form discrimination at high levels. However, as it involved only two stimuli (square and rectangle), even an impoverished representation of the stimuli might suffice for accurate discrimination, and so the above hypothesis would not predict an impairment in this base task, and none was found. At this point, more difficult probe trials were introduced in which the dimensions of the rectangle were progressively adjusted so that the rectangle increasingly resembled the square. The most difficult probe trials were at the limits of the abilities of the control rats to discriminate between the square and rectangle. Even so, rats with perirhinal cortex lesions performed entirely normally on this task, despite having functional perirhinal lesions as evidenced by their performance on another task (to be discussed later). This was the case whether the simple square/rectangle discrimination was learned before or after the perirhinal lesion, and thus it cannot be attributed to a difference in strategies in learning the basic task in the lesioned state. Thus we concluded that after functional perirhinal cortex lesions rats may represent at least some visual stimuli in a manner that allows very-fine-grained discriminations. This finding is not in accordance with a simple view that after perirhinal cortex lesions animals represent all stimuli in an imprecise or impoverished manner. It is clear that after perirhinal cortex lesions rats can maintain a highly specified representation of at least some visual stimuli.

The stimuli used by Eacott et al. (2001) were, by design, very simple. In fact, the rat did not need to represent every aspect of each stimulus to succeed in this task as the square always had a shorter top edge than the rectangle. Therefore this discrimination could simply be solved on the basis of this single feature: edge length. There is accumulating evidence that learning of discriminations that can be performed on the basis of a single feature, such as length or colour, may proceed entirely normally after perirhinal cortex lesions. For example, Buckley, Gaffan, and Murray (1997) and Buckley, Booth, Rolls, and Gaffan (2001) showed that monkeys with perirhinal cortex lesions had normal thresholds for discriminating simple perceptual features such as colour. Such single-feature processing is likely to take place in earlier visual areas such as monkeys inferotemporal cortex (TE), and, consistent with this idea, Buckley et al. (1997) found that monkeys with lesions of the middle temporal gyrus were impaired on a colour discrimination. Therefore the good discrimination of single features seen in the perirhinal group, in both Buckley et al.'s (1997) study and that of Eacott et al. (2001), may have been mediated by such processing.

An alternative hypothesis as to the role of perirhinal cortex in object processing is that perirhinal cortex may only become important where whole objects need to be represented. The difference between an object, as the term is used here, and a single feature is that an object has a number of features, some of which may be shared with other objects. The object is defined by a unique set of features. The role of perirhinal cortex in perception and memory of objects may be in associating together the individual features that together represent the whole object. According to this view, in the absence of the perirhinal cortex, discriminations that rely on single features, such as the above task, will proceed normally, but those that rely on associating individual features or elements will be impaired.

In order to test whether perirhinal cortex is involved generally in associating together multiple stimuli, we employed a task where rats learned associations between visual stimuli that were successively presented. We (Eacott, Norman, & Gaffan, 2003) taught rats to use a secondary visual reinforcer in order to learn visual discriminations. In this task, rats learn a number of two-choice discriminations between stimuli (2-D shapes appearing on screens) in the same computer-controlled Y-maze as that used by Eacott et al. (2001). However, in this task, approaching the correct shape was not directly rewarded by food reward, but by the appearance on the screen of the visual secondary reinforcer, a horizontal grey bar, which appeared directly over the correct shape. The appearance of the secondary reinforcer signalled that a correct response had been made. Primary reinforcement (food pellets) was delivered only after five correct responses (not including the first trial, see Eacott et al., 2003, for details). As a result, no primary reinforcement could be received before Trial 6, and any learning that was evident before this point could only have been mediated by the visual–visual association between the visual discriminandum and the visual secondary reinforcer. Although such learning proved to be very difficult even for intact rats, subsequent lesions of the perirhinal cortex had no effect on learning of new discriminations using the same secondary reinforcer. Indeed, perirhinal lesioned rats showed entirely normal learning of the visual discriminations even over the first six trials during which no primary reinforcement was given. In other words, learning of new visual–visual associations between the discriminanda and the secondary reinforcer could proceed normally in the absence of the perirhinal cortex.

The absence of a perirhinal lesion effect in the secondary visual reinforcer task contrasts with evidence from both monkeys (Buckley & Gaffan, 1998a) and rats (Eacott et al., 2001) that, in other circumstances, visual–visual association is impaired after perirhinal lesions. Our own study (Eacott et al., 2001) was designed to investigate visual–visual associations between elements of a discriminandum. The rats learned a two-choice discrimination in which there were four highly similar visual discriminanda. The compound discriminanda consisted of two separable components: a base stimulus (background shape) and an overlying stimulus (superimposed lines). However, there were only two base stimuli (e.g., A and B) and two overlay stimuli (e.g., x and y), and these were recombined to form four unique discriminanda (e.g., Ax, Ay, Bx, By) each of which shared its base component with one other discriminandum and its overlay component with a different single discriminandum. The reward values of the discriminanda were such that neither base nor overlay component alone predicted the reward value of a compound discriminandum, but the combination of base and overlay defined a unique discriminandum, which had a consistent reward value (e.g., Ax^+, Ay^-, Bx^-, By^+, where $^+$ indicates a rewarded stimulus, and $^-$ indicates an unrewarded stimulus). In order to succeed in learning this complex discrimination, it was necessary for the rats to learn to associate two visual components to form a gestalt or whole stimulus representation of the discriminanda. In this case, we found that the rats with perirhinal cortex lesions were impaired only during a late phase of training where they had to distinguish discriminanda that shared features (Eacott et al., 2001). In this task, it is not possible to succeed on the basis of a single component or element of the discriminanda, but the components of the stimuli must be associated to form a whole-object representation. Thus this study supported the view that perirhinal cortex is important in forming at least some types of visual–visual association.

The contrast between the success of the perirhinal lesioned rats in making visual–visual associations in the secondary reinforcer task, even over as few as six trials, and the failure to efficiently form such associations in the compound discriminanda task, even over many hundreds of trials, is striking. Both tasks appear to call for learning visual–visual associations. However, there are also some important differences between the tasks, which may shed light on the role of the perirhinal cortex. First, although the secondary reinforcer task (Eacott et al., 2003) involved a large number of stimuli in total (summed across all the two-choice discriminations that the rats learned), each individual discrimination only involved three stimuli—the correct shape, the nonrewarded shape, and the secondary reinforcer—and one of these, namely the secondary reinforcer, was highly familiar having occurred in many previous discriminations. Therefore each individual problem might perhaps be learned by finding and using a simple feature that differentiated the correct and incorrect shapes, and associating that feature of the correct shape with a feature of the well-learned secondary reinforcer. We previously argued that single-feature learning does not require the perirhinal cortex. In contrast, in the compound discrimination task of Eacott et al. (2001), each component feature (base or overlay) had no meaning independent of the other parts of the discriminandum, and an association between features was required to form a gestalt representation of an object (a rewarded discriminandum), which had a consistent reward value. In sum, the former task requires a visual association between stimuli, which may rely on single features, while the latter requires a within-object association of features or components to form an object representation. Therefore comparison of these two tasks strengthens the evidence in favour of the proposals mentioned earlier that the role of the perirhinal cortex lies in object representation and not in representing all possible associations between stimuli.

In the above study (Eacott et al., 2001), the discriminanda contained feature ambiguity in that they shared some features and therefore could not be solved by processing individual elements or features in isolation. Resolving such feature ambiguity has been suggested to be a crucial part of forming object representations in perirhinal cortex (Bussey, Saksida, & Murray 2002, 2003, this issue). Bussey and colleagues argue that the rostral portions of the ventral processing stream are important for processing relationships between features of an object, with increasing levels of feature integration at successive levels of the ventral processing stream. Bussey et al. (2002, 2003) have shown that monkeys with perirhinal lesions are specifically impaired at discriminating between stimuli that are designed to contain similar features and have backed up their interpretation of the lesion data with connectionist simulations (Bussey & Saksida, 2002). Thus Bussey claims (Bussey et al., 2002, 2003, this issue) that the perirhinal cortex is crucial for disambiguating discriminanda that share features, and he calls this feature ambiguity (see also Lee, Barense, & Graham, this issue).

Our own studies have examined this view in the rat (Norman & Eacott, 2004) using a spontaneous recognition paradigm. The spontaneous recognition task is a one-trial test of memory, which exploits rats' natural tendency to explore novelty; when freely exploring in an open field containing a familiar object and a novel object, intact rats show a strong preference for exploring the novel object (Ennaceur & Delacour, 1988). In a variation of this task, Norman and Eacott (2004) manipulated the novel object so that feature overlap or ambiguity could be investigated. Sham-operated and perirhinal-lesioned rats were compared on versions of the task in which the novel objects had differing levels of feature overlap with the familiar objects and so introduced feature ambiguity. Feature ambiguity was manipulated by using objects made entirely from children's play blocks (Duplo^{TM}). To produce stimuli with a relatively low level of feature ambiguity, the novel and the familiar object were constructed from different numbers and types of blocks. As a result they shared few features, but as all were constructed from similar material they shared their surface texture and the presence of 90-degree corners, and so there was inevitably some degree of feature overlap. To achieve a high level of feature ambiguity, novel objects were constructed by reconfiguring exactly the same number and types of blocks that made up the familiar stimulus. Thus in the high feature ambiguity task, the two stimuli deliberately contained many overlapping features, although they appeared in a modified spatial arrangement. These two levels of feature overlap or ambiguity were compared with performance in the standard version of the task, which uses standard junk objects such as bottles or candlesticks, which are unlikely to share many features and so have very little or no feature ambiguity. Although perirhinal animals were impaired on all versions of this object recognition task at longer delays, the severity of the deficit varied according to the level of feature ambiguity present; in fact where feature ambiguity was high, there was no evidence that the perirhinal-lesioned animals could discriminate between the two objects at all, even over delays as short as one minute (see Figure 1). Although the high feature overlap condition of the task (in which the novel stimulus was a reconfigured version of the familiar stimulus) was difficult even for sham-operated rats, other difficult control tasks were run in order to demonstrate that the perirhinal impairment varied with the level of feature ambiguity and not with task difficulty. So we concluded that a deficit in resolving feature ambiguity after perirhinal lesions was the primary impairment.

However, some studies report impairments on tasks where there is no explicit feature ambiguity. For example, monkeys with rhinal (perirhinal + entorhinal) lesions (Eacott et al., 1994) or perirhinal lesions (Buckley et al., 1997; Meunier, Bachevalier, Mishkin, & Murray, 1993) are impaired in trial-unique match- or nonmatch-to-sample, tasks in which the stimuli have no explicit feature overlap or ambiguity and thus do not appear at first glance to demand perirhinal involvement. However, as the number of stimuli in a task

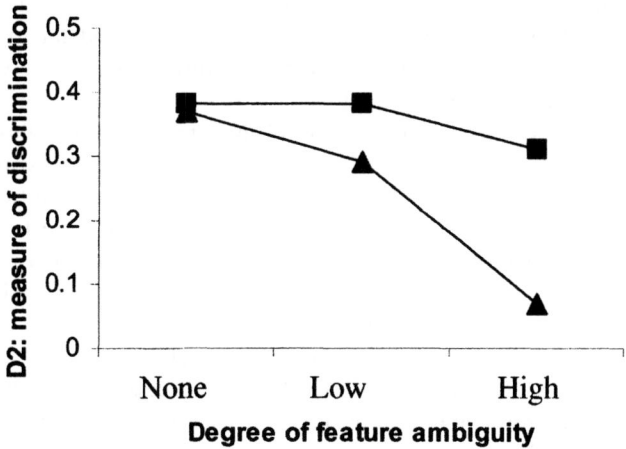

Figure 1.   The figure shows the performance of sham-operated (squares) and perirhinal-lesioned (triangles) rats at one-minute delay on three versions of the spontaneous recognition task of Norman and Eacott (2004).

increases, the potential problem of discriminating between them rises, and the greater the likelihood is that a given stimulus will share features with another. Therefore a task in which there are a large number of different stimuli, such as trial-unique matching or nonmatching, will require that the stimuli are encoded using multiple features bound into a whole-object representation, even though no explicit feature ambiguity is introduced into the task. This analysis can also explain other findings where the severity of the perirhinal lesion effect increases with the number of stimuli concurrently being learned about (Buckley & Gaffan, 1998b).

The central role of perirhinal cortex in representing objects has recently been contrasted with that of the postrhinal cortex. As previously discussed, the perirhinal cortex and postrhinal cortex project in parallel to the hippocampus, have heavy interconnections, and yet have distinct cortical inputs and neural responsivity. This pattern of similarities and differences might suggest that the perirhinal cortex and postrhinal cortex have different yet complementary roles in perception and memory. Yet there is very much less lesion evidence concerning the role of the postrhinal (or in monkeys, parahippocampal) cortex especially as many studies (e.g., Eacott et al., 1994; Mumby & Pinel, 1994) have combined perirhinal and postrhinal (or parahippocampal) lesions. Although, as discussed above, the responses of neurons in postrhinal cortex suggest a role in spatial processing, the lesion evidence is less clear. While lesions of monkey parahippocampal cortex have been reported to cause deficits in object–place associations (Malkova & Mishkin, 2003) and spatial reversal (Teng, Squire, & Zola, 1997), spatial learning deficits are not universally found (Murray, Baxter, & Gaffan, 1998). In the rat, neither PeRh nor PoRh lesions impair place learning in the water maze (e.g., Burwell et al., 2004), which again suggests that PoRh is not always critical to successful performance of spatial tasks. So, in a recent set of studies, we (E. A. Gaffan, Eacott, & Simpson, 2000; E. A. Gaffan, Healey, & Eacott, 2004; Norman & Eacott, 2005) have contrasted the effects of postrhinal and perirhinal lesions in two different visual learning paradigms, which allow us to investigate visuospatial representation—that is, how rats represent both objects and their positions or locations.

The first of these paradigms (E. A. Gaffan et al., 2000, 2004) used the same Y-maze as did Eacott et al. (2001) to train rats in a visual discrimination paradigm, called "constant-negative", which allows us to investigate how multifeatured objects, or scenes (arrays of objects), are encoded. In the constant-negative task, rats choose on each trial between two stimuli (objects or scenes) displayed in two arms of the Y-maze. Any stimulus can appear in any of the three arms. One of the stimuli, the *constant*, is highly familiar, having appeared on many previous trials. The other stimulus, the *variable*, changes from trial to trial. Variables may differ from the constant in several ways—for example, replacing objects, changing their features, on changing their position on the screen. Note that, because a given scene can appear in different arms, positions of objects are defined egocentrically, relative to a rat's view of the screen, and not allocentrically with respect to fixed room cues.

The rats were rewarded with food for choosing the variable on any trial. Their ability to encode aspects of objects or scenes was assessed by their success in discriminating a particular class of variables; for example, a rat who reliably chooses variables that differ from the constant only in what objects they contain is able to encode objects. This task does not require memory for the constant stimulus, which was always present. Rather, it taxes rats' ability to precisely encode or represent aspects of the constant.

Gaffan et al. (2000) trained rats with a constant scene that comprised three objects in three different screen positions. Three types of test trial were compared: type O where the variable scene contained new objects at the same positions as those occupied in the constant; type P where the objects were the same as in the constant, but were moved to new positions; and type X where constant objects and positions were interchanged. Rats with perirhinal lesions were impaired only on tests of type O, implying that they had difficulty in representing the objects in the constant. This object-related impairment is specific to perirhinal cortex, because it was not found in rats with lesions of the fornix or other diencephalic projections of the hippocampus, or of entorhinal cortex (E. A. Gaffan, Bannerman, Warburton, & Aggleton, 2001).

E. A. Gaffan et al. (2004, Exp. 1) compared rats with perirhinal and postrhinal lesions in a similar task and varied the discriminability of the stimuli so as to increase the sensitivity of the tests. Once again the perirhinal-lesioned group showed an impairment on type O tests. Rats with postrhinal lesions did as well as controls on type O, but were impaired on type P, suggesting some difficulty in precisely representing egocentric positions of objects—consistent with the hypothesis of greater involvement in spatial encoding. The dissociation was not complete, however, because the perirhinal-lesioned group, too, showed some impairment on difficult type P trials. To follow up these findings, E. A. Gaffan et al. (2004) carried out a second experiment in which the constants were not scenes but single large objects. They gave test trials of type P where variables consisted of a constant object in a new position, type O where a constant object was replaced by a new one in the same position, and two types where only one aspect of a constant object was changed, either its outline shape (type S) or its internal fill pattern (type F). With these stimuli, a clear dissociation emerged. The postrhinal-lesioned group were impaired only on type P trials, never on any type where objects or their features were changed. The perirhinal-lesioned group, conversely, performed normally on these simple type P trials, but showed some impairment when a whole object (type O) or its outline shape (type S) were changed. These effects were mild and were most apparent early in learning, suggesting that the two cortical areas are involved in initial formation of object or position representations.

We emphasize again that, in the constant-negative task, the constant stimulus need not be retrieved from memory, but is perceptually compared with another stimulus. These experiments, then, provide the first evidence that rats' perirhinal and postrhinal cortex are differentially involved in visuospatial representation—perirhinal being more specialized for representing objects, postrhinal for egocentric locations. Presumably, lesions of perirhinal and postrhinal cortex should also have a different impact in tasks where memory for these aspects was probed.

The dissociation is not completely clear-cut, because other studies (in addition to E. A. Gaffan et al., 2004, Exp. 1) have implicated rats' perirhinal cortex in various tasks that have a spatial element—for example, associating 3-D objects with positions in space (Bussey, Dias, Amin, Muir, & Aggleton, 2001; Liu & Bilkey, 2001) or associating shock with a multifeatured context, where Bucci and colleagues (Bucci, Phillips, & Burwell, 2000; Bucci, Saddoris, & Burwell, 2002) found that perirhinal and postrhinal lesions produced similar impairments. However, our experiments suggest that the two cortical areas can, under certain circumstances, play contrasting roles in visuospatial representation.

In a subsequent series of experiments, Norman and Eacott (2005) have followed up this suggestion of a postrhinal role in visuospatial representations using the spontaneous recognition paradigm. Given evidence that either peri- or postrhinal lesions may impair contextual conditioning (Bucci et al., 2000, 2002), we investigated the role of these areas in forming associations between an object and the context in which it appears (context task, see Figure 2a). This was contrasted with a similar noncontextual task in which the association was between two objects (object task, see Figure 2b). In the first of these tasks (Figure 2a) rats were exposed to objects in the open field in two presentation phases. In the first, the rats were given time to explore two objects in Context 1 (here represented as two cylinders in a grey context). After a brief delay to remove the rats and rearrange the context and objects, they then explored two different objects in a different context (floor and wall inserts of differing textures and contrasts were used to make the different contexts). After a delay, the test phase took place. For the test phase, the rats were replaced in the open field, which contained two objects in one of the contexts (counterbalanced across rats). One of the objects was in the same context as it had previously appeared in (in this example, the cylinder)—these were called congruent configurations. The second object was in an incongruent configuration

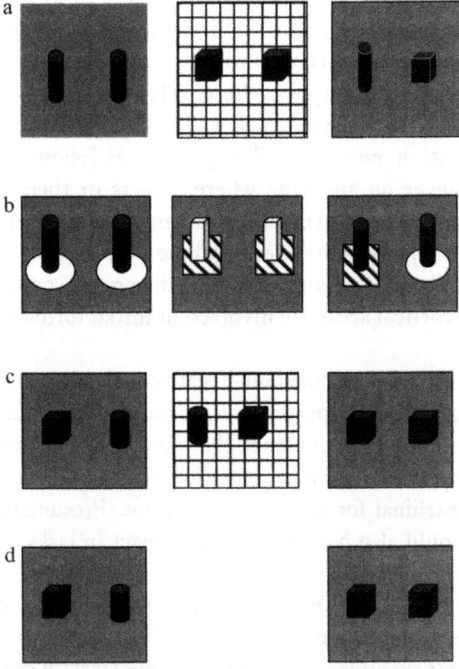

Figure 2.   The figure shows a schematic representation of an example of the experimental procedure from four experiments; (a) shows two exposure phases and a test phase from the *what, which* experiment from Norman and Eacott (2005, Exp. 1); (b) similarly shows the three phases from Experiment 2 of Norman and Eacott (2005); (c) shows two exposure phases and a test phase from the *what, where, which* experiment (Eacott & Norman, 2004, Exps. 1 and 2); (d) shows the single exposure phase and a test phase from the *what, where* experiment of Eacott and Norman (2004, Exp. 3).

(here the cube). In the spontaneous recognition paradigm, a change in context between two presentations of an object results in increased exploration of the object, even when all objects and contexts are themselves familiar (Dellu, Fauchey, LeMoal, & Simon, 1997; Dix & Aggleton, 1999). This increased exploration of the object in the incongruent context reflects memory for the context in which the object was originally experienced. Rats with perirhinal lesions show an intact memory for the object–context association at short delays (see Figure 3), as evidenced by the intact preference for exploring the object that appears in an incongruent context. The results of this study were contrasted with those in a similar, but non-contextual, task in which the association was between two objects rather than between an object and a context (object task, see Figure 2b). As before, the task consisted of two presentation phases and a test phase. In the first presentation phase the rat explored two copies of an object, but in this study the objects were placed in or on another object. For example, as shown in Figure 2b, the first presentation phase consists of the rat exploring two cylinders each placed on a white disc. In the second presentation phase, two novel objects, on or in another object, are explored. As before, the test phase consists of two objects (in this example, cylinders) but one is in an incongruent configuration with the associated object. For example, in Figure 2b, the cylinder on the left is in an incongruent association with the striped square on which it sits, while the cylinder on the right is in a congruent pairing with the white disc. Note that only the pairing of objects is novel, as all objects are themselves equally familiar to the rats. In this object task, in contrast to the previously described contextual task, the perirhinal lesioned rats were severely impaired (see Figure 3). Rats with lesions of postrhinal cortex, however, showed the converse dissociation; they showed no memory for the object–context association, while being unimpaired at the object version of the task (see Figure 3). It is interesting to compare these findings with those of Bucci et al. (2000, 2002) who showed that both perirhinal and postrhinal lesions disrupt rats' ability to associate a context (multifeatured chamber) with foot shock. However, Norman and Eacott's results suggest that, when objects and contexts are experienced together, it is postrhinal cortex that has the predominant role in representing context.

Our two sets of studies discussed here used very different apparatuses (automated Y-maze vs. open field) and markedly different paradigms (motivated discrimination vs. spontaneous recognition), yet both suggest that in the rat the roles of the perirhinal cortex and postrhinal cortex are dissociable, but may be complementary. Postrhinal cortex does not play a crucial role in object recognition, whereas the role of perirhinal cortex here is clear. In contrast, postrhinal cortex is involved in memory for position of object, whether that is egocentric position within a scene or its presence within a particular context.

The above discussion suggests that memory for objects and the locations or context (scenes) in which they are experienced are processed by the perirhinal and postrhinal cortex, respectively. However, in much of everyday life our memory for objects does not consist merely of objects existing in scenes. Instead we have a rich memory for events that we have experienced. Memory for such discrete and personally experienced episodes is termed episodic memory, and a key characteristic of such memory in humans is that retrieval of such memory involves a conscious experience of recall, or of reexperiencing the past (Tulving, 1983). Examination of episodic memory in animals has proved controversial (e.g., Suddendorf & Busby, 2003), but many accept that while the difficulty of showing conscious recall excludes using the term episodic, an ability which is akin to episodic memory exists in animals and has been termed episodic like. Episodic-like memory has many of the character-istics of episodic memory in humans, although there is no necessity of demonstrating con-scious experience of recall. Tulving (1983) defined episodic memory as that which receives and stores information about temporally dated episodes or events and temporal–spatial relations between them, and thus the central characteristics of episodic-like memory are a memory for objects (what), their spatial location (where) and the temporal context in which they were experienced (when). This definition, focusing on memory for what, where, and when, has successfully been used to develop models of episodic-like memory in birds (e.g., Clayton, Bussey, Emery, & Dickinson, 2003; Clayton & Dickinson, 1998; Clayton, Griffiths, Emory, & Dickinson, 2001). Recently this approach has been developed to examine the role of perirhinal and postrhinal cortex in memory for these components of episodic-like memory in rats (Eacott & Norman, 2004). In addition, given that both these areas project heavily to the hippocampus, long linked to episodic memory in humans (Cummings, Tomiyasu, Read, & Benson, 1984; Rempel-Clower, Zola-Morgan, Squire, & Amaral, 1996; Scoville & Milner, 1957; Zola-Morgan, Squire, & Amaral 1986), we also examined the effects of disruption of the hippocampal output by way of a fornix lesion.

Given our previous work demonstrating the importance of context or scenes in postrhi-nal function (E. A. Gaffan et al., 2000, 2004; Norman & Eacott, 2005) and the work of D. Gaffan (1994) suggesting that memory for complex scenes may be analogous to episodic memory in animals, we developed a task to study memory for objects, their spatial location, and the context (or scene) in which they appeared. This approach diverges from the *what, where, when* model, which has been used with success in other laboratories with birds (Clayton et al., 2003; Clayton & Dickinson, 1998; Clayton et al., 2001) but which has proved more difficult to apply to rats (Bird, Roberts, Abroms, Kit, & Crupi, 2003; Ergorul & Eichenbaum, 2004; Hampton & Schwartz, 2004). However, termed *what, where, which*, our model expands the notion of a temporal context, to include any contextual cue that may define the occasion on which an event occurred. While time may serve as such a contextual cue, other cues may also mark such changes of occasion, including changes in context. Using a spontaneous recognition paradigm, we have found that intact rats show increased explo-ration of an object that is appearing in a novel location, as defined by its spatial position within a visually defined context, even though the object, its location, and the context were all equally familiar. In this study, rats were exposed to objects in the open field in two presentation phases (see Figure 2c). After a delay, they were then replaced in the open field, which contained two copies of one of the objects. One of the objects was in a novel location for this object in this context (in Figure 2c the cube on the right is in an incongruent configuration with position and context) although all objects, positions, and contexts were themselves familiar, as were configurations of any two of these aspects. Only the configura-tion of object, position, and context was novel. Because the design required that the rats were exposed to two different contexts, each containing objects, and could be tested for memory in either context, the rats were able to demonstrate that they possessed a detailed memory for what they had seen and where in each of the two contexts. Moreover, this memory for *what, where, which* was quickly and spontaneously acquired and was relatively long lasting, as memory for two highly similar events could be demonstrated over periods as long as one hour. Crucially, we found that neither perirhinal nor postrhinal cortex lesions disrupted the ability to form *what, where, which* memories, whereas rats with fornix lesions could not perform above chance in this task, even after delays as short as 2 minutes (see Figure 3, *what, where, which* task). Fornix-lesioned rats have previously been found to be impaired on an object–place (*what, where*) task (Ennaceur & Meliani, 1992; Ennaceur, Neave, & Aggleton, 1997) in which objects are found in novel positions within a single context (see Figure 2d). Therefore we also considered the possibility that this was the source of the fornix-lesioned group's difficulty with the *what, where, which* task. Yet, we found no object–place impairments in this group of rats even at delays of 5 minutes, considerably longer than the delays in which they showed a severe *what, where, which* deficit (see Figure 3). The delays we used were relatively brief compared with delays used by Ennaceur and colleagues (Ennaceur & Meliani, 1992; Ennaceur et al., 1997), so our failure to find an object–place deficit in our rats is perhaps not surprising; an impairment at longer delays might reasonably be predicted. However, the contrast between the fornix group's normal performance of *what, where* at 5 minutes and their complete failure on *what, where, which* at only 2 minutes is striking and emphasizes the importance of the fornix in the *what, where, which* task. Moreover, these results demonstrate a further dissociation among the rodent equivalents of primate medial temporal structures, between the effects of fornix lesions and lesions of the peri- and postrhinal cortices (see Figure 3).

The series of studies outlined above emphasize that the regions within the medial temporal lobe cannot be see as a unified and indivisible memory system as envisaged by some (Zola-Morgan, Squire, Amaral, & Suzuki, 1994). Instead each of the contributing regions serves an independent and doubly dissociable function in memory involving objects. In some ways, however, our data raise further questions about the interrelationship of these areas. For example, as detailed above, perirhinal cortex and postrhinal cortex in the rat provide heavy parallel inputs to the hippocampus. In our studies, we do not directly lesion the

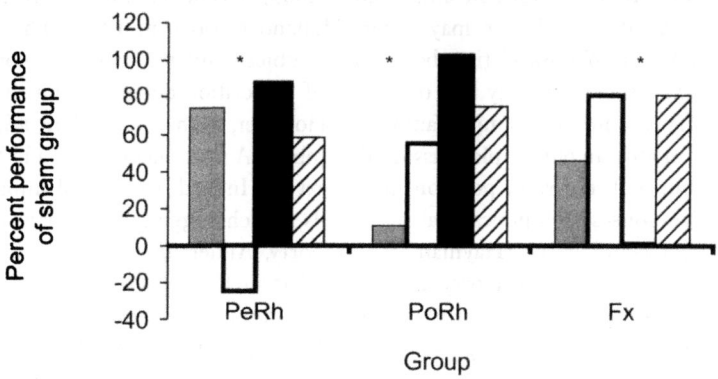

Figure 3.   The figure shows the performance of three groups of lesioned rats expressed as a percentage of the mean performance of the sham-operated group on the same task. Grey bars show performance on the object–context (*what, which*) task (Exp. 1) of Norman and Eacott (2005); white bars show performance on the object task (Exp. 2) of the same paper; black bars show the performance of the same animals on the *what, where, which* task (Exp. 2) of Eacott and Norman (2004). Striped bars show performance on the *what, where* task (Exp. 3) of Eacott and Norman (2004). Asterisked bars are those tasks on which the relevant group showed a significant impairment relative to the sham-operated group at this delay (for full details see relevant paper). All data comes from the 2-minute delay condition of the relevant task.

hippocampus but disrupt its function by way of fornix lesions. Nonetheless, perirhinal lesions and postrhinal lesions do not impair performance on the *what, where, which* task of Eacott and Norman (2004), which is so impaired by fornix lesions. This result, in fact, is somewhat counterintuitive, as the task involves elements of the object–context (*what, which*) task in which postrhinal animals were severely impaired (Norman & Eacott, 2005). The question arises as to how animals who cannot judge the incongruency of an object in a context (Norman & Eacott, 2005), can nonetheless judge the incongruency of an object in a place in a context (Eacott & Norman, 2004). This has led us to consider further the nature of the memory processing in the two tasks. Recent reports (Fortin, Wright, & Eichenbaum, 2004) have emphasized that a single memory task might have contributions from more than one memory process. In a food-rewarded old–new odour recognition paradigm, Eichenbaum and colleagues presented evidence that claimed to show that both familiarity and recall processes contribute to performance in the rat. Moreover, the contribution to performance from recall processes was entirely removed by hippocampal lesions. In our complex versions of the spontaneous recognition paradigm we have no reason to doubt that similar processes are also being engaged. But one can speculate that relatively minor changes in the task might emphasize one or other of the memory processes and thus change the relative contributions of different contributing brain regions. Winters, Forwood, Cowell, Saksida, and Bussey (2004) have suggested that the role of the hippocampus in such recognition tasks becomes important when spatial or contextual factors are involved. Thus we speculate that the *what, where, which* task of Eacott and Norman (2004), despite relying on a recognition paradigm, might be calling on the recall processes of the hippocampus much more heavily than does the object–context (*what, which*) task of Norman and Eacott (2004). In other words, in considering the effects of lesions on tasks, we should not only consider the physical elements to be associated, but should equally consider the nature of the underlying processes. Indeed, Norman and Eacott (2005) speculated that the role of context in the two tasks discussed here may differ (Holland & Bouton, 1999). The *what, which* task relies on a simple association between the object and its associated context. Such object–context associations may rely on a type of association dependent on the postrhinal cortex and may require only simple familiarity. However, in the *what, where, which* task, the context may modify an object–place association (object A is associated with a left position in context X, but with the right position in context Y). Indeed, contextual information can modify the responses of hippocampal place cells, which suggests a mechanism by which such an effect could occur (Hayman, Chakraborty, Anderson, & Jeffery, 2003; Jeffery, Anderson, Hayman, & Chakraborty, 2004) and that such a role is dependent on the hippocampus. Thus the postrhinal cortex may only be important for detection of incongruent context associations that are recognized by familiarity processes alone. Detection of incongruency in the *what, where, which* task, however, may be not be possible via mere familiarity of the associations and may instead rely far more heavily on the recall circuits (Aggleton & Brown, 1999, this issue) and thus be reliant in this case on the hippocampus.

We suggested at the outset that perirhinal and postrhinal cortex may have differing yet complementary roles in memory. Analysis of the patterns of impairments and spared abilities in the series of experiments from our laboratories supports this assertion and, in addition, suggests that their respective functions differ from, yet are complementary to, that of the hippocampus. The perirhinal cortex is important for learning about features that define objects, while the postrhinal cortex is important in learning about within-scene position and contexts. However, in both the process may be one of familiarity, not recall. In contrast, the hippocampus may be critical in tasks that call for memory of objects, their positions, and the contexts in which the occur. Termed *what, where, which* memory, this type of association may require a different type of memory process, one that is more akin to human episodic memory.

REFERENCES

Aggleton, J. P., & Brown, M. W. (1999). Episodic memory, amnesia and the hippocampal-anterior thalamic axis. *Behavioural and Brain Sciences, 22*, 425–489.

Aggleton, J. P., & Brown, M. W. (this issue). Contrasting hippocampal and perirhinal cortex function using immediate early gene imaging. *Quarterly Journal of Experimental Psychology, 58B*, 218–233.

Bird, L. R., Roberts, W. A., Abroms, B., Kit, K. A., & Crupi, C. (2003). Spatial memory for food hidden by rats (Rattus norvegicus) on the radial maze. Studies of memory for where, what, and when. *Journal of Comparative Psychology, 117*, 176–187.

Bucci, D. J., Phillips, R. G., & Burwell, R. D. (2000). Contributions of postrhinal and perirhinal cortex to contextual information processing. *Behavioral Neuroscience, 114*, 882–894.

Bucci, D. J., Saddoris, M. P., & Burwell, R. D. (2002). Contextual fear discrimination is impaired by damage to the postrhinal or perirhinal cortex. *Behavioral Neuroscience, 116*, 479–488.

Buckley, M. J. (this issue). The role of the perirhinal cortex and hippocampus in learning, memory, and perception. *Quarterly Journal of Experimental Psychology, 58B*, 264–268.

Buckley, M. J., Booth, M. C. A., Rolls, E. T., & Gaffan, D. (2001). Selective perceptual impairments after perirhinal cortex ablation. *Journal of Neuroscience, 21*, 9824–9836.

Buckley, M. J., & Gaffan, D. (1998a). Perirhinal cortex ablation impairs configural learning and paired-associate learning equally. *Neuropsychologia, 36*, 535–546.

Buckley, M. J., & Gaffan, D. (1998b). Perirhinal cortex ablation impairs visual object identification. *Journal of Neuroscience, 18*, 2268–2275.

Buckley, M. J., Gaffan, D., & Murray, E. A. (1997). Functional double dissociation between two inferior temporal cortical areas. Perirhinal cortex versus middle temporal gyrus. *Journal of Neurophysiology, 77*, 587–598.

Burwell, R. D., & Amaral, D. G. (1998a). Cortical afferents of the perirhinal, postrhinal and entorhinal cortices of the rat. *Journal of Comparative Neurology, 398*, 179–205.

Burwell, R. D., & Amaral, D. G. (1998b). Perirhinal and postrhinal cortices of the rat. Interconnectivity and connections with the entorhinal cortex. *Journal of Comparative Neurology, 391*, 293–321.

Burwell, R. D., & Hafeman, D. M. (2003). Positional firing properties of postrhinal cortex neurons. *Neuroscience, 119*, 577–588.

Burwell, R. D., Saddoris, M. P., Bucci, D. J., & Wiig, K. A. (2004). Corticohippocampal contributions to spatial and contextual learning. *Journal of Neuroscience, 24*, 3826–3836.

Burwell, R. D., Witter, M. P., & Amaral, D. G. (1995). The perirhinal and postrhinal cortices of the rat: A review of the neuroanatomical literature and comparison with findings from the monkey brain. *Hippocampus, 5*, 390–408.

Bussey, T. J., Dias, R., Amin, E., Muir, J. L., & Aggleton, J. P. (2001). Perirhinal cortex and place–object conditional learning in the rat. *Behavioral Neuroscience, 115*, 776–785.

Bussey, T. J., & Saksida, L. M. (2002). The organization of visual object representations; a connectionist model of effects of lesions in perirhinal cortex. *European Journal of Neuroscience, 15*, 355–364.

Bussey, T. J., Saksida, L. M., & Murray, E. A. (2002). Perirhinal cortex resolves feature ambiguity in complex visual discriminations. *European Journal of Neuroscience, 15*, 365–374.

Bussey, T. J., Saksida, L. M., & Murray, E. A. (2003). Impairments in visual discrimination after perirhinal cortex lesions, testing 'declarative' vs. 'perceptual-mnemonic' views of perirhinal cortex function. *European Journal of Neuroscience, 17*, 649–660.

Bussey T. J., Saksida, L. M., & Murray, E. A. (this issue). The perceptual-mnemonic/feature conjunction model of perirhinal cortex function. *Quarterly Journal of Experimental Psychology, 58B*, 269–282.

Clayton, N. S., Bussey, T. J., Emery, N. J., & Dickinson, A. (2003). Prometheus to Proust, the case for behavioural criteria for mental time travel. *Trends in Cognitive Science*, 7, 346–347.

Clayton, N. S., & Dickinson, A. (1998). What, where and when, episodic like memory during cache recovery by scrub jays. *Nature*, 395, 272–274.

Clayton, N. S., Griffiths, D. P., Emory, N. J., & Dickinson, A. (2001). Elements of episodic-like memory in animals. *Philosophical Transactions of the Royal Society of London. Series B: Biological Sciences*, 356, 1483–1491.

Cummings, J. L., Tomiyasu, S., Read, S., & Benson, D. F. (1984). Amnesia with hippocampal lesions after cardiopulmonary arrest. *Neurology*, 34, 679–681.

Dellu, F., Fauchey, V., LeMoal, M., & Simon, H. (1997). Extension of a new two-trial memory task in the rat. Influence of environmental context on recognition processes. *Neurobiology of Learning and Memory*, 67, 112–120.

Dix, S. L., & Aggleton, J. P. (1999). Extending the spontaneous preference test of recognition; evidence of object-location and object-context recognition. *Behavioural Brain Research*, 99, 191–200.

Eacott, M. J., Gaffan, D., & Murray, E. A. (1994). Preserved recognition memory for small sets and impaired stimulus identification for large sets following rhinal cortex ablations in monkeys. *European Journal of Neuroscience*, 6, 1466–1478.

Eacott, M. J., & Heywood, C. A. (1995). Perception and memory, action and interaction. *Critical Reviews in Neurobiology*, 9, 311–320.

Eacott, M. J., & Norman, G. (2004). Integrated memory for object, place and context in rats: A possible model of episodic-like memory in rats? *Journal of Neuroscience*, 24, 1948–1953.

Eacott, M. J., Machin, P. E., & Gaffan, E. A. (2001). Elemental and configural visual discrimination learning following lesions to perirhinal cortex in the rat. *Behavioural Brain Research*, 124, 55–70.

Eacott, M. J., Norman, G., & Gaffan, E. A. (2003). The role of perirhinal cortex in visual discrimination learning for visual secondary reinforcement in rats. *Behavioral Neuroscience*, 117, 1318–1325.

Ennaceur, A., & Delacour, J. (1988). A new one-trial test for neurobiological studies of memory in rats. 1. Behavioral-data. *Behavioural Brain Research*, 31, 47–59.

Ennaceur, A., & Meliani, K. (1992). A new one-trial test for neurobiological studies of memory in rats. 3. Spatial vs. nonspatial working memory. *Behavioural Brain Research*, 51, 83–92.

Ennaceur, A., Neave, N., & Aggleton, J. P. (1997). Spontaneous object recognition and object location memory in rats; the effects of lesions in the cingulate cortices, the medial prefrontal cortex, the cingulum bundle and the fornix. *Experimental Brain Research*, 113, 509–519.

Ergorul, C., & Eichenbaum, H. (2004). The hippocampus and memory for "What", "Where", and "When". *Learning and Memory*, 11, 397–405.

Fortin, N. J., Wright, S. P., & Eichenbaum, H. (2004). Recollection-like memory retrieval in rats is dependent on the hippocampus. *Nature*, 431, 188–191.

Gaffan, D. (1994). Scene-specific memory for objects: A model of episodic memory impairment in monkeys with fornix transection. *Journal of Cognitive Neuroscience*, 6, 305–320.

Gaffan, E. A., Bannerman, D. M., Warburton, E. C., & Aggleton, J. P. (2001). Rats' processing of visual scenes, effects of lesions to fornix, anterior thalamus, mamillary nuclei or the retrohippocampal region. *Behavioural Brain Research*, 121, 103–117.

Gaffan, E. A., & Eacott, M. J. (1995). A computer controlled maze environment for testing visual memory in the rat. *Journal of Neuroscience Methods*, 60, 23–27.

Gaffan, E. A., Eacott, M. J., & Simpson, E. L. (2000). Perirhinal ablation in rats selectively impairs object identification in a simultaneous visual comparison task. *Behavioral Neuroscience*, 114, 18–31.

Gaffan, E. A., Healey, A. N., & Eacott, M. J. (2004). Objects and positions in visual scenes; effects of perirhinal and postrhinal cortex lesions in the rat. *Behavioral Neuroscience*, 118, 992–1010.

Hampton, R. R. (this issue). Monkey perirhinal cortex is critical for visual memory, but not for visual perception: Reexamination of the behavioural evidence from monkeys. *Quarterly Journal of Experimental Psychology*, 58B, 283–299.

Hampton, R. R., & Schwartz, B. L. (2004). Episodic memory in nonhumans, what, and where, is when? *Current Opinion in Neurobiology*, 14, 192–197.

Hayman, R. M. A., Chakraborty, S., Anderson, M. I., & Jeffery, K. J. (2003). Context-specific acquisition of location discrimination by hippocampal place cells. *European Journal of Neuroscience*, 18, 2825–2834.

Holdstock, J. S. (this issue). The role of the human medial temporal lobe in object recognition and object discrimination. *Quarterly Journal of Experimental Psychology*, 58B, 326–339.

Holland, P. C., & Bouton, M. E. (1999). *Current Opinion in Neurobiology, 9*, 195–202.

Jeffery, K. J., Anderson, M. I., Hayman, R., & Chakraborty, S. (2004). A proposed architecture for the neural representation of spatial context. *Neuroscience and Biobehavioral Reviews, 28*, 201–218.

Lee, A. C. H., Barense, M. D., & Graham, K. S. (this issue). The contribution of the human medial temporal lobe to perception: Bridging the gap between animal and human studies. *Quarterly Journal of Experimental Psychology, 58B*, 300–325.

Liu, P., & Bilkey, D. K. (2001). The effect of excitotoxic lesions centered on the hippocampus or perirhinal cortex in object recognition and spatial memory tasks. *Behavioral Neuroscience, 115*, 94–111.

Malkova, L., & Mishkin, M. (2003). One-trial memory for object–place associations after separate lesions of hippocampus and posterior parahippocampal region in the monkey. *Journal of Neuroscience, 23*, 1956–1965.

Meunier, M., Bachevalier, J., Mishkin, M., & Murray, E. A. (1993). Effects on visual recognition of combined and separate ablations of the entorhinal and perirhinal cortex in rhesus-monkeys. *Journal of Neuroscience, 13*, 5418–5432.

Mumby, D. G., & Pinel, J. P. J. (1994). Rhinal cortex lesions and object recognition in rats. *Behavioral Neuroscience, 108*, 1–8.

Murray, E. A., Baxter, M. G., & Gaffan, D. (1998). Monkeys with rhinal cortex damage or neurotoxic hippocampal lesions are impaired on spatial scene learning and object reversals. *Behavioral Neuroscience, 112*, 1291–1303.

Murray, E. A., & Bussey, T. J. (1999). Perceptual-mnemonic functions of the perirhinal cortex. *Trends In Cognitive Sciences, 3*, 142–151.

Murray, E. A., Bussey, T. J., Hampton, R. R., & Saksida, L. M. (2000). The parahippocampal region and object identification. *Annals of the New York Academy of Sciences, 911*, 166–174.

Norman, G., & Eacott, M. J. (2004). Impaired object recognition with increasing levels of feature ambiguity in rats with perirhinal cortex lesions. *Behavioural Brain Research, 148*, 79–91.

Norman, G., & Eacott, M. J. (2005). Dissociable effects of lesions to the perirhinal cortex and the postrhinal cortex on memory for context and objects in rats. *Behavioral Neuroscience, 119*, 557–566.

Rempel-Clower, N. L., Zola-Morgan, S., Squire, L. R., & Amaral, D. G. (1996). Three cases of enduring memory impairment following bilateral damage limited to the hippocampal formation. *Journal of Neuroscience, 16*, 5233–5255.

Scoville, W. B., & Milner, B. (1957). Loss of recent memory after bilateral hippocampal lesions. *Journal of Neurology, Neurosurgery and Psychiatry, 20*, 11–21.

Suddendorf, T., & Busby, J. (2003). Mental time travel in animals? *Trends in Cognitive Science, 7*, 391–396.

Suzuki, W. A., & Amaral, D. G. (1994). The perirhinal and parahippocampal cortices of the macaque monkey: Cortical afferents. *Journal of Comparative Neurology, 350*, 497–533.

Teng, E., Squire, L. R., & Zola, S. (1997). Different roles for the parahippocampal and perirhinal cortices in spatial reversal. *Society for Neuroscience Abstracts, 23*, 12.7.

Tulving, E. (1983). *Elements of episodic memory.* Oxford, UK: Clarendon Press.

Vann, S. D., Brown, M. W., Erichsen, J. T., & Aggleton, J. P. (2000). Fos imaging reveals differential patterns of hippocampal and parahippocampal subfield activation in rats in response to different spatial memory tests. *Journal of Neuroscience, 20*, 2711–2718.

Wan, H. M., Aggleton, J. P., & Brown, M. W. (1999). Different contributions of the hippocampus and perirhinal cortex to recognition memory. *Journal of Neuroscience, 19*, 1142–1148.

Winters, B. D., Forwood, S. E., Cowell, R. A., Saksida, L. M., & Bussey, T. J. (2004). Double dissociation between the effects of peri-postrhinal cortex and hippocampal lesions on tests of object recognition and spatial memory: Heterogeneity of function within the temporal lobe. *Journal of Neuroscience, 24*, 5901–5908.

Zhu, X. O., Brown, M. W., & Aggleton, J. P. (1995a). Neuronal signalling of information important to visual recognition memory in rat rhinal and neighbouring cortices. *European Journal of Neuroscience, 7*, 753–765.

Zhu, X. O., Brown, M. W., McCabe, B. J., & Aggleton, J. P. (1995b). Effects of novelty or familiarity of visual stimuli on the expression of the immediate early gene c-fos in the rat brain. *Neuroscience, 69*, 821–829.

Zola-Morgan, S., Squire, L. R., & Amaral, D. G. (1986). Human amnesia and the medial temporal region, enduring memory impairment following a bilateral lesion limited to field CA1 of the hippocampus. *Journal of Neuroscience, 6*, 2950–2967.

Zola-Morgan, S., Squire, L. R., Amaral, D. G., & Suzuki, W. A. (1994). Lesions of perirhinal cortex that spare the amygdala and hippocampal formation produce severe memory impairment. *Journal of Neuroscience, 9*, 4355–4370.

THE QUARTERLY JOURNAL OF EXPERIMENTAL PSYCHOLOGY
2005, 58B (3/4), 218–233

Contrasting hippocampal and perirhinal cortex function using immediate early gene imaging

John P. Aggleton

Cardiff University, Wales, UK

Malcolm W. Brown

MRC Centre for Synaptic Plasticity, University of Bristol, UK

The perirhinal cortex and hippocampus have close anatomical links, and it might, therefore, be predicted that they have close, interlinked roles in memory. Lesion studies have, however, often failed to support this prediction, providing dissociations and double dissociations between the two regions on tests of object recognition and spatial memory. In a series of rat studies we have compared these two regions using the expression of the immediate early gene c-*fos* as a marker of neuronal activity. This gene imaging approach makes it possible to assess the relative involvement of different brain regions and avoids many of the limitations of the lesion approach. A very consistent pattern of results was found as the various hippocampal subfields but not the perirhinal cortex show increased c-*fos* activity following tests of spatial learning. In contrast, the perirhinal cortex but none of the hippocampal subfields show increased c-*fos* activity when presented with novel rather than familiar visual objects. When novel scenes are created by the spatial rearrangement of familiar objects it is the hippocampus and not the perirhinal cortex that shows c-*fos* changes. This double dissociation for gene expression accords with that found from lesion studies and highlights the different contributions of the perirhinal cortex and hippocampus to memory.

As a component part of the parahippocampal region, the perirhinal cortex (areas 35 and 36) is situated in close proximity to the hippocampus. The two structures are reciprocally linked both directly and indirectly though their dense connections with the entorhinal cortex (Figure 1a). Via these connections, which are found in both the primate and the rodent brain (Burwell, Witter, & Amaral, 1995; Witter, Groenewegen, Lopes da Silva, & Lohman, 1989), the perirhinal cortex is assumed to exert an important influence over the hippocampus and vice versa (Burwell et al., 1995; Burwell & Witter, 2002; Squire & Zola, 1996). The functional

Correspondence should be addressed to John Aggleton, School of Psychology, Park Place, Cardiff University, Cardiff, Wales, CF10 3AT, UK. Email: aggleton@cf.ac.uk

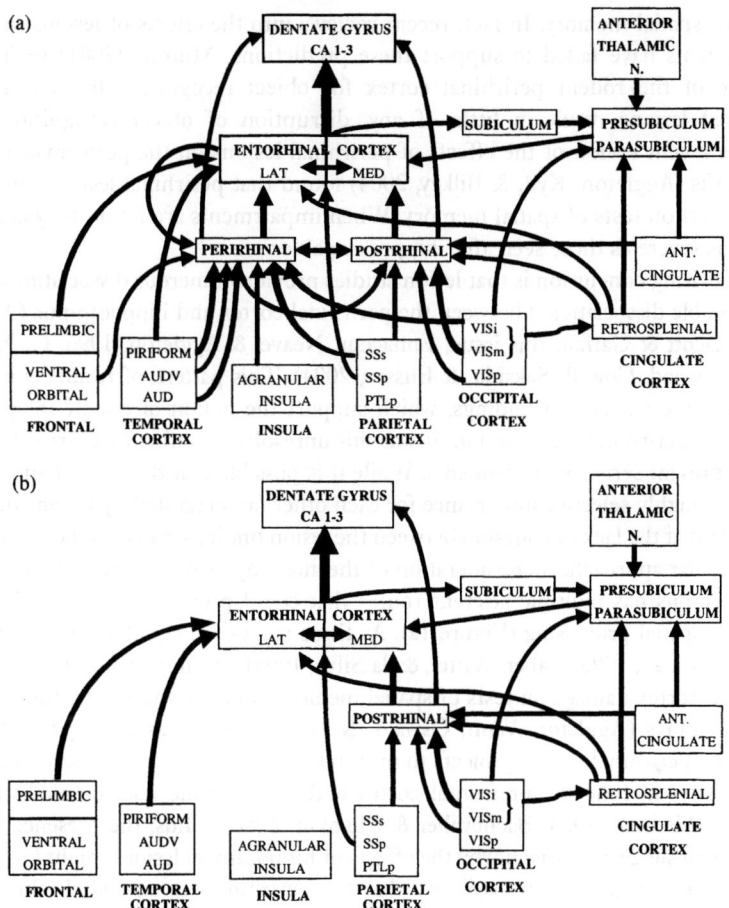

Figure 1.   (a) Schematic diagram of cortical inputs to the hippocampus via the parahippocampal region. The thickness of the arrows indicates the relative sizes of the projections. (b) Schematic diagram showing the many routes to the hippocampus that are preserved after removal of the perirhinal cortex.

importance of these influences has, however, proved more difficult to discern, and they are the focus of this review.

One obvious clue to the possible nature of these interactions lies in the fact that both the perirhinal cortex and the hippocampus are key regions for specific aspects of memory. It has long been appreciated that the hippocampus is vital for certain forms of spatial memory— for example, tasks that tax allocentric processing (Morris, Garrud, Rawlins, & O'Keefe, 1982; O'Keefe & Nadel, 1978). In more recent years it has become evident that the perirhinal cortex is required for normal object recognition memory (Brown & Aggleton, 2001; Mumby & Pinel, 1994; Murray, 1992). In view of their interconnections a more parsimonious account of their mnemonic roles would be that they are tightly linked—that is, that the hippocampus is also important for object recognition while the perirhinal cortex is also required for spatial memory. In fact, recent reviews into the effects of lesions in these two areas in rodents have failed to support these predictions. Mumby (2001) reaffirmed the importance of the rodent perirhinal cortex for object recognition but concluded that hippocampal lesions result in little, if any, disruption of object recognition memory. Similarly, a recent review of the effects of perirhinal lesions on the performance of spatial memory tasks (Aggleton, Kyd, & Bilkey, 2004) found that perirhinal lesions often have no apparent effect on tests of spatial memory. When impairments are found they are transient and never as severe as those seen after hippocampal lesions.

The surprising conclusion is that lesion studies provide numerous dissociations (and, collectively, double dissociations) between the perirhinal cortex and hippocampus (Aggleton et al., 2004; Eacott & Gaffan, this issue, Ennaceur, Neave, & Aggleton, 1996; Gaffan, 1994a; Winters, Forwood, Cowell, Saksida, & Bussey, 2004). This pattern of results is in apparent conflict with the anatomical findings, which support the notion of closely integrated hippocampal and perirhinal cortex action. It was this unresolved conflict that formed the rationale for the present series of experiments. While it is possible that these two regions are only of limited or highly selective importance for each other (as suggested by lesion studies), it is also possible that the lack of cohesion between the lesion findings reflects inherent limitations in the particular approach. A consideration of the anatomy of the temporal lobe reveals that the hippocampus has multiple afferent routes that could provide the sensory information required for spatial processing (Figure 1a). As these routes are parallel (Burwell & Amaral, 1998; Burwell et al., 1995; Naber, Witter, & da Silva, 1999) it is quite plausible that the effects of perirhinal cortex damage on tests of spatial memory could be masked by the recruitment of other pathways (Aggleton, Vann, Oswald, & Good, 2000b), as depicted in Figure 1b. Likewise, the perirhinal cortex projects to multiple regions implicated in recognition memory. These sites include the prefrontal cortex and the thalamic nucleus medialis dorsalis (Aggleton & Mishkin, 1983; Bachevalier & Mishkin, 1986). Thus, the presence of parallel pathways again suggests a reason why the effects of hippocampal lesions might be attenuated.

This review considers the possibility that the perirhinal cortex and the hippocampus closely cooperate for spatial memory and object recognition memory, but that the insensitivity of the lesion technique masks this cooperation. For this reason, we compared the contributions of the hippocampus and perirhinal cortex to these forms of memory using the expression of the immediate early gene (IEG) c-*fos*. Gene expression studies were selected as they make it possible to compare levels of neuronal activation in multiple sites (e.g., in both the hippocampus and the perirhinal cortex) in the same tissue in normal brains. The use of IEGs as markers of functional involvement stems from findings that, (a) IEGs have the distinctive property of being induced as a direct result of neuronal activation rather than via other genes (Herrera & Robertson, 1996), which means that they may be more directly coupled to neuronal activity; (b) some IEGs, including c-*fos*, are inducible transcription factors so that they control the downstream expression of many other genes. These genes are likely to include some of those that are important for neuronal plasticity (Herdegen & Leah, 1998; Tischmeyer & Grimm, 1999). Although IEGs such as c-*fos* can provide an index of neuronal activity this is not true for all neuronal systems (Herdegen, 1996), and so it is important to confirm the sensitivity of any given system under investigation.

This review describes the outcome of a series of studies that have examined the expression of c-*fos*. This IEG was selected as it is expressed throughout the temporal lobe. A feature of this IEG is that it has a relatively low basal rate of expression, making it an especially good marker when there are environmental changes (Zengenehpour & Chaudhari, 2002). Furthermore, activation of c-*fos* has repeatedly been linked to learning processes (Fleischmann et al., 2003; Guzowski, Setlow, Wagner, & McGaugh, 2001; Hall, Thomas, & Everitt, 2001; Mileusnic, Anoknin, & Rose, 1996; Swank, Ellis, & Cochran, 1996; Tischmeyer & Grimm, 1999). This evidence includes the results of antisense experiments, which make it possible to block the ability of the mRNA to produce the Fos protein. The outcome of infusions of antisense for Fos mRNA into the dorsal hippocampus support the importance of Fos production for normal spatial learning in the radial-arm maze (He, Yamada, & Nabeshima, 2002).

Our strategy has been to compare c-*fos* activity in rats performing tests of memory with that in control rats that are performing tasks closely matched for motor and sensory demands but which lack the critical learning component. In some cases, such as the paired-viewing procedure described below, it is possible to make within-animal comparisons. In all cases c-*fos* activity was estimated by measuring levels of Fos, the protein product of the c-*fos* gene. Fos was visualized using immunohistochemical methods in rats that were perfused 90 min after completion of the learning (or control) task. The 90-min interval reflects the time needed to allow Fos levels to peak following initial activation (Zengenehpour & Chaudhari, 2002). This interval does, however, highlight the fact that while this approach has exceptionally high anatomical resolution it has poor temporal resolution. This interval also shows that the initiation of Fos production is unlikely to be directly involved in the performance of those memory tasks, such as tests of working memory, where the interval between an event and an appropriate change of behaviour is often much shorter than 90 minutes.

c-*fos* activity and the performance of spatial tasks sensitive to hippocampal lesions

The first study (Figure 2a) examined Fos levels in the brains of rats shortly after performing a spatial working memory task in an eight-arm radial maze (Vann, Brown, & Aggleton, 2000a; Vann, Brown, Erichsen, & Aggleton, 2000b). A working memory test was used for the experimental group as this made it possible to demonstrate spatial learning during the critical (final) test session. In the standard version of this task rats are trained to select different arms of a radial maze on every choice, thereby not running down a previously visited arm that now has no food at the end. Although rats can use a variety of strategies to solve this task, there is much evidence that they will typically use allocentric cues to distinguish the choice of arms (Olton & Samuelson, 1976). While damage to the hippocampal system severely impairs this task (Olton, Becker, & Handelman, 1979; Warburton, Morgan, Baird, Muir, & Aggleton, 2001), rats with perirhinal cortex lesions either perform at normal levels or show only transient impairments (Aggleton et al., 2004).

Rather than give each rat a single trial (terminated when all eight arms had been visited), the rats were trained on a series of radial-arm maze trials within a single session. This modification extended the training session so as to increase the IEG signal strength. During each session, which lasted for 30 min, the rats completed five or six trials (each trial consisted of visiting eight different arms). The control group were in a "yoked" regime so that they received food rewards at the same frequency as that for the experimental animals.

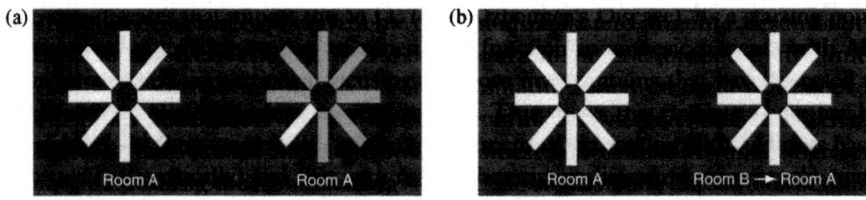

Figure 2. Testing procedures to examine c-*fos* expression following spatial memory tasks (Vann et al., 2000a, 2000b). (a) Working memory in a radial-arm maze (left) versus single-arm control group (right). (b) Novel room study in which the control group are trained in Room A, while the experimental group are trained in Room B but tested in Room A on the final test session.

The difference was that these control rats just ran up and down one arm of the radial maze, and so there was no explicit memory load (Figure 2a).

Counts of Fos-positive cells revealed two quite different patterns of activity change in the parahippocampal region. Significantly higher Fos counts were found in those animals performing the spatial working memory task not only in all hippocampal subfields that were counted (dentate gyrus, CA3, CA1) but also the dorsal subiculum, ventral subiculum, presubiculum, postsubiculum, parasubiculum, lateral and medial entorhinal cortices, and the postrhinal cortex (Vann et al., 2000a, 2000b). In contrast, the perirhinal cortex did not show an increase in Fos counts. The perirhinal cortex was, therefore, the only parahippocampal region not to show a significant Fos increase. This differential pattern of activation among the parahippocampal cortices was reflected by significant interactions between the pattern in the perirhinal cortex and that in the three other regions (lateral entorhinal, medial entorhinal, postrhinal cortices). These results fail to support the view that the perirhinal cortex is normally important for spatial memory tasks but that parallel inputs mask the effects of its removal. Additionally, other studies have also shown that increases in hippocampal c-*fos* activity are associated with spatial learning in the radial-arm maze (Touzani, Marighetto, & Jaffard, 2003) and the water maze (Guzowski et al., 2001).

A quite different approach was used in a study that compared Fos levels in rats trained in the Morris water maze on one of two spatial tasks (Jenkins, Amin, Harold, Pearce, & Aggleton, 2003). Rats were either trained on a working memory task that taxed the ability to learn a location within a session, or trained to find a submerged platform that was always a fixed distance and direction from a visible landmark in the pool ("landmark" task). Structural equation modelling (SEM) revealed that the tasks resulted in two quite different patterns of relative Fos change in the structures comprising the hippocampal and parahippocampal regions (Figure 3). For the working memory task, which is highly sensitive to hippocampal damage, the best fit model indicated a continuous anatomical route linking both the postrhinal and perirhinal cortices to the entorhinal cortex and thence to the dentate gyrus and subfield CA1 (Figure 3a). Interestingly, the perirhinal cortex was linked to the lateral entorhinal cortex while the postrhinal cortex was linked to the medial entorhinal cortex, in accordance with their anatomical connections (Burwell & Witter, 2002; Naber et al., 1999). For the landmark task, which is less sensitive to hippocampal damage (Pearce, Roberts, & Good, 1998), the best fit model indicated a route from both the postrhinal and perirhinal cortices to the entorhinal cortex, but from there the path moved to the subiculum,

'Place' task   'Landmark' task

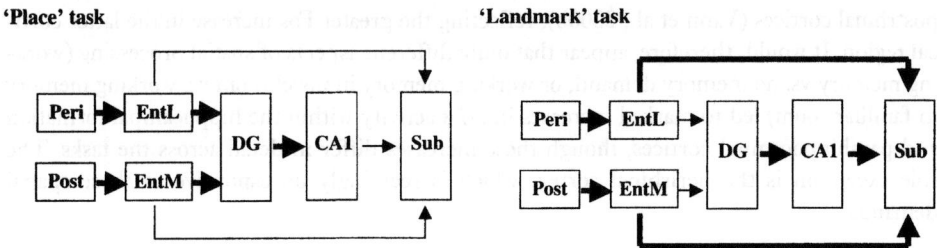

Figure 3. Pathways derived from structural equation modelling of Fos levels following two different spatial memory tasks (Jenkins et al., 2004). The thick lines are those that show significant linkages. (a) "Place" task—working memory in the water maze, selected as it is sensitive to hippocampal lesions; (b) "Landmark" task—vector heading task in the water maze that is insensitive to lesions of hippocampus proper. DG, dentate gyrus; EntL, lateral entorhinal cortex; EntM, medial entorhinal cortex; Peri, perirhinal cortex; Post, postrhinal cortex; Sub, subiculum.

so side-stepping the dentate gyrus and CA fields (Figure 3b). In summary, this experiment (Jenkins et al., 2003) revealed two different modes of temporal lobe activity associated with spatial processing. Only for the task (working memory) that is most sensitive to hippocampal damage did the pattern of IEG activity link across the parahippocampal region and across the hippocampus proper (Figure 3a). A feature of both tasks, however, was the link between the perirhinal cortex and postrhinal cortex to different parts of the entorhinal cortex, consistent with their respective connection patterns. This suggests that there is an ongoing interaction between these parahippocampal regions, irrespective of task demands. It should be added that this SEM procedure could not be used to compare all hippocampal inputs, and so the importance of other routes (Figure 1) will need to be examined before a more comprehensive picture of the active pathways during spatial memory can be determined. Furthermore, the SEM analysis is based around correlations between areas, and so the incorporation of the perirhinal cortex need not conflict with other studies (e.g., Vann et al., 2000a; 2000b) that failed to find a distinct perirhinal activation when comparing an allocentric spatial memory task with its own control task.

c-*fos* activation following exposure to a novel location

A variation of the radial-arm maze procedure was used to look at c-*fos* responses after a rat was placed in a novel room (Vann et al., 2000a, 2000b). The radial-arm maze task was used to encourage the rats to learn the new spatial landmarks. For this study, two groups of rats were trained on the radial-arm maze task, one group in Room A and the other in Room B. On the critical, final day the rats in Room B were moved into Room A to perform the radial-arm maze task (Figure 2b). Now both groups were in the same location (Room B), but for one group the room layout was quite novel, and the rats would be expected to learn this new layout to help solve the task. Although the two groups did not differ on behavioural measures (e.g., number of errors), the novel room group showed significantly higher Fos levels than did the familiar room group in the hippocampus proper (dentate gyrus, CA1, CA3), dorsal subiculum, presubiculum, postsubiculum, and parasubiculum, as well as in the lateral and medial entorhinal cortices. While there was an overall rise in perirhinal counts this change was not significant. Furthermore, there was a significant interaction between the perirhinal and postrhinal cortices (Vann et al., 2000b), reflecting the greater Fos increase in the latter corti-cal region. It would, therefore, appear that quite different aspects of spatial processing (work-ing memory vs. no memory demand, or working memory in novel room vs. working memory in familiar room) led to marked increases in c-*fos* activity within the hippocampal formation and parahippocampal cortices, though these increases differ in detail across the tasks. The sole exception is the perirhinal cortex, which is seemingly unresponsive to these spatial demands.

c-*fos* activation following exposure to novel visual objects

Two very different methods have been used to examine c-*fos* expression following exposure to novel visual stimuli, although the pattern of results has been consistent. For both methods levels of activation were compared between animals (or hemispheres) that had been shown either novel or familiar objects. Unlike the spatial tasks, which tax working memory, the procedures were passive in that there was no independent, behavioural measure of novelty detection. In spite of the different features of the two methods a consistent pattern emerged: Exposure to novel compared to familiar objects resulted in increased Fos levels in the perirhinal cortex, while the hippocampus appeared unresponsive.

In the first study (Zhu, Brown, McCabe, & Aggleton, 1995b) rats were placed in a cham-ber ($45 \times 30 \times 35$ cm) with opaque sides except for a Perspex front wall with a half-silvered glass wall. Through this wall objects could be seen when illuminated. Rats were trained to poke their muzzles though an observing hole to gain a sweet drink, so placing the animal in front of the stimuli to be presented. This arrangement ensured that the experimenter could control when and for how long novel or familiar objects could be observed. Comparisons were made between rats shown all novel objects or, all familiar objects, or those for which the light was switched on and off but no objects were presented. Fos levels were found to be significantly higher for novel than for familiar objects in a group of sites involved in visual processing (Zhu et al., 1995b). These sites were the occipital cortex, area TE (Te2), and the perirhinal cortex. No evidence was found for a Fos increase in the hippocampus (dentate gyrus, CA1, CA3 combined) or entorhinal cortex.

Using the same apparatus Zhu, McCabe, Aggleton, and Brown (1997) provided more evidence that the perirhinal cortex and hippocampus can respond independently. Two groups of rats were shown novel objects, exactly as in the study described above, except that for one group the chamber was familiar and for the other it was novel. Comparisons between the ratio of perirhinal to hippocampal Fos counts in the two conditions showed a significant difference, as only hippocampal counts (dentate gyrus, CA1, CA3 combined) were increased in the novel environment (Zhu et al., 1997).

A limitation of this apparatus for presenting novel stimuli is that the behaviours of the experimental and control groups may be difficult to equate. It might be, for example, that the rats shown novel objects are more active, so leading to Fos changes that are not a direct reflection of distinguishing novel from familiar. In order to eliminate this problem we intro-duced the "paired-viewing" procedure (Zhu, McCabe, Aggleton, & Brown, 1996). In this procedure, rats are first trained to lick at a spout that can be reached through an observing hole in the transparent wall of a test chamber. Once again, the licking ensures that the

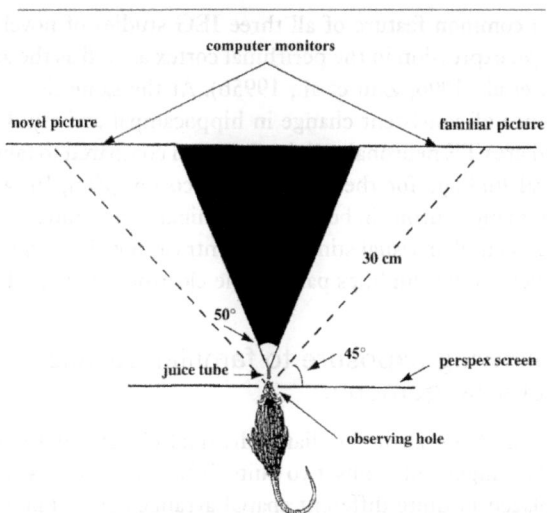

Figure 4.   Plan of paired-viewing apparatus. When the rat's head was in the observing hole in the Perspex screen the right eye could only see the right monitor, and the left eye could only see the left monitor. Stimuli were presented on both monitors simultaneously. (In some studies stimuli were presented behind a one-way mirror rather than on a monitor.)

position and orientation of the rat are fixed. The rat is then shown different stimuli to the left and right visual fields (Figure 4). The optic pathways in the rat show cross-over for stimuli appearing in the monocular visual field of each eye, so that stimuli shown to the left monocular visual field (left eye) will first be processed in the right hemisphere. The opposite is true for the right monocular visual field (left hemisphere first). By simultaneously presenting novel stimuli to one eye and only familiar stimuli to the other eye of a rat, it is possible to compare Fos levels within a single animal. This method ensures optimal control conditions, as factors such as motor activity and levels of arousal and hormones are matched. Moreover, sensory stimulation can also be matched by counterbalancing the particular stimuli used as novel and familiar across the animals. A limitation is that some information from the two eyes is likely eventually to cross over, and so areas that would show differential responses to novel or familiar stimuli may fail to do so, especially if they are relatively far down the processing pathways.

Initial paired-viewing procedure experiments used three-dimensional objects (novel or familiar) set behind one-way mirrors. Increased Fos levels were found in the "novel object" hemisphere in area TE and the perirhinal cortex, but not in the hippocampus or entorhinal cortex (Zhu et al., 1996). A later refinement of the paired-viewing procedure (Figure 4) has been to use two-dimensional stimuli displayed on computer monitors (Wan, Aggleton, & Brown, 1999). When rats were shown novel two-dimensional pictures to one eye and familiar pictures to the other eye, differential levels of Fos were once again found in the perirhinal cortex and area TE (Wan et al., 1999). In this study separate counts were made in the dentate gryus, CA1, CA3, ventral subiculum, and entorhinal cortex. None of these hippocampal regions showed a Fos change in response to the novel stimuli.

It can be seen that a common feature of all three IEG studies of novel versus familiar objects is a change in c-*fos* expression in the perirhinal cortex as well as the adjacent area TE (Wan et al., 1999; Zhu et al., 1996; Zhu et al., 1995b). At the same time the IEG studies failed to find any evidence of consistent change in hippocampal activity. The evidence for increased perirhinal and area TE neuronal activity for novel compared to familiar stimuli fits with electrophysiological findings for the rat temporal cortex (Zhu, Brown, & Aggleton, 1995a), which show that many neurons in both the perirhinal cortex and area TE show larger responses to novel than to familiar visual stimuli. In contrast, few do so in the hippocampus (Zhu et al., 1995a). Hence the Fos findings parallel the electrophysiological results.

c-*fos* activation following exposure to familiar objects set in novel spatial arrangements

Novelty can arise not only from an unfamiliar individual object but from familiar items placed in a novel spatial arrangement. Thus, two quite different scenes may contain the same items but have them placed in quite different spatial arrangements. The next two studies used c-*fos* expression to compare how different brain regions respond to the novelty that arises when familiar items are placed in unfamiliar spatial arrays. The principal question is whether this form of novelty would activate the hippocampus, the perirhinal cortex, or both.

The first study used the paired-viewing procedure (Wan et al., 1999). During training rats were shown separate visual scenes to the two eyes, each scene comprising three individual stimuli set in fixed locations on the computer screen. The critical test involved presenting scenes that were composed of novel spatial configurations of familiar, individual stimuli to one eye while familiar configurations were presented to the other eye (Figure 5). The same locations on the computer screen were occupied for both the familiar and novel arrays of three stimuli, the critical difference being that the identity at each location had changed in the novel arrangement (Figure 5). Within-subject comparisons showed that there were significant differences between the two hemispheres in the dentate gyrus, CA1, ventral subiculum, and postrhinal cortex. For the postrhinal cortex and for CA1 the Fos counts were higher in the hemisphere first processing the novel arrays. In contrast, for the dentate gyrus and the ventral subiculum lower Fos counts were found in the "novel" hemisphere. No significant hemisphere difference was found in either the perirhinal cortex or area TE. This result therefore establishes a double dissociation between the processing of novel and familiar arrangements and of novel and familiar individual items: The relative familiarity of the arrangements differentially activates hippocampal and postrhinal areas, whereas for the individual items perirhinal cortex and TE are differentially activated.

A very different procedure was used by Jenkins, Amin, Pearce, Brown, and Aggleton (2004) to address a similar question. In this study two groups of rats ("novel" and "familiar") were trained on the same working memory task in the radial-arm maze. The Novel group was trained in a distinctive cue-controlled environment created by closing a curtain surrounded the maze. Eight distinctive figures (each 40×60 cm) were attached to the inside of the curtain at positions between the arms of the maze (Figure 6). The positions of these figures remained fixed until the final session. On the final test day the positions of these familiar, extramaze cues were rearranged for group novel. The new spatial configuration of the cues now matched the arrangement of the same figures that had been used throughout training

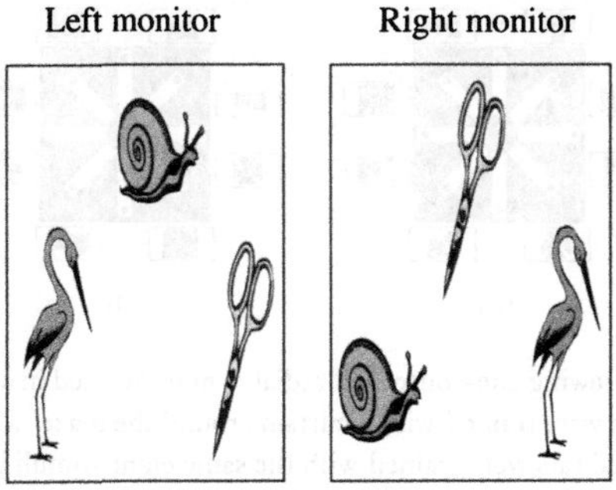

Left monitor   Right monitor

Figure 5. Example of stimuli used in spatial configuration task (Wan et al., 1999). Rats are repeatedly exposed to specific stimuli in fixed spatial arrangements, but on the test day one eye is presented with novel spatial arrangements of the same individual stimuli.

for group familiar (including the final day). Rearrangement of visual stimuli (group novel) led to significant increases in Fos-positive cells in various hippocampal subfields (rostral CA1, rostral CA3, and rostral dentate gyrus) as well as in the parietal cortex and the post-subiculum. In contrast, no changes were observed in parahippocampal sites (the perirhinal cortex, postrhinal cortex, and lateral and medial entorhinal cortices). The lack of any Fos differences in sites conveying afferent visual information to the hippocampus suggests that the signal for spatial rearrangement is generated in the hippocampus itself. These findings are consistent with evidence that fornix lesions are especially disruptive when distal cues attached to the inside of a curtain around a radial arm maze are spatially rearranged between sessions (Hudon, Dore, & Goulet, 2003).

The two different Fos procedures have quite different merits. In the paired-viewing procedure (Wan et al., 1999) the control conditions are unusually well matched and the statistical analyses are within subject, so reducing the impact of individual variation. The novel spatial configuration task in the radial-arm maze (Jenkins et al., 2004) has the advantage that the rats' accurate choice behaviour shows that they are actively processing the rearranged stimuli (i.e., it is not passive). In addition, this design avoids the potential problem that interhemispheric pathways could reduce any hemispheric differences (as in the paired-viewing procedure). The fact that these very different procedures give similar results adds weight to the general conclusions that can be drawn. The first is that subfields of the hippocampus are sensitive to the novel spatial rearrangement of familiar stimuli, although this may be reflected as either increased or decreased Fos levels. The second is that the perirhinal cortex appears insensitive to this form of novelty.

Conclusions

The overall pattern of IEG results is remarkably clear, despite the use of different behavioural methods (Table 1). The presentation of novel, individual visual stimuli evokes increased Fos levels in the perirhinal cortex but not in the hippocampus (dentate gyrus, CA subfields, subicular cortices) or the entorhinal cortex. In contrast, performing spatial tasks

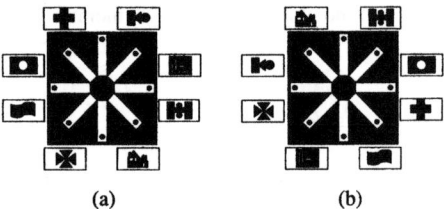

(a)   (b)

Figure 6.   Schematic diagram showing cue-controlled radial-arm maze used in spatial rearrangement experiment (Jenkins et al., 2004). Control rats were trained with a curtain around the maze, with eight distinctive stimuli in set places (Condition b). Experimental rats were trained with the same eight stimuli but in a different spatial arrangement (Condition a). On the final test day the experimental rats were switched to Condition b.

or being exposed to novel spatial arrays of visual stimuli does not increase perirhinal Fos levels but does alter Fos levels throughout much of the hippocampal formation (Table 1). These findings provide a clear answer to the questions posed at the beginning of this review. Just as lesion studies in rats reveal dissociations between the effects of hippocampal lesions and perirhinal lesions on tests of object recognition and spatial memory, so IEG studies provide exactly the same pattern of dissociations (and double dissociations). As a consequence, there is no support for the idea that the results of lesion studies on these functions are misleading because the importance of the perirhinal cortex (or hippocampus) is masked by parallel pathways that support the function when one route is surgically removed.

Before accepting this conclusion a number of minor issues should be addressed. The first is that the experiments all concern the IEG c-*fos*. Although a more comprehensive analysis would have included other IEGs, the striking double dissociations that are found for c-*fos* provide strong evidence of a qualitative difference in function, irrespective of what may yet be found for other IEGs. Furthermore, the Fos dissociations match those found for lesion studies, which establish the necessity of different brain regions for recognition or spatial memory. A second issue is that all of the studies of item novelty used visual stimuli. It is possible that the use of another modality might produce a different set of results. While there are no behavioural studies in the rat that have tested the effects of lesions on auditory recognition memory, one IEG study has examined this issue (Wan et al., 2001). Novel sounds were associated with increased Fos levels compared to familiar sounds in the auditory association cortex but comparable changes were not found in the perirhinal cortex nor in the hippocampal region. Data emerging from lesion studies with monkeys have also failed to show that the perirhinal cortex is important for auditory recognition memory (Saunders, Fritz, & Mishkin, 1998), in contrast to the effects of similar lesions on visual recognition memory (Murray, 1992; Suzuki, 1996). This apparent difference between sensory modalities clearly warrants further investigation. A third factor is that Fos levels were measured in rats perfused 90 min after completion of the learning (or control) task. This lengthy period is likely to increase the noise-to-signal ratio in the IEG method. In spite of this limitation, consistent Fos differences were found across different classes of memory task.

An underlying issue concerns the apparent mismatch between the close anatomical relationships between the perirhinal cortex and the hippocampus and the strikingly different

TABLE 1
Summary table of c-*fos* expression studies

	Ref.	Hippocampus	Subiculum	Entorhinal	Postrhinal	Perirhinal
Spatial working memory–RAM	1	↑	↑	↑	↑	–
Novel room working memory	1	↑	↑	↑	↑	–
Novel object	2	–		–		↑
Novel object–paired viewing	3	–		–		↑
Novel object–paired viewing	4	–		–	–	↑
Novel spatial arrangement– paired viewing	4	↑↓	↓	–	↑	–
Novel spatial arrangement – RAM	5	↑	–	–	–	–

Note: The table shows the very different patterns of Fos changes found in spatial tasks compared with tasks involving the presentation of novel objects. ↑ increased Fos counts; ↓ decreased Fos counts; – no significant change in Fos levels. When Fos counts were not made the column is left blank. In one study (4) different subfields of the hippocampus showed increased and decreased counts. RAM–radial-arm maze. References: 1. Vann et al. (2000a, 2000b); 2. Zhu et al. (1995b); 3. Zhu et al. (1996); 4. Wan et al. (1999); 5. Jenkins et al. (2004).

functional properties of these two areas as measured both by lesions and by IEG expression. It is not appropriate simply to ignore these connections; the task is to identify under what special conditions they become important. Clues are provided when demands are simultaneously placed on both stimulus identity and stimulus location as "object-in-place" tasks appear to tax the interactions between the perirhinal cortex and the hippocampus. Evidence for this view comes from studies that have compared the effects of selective lesions on a discrimination task in which learning the combination of item identity and the scene within which it is placed should provide optimal rates of learning (Gaffan, 1994b). Deficits are found on this task after bilateral lesions of the fornix, which supplies critical connections to and from the hippocampus, in both monkeys (Gaffan, 1994b) and humans (Aggleton et al., 2000a). Furthermore, crossed lesion studies in monkeys indicate that the perirhinal cortex works in concert with the hippocampus to solve these tasks (Gaffan & Parker, 1996). Other supporting data have come from lesion studies of rats performing other behavioural tasks that involve the combination of item identification and item location. Here, there is also growing evidence for a perirhinal involvement (Aggleton et al., 2004; Bussey & Aggleton, 2002).

This form of object and place interaction might, at first, seem at odds with the two IEG results in the previous section where the combination of identity and location affected hippocampal but not perirhinal Fos levels. In fact, almost all of the preceding IEG results can be explained if it is assumed that (a) object identity is constantly being processed by the perirhinal cortex (irrespective of whether it is familiar or not) and that some of this information is relayed to the hippocampus, (b) spatial information is processed in the hippocampus but not the perirhinal cortex, and (c) differential responses to novel objects are found in the perirhinal cortex but not the hippocampus. Thus, as neither experiment examining object–place combinations used novel individual stimuli the lack of a perirhinal Fos increase is to be predicted. The only exception to this account concerns the radial-arm maze task in which rats were tested in a novel room (Vann et al., 2000a, 2000b). Here, the animals were exposed to novel items yet the Fos increases in the perirhinal cortex failed to reach significance and were much less prominent than those found in the hippocampus. One possibility is that the nonsignificant result for the perirhinal cortex reflects the need for more power—that is, either larger group sizes or perhaps more novel objects in the training room. This is possible because the Fos changes, although smaller than those in the hippocampus, did approach significance (Vann et al., 2000b). Another possibility is that this difference is because the novel stimuli used in the radial-arm maze IEG experiment were treated as spatial arrays and not individual objects. This explanation is circular and ultimately requires an independent definition of when objects become spatial arrays and vice versa. Nevertheless, there is evidence that object size and appearance are important factors as to whether the hippocampus is necessary for detecting novelty, with increasing importance being found for objects of increasing size and plainness (Cassaday & Rawlins, 1997). Linked to this finding is evidence of greater recruitment of the hippocampal system by distal than by proximal cues (Hudon et al., 2003). While the precise parameters underlying these important differences have yet to be determined, it seems possible that certain stimuli could be viewed as an individual object or as part of a spatial array given particular experimental conditions, with potentially different IEG outcomes.

In conclusion, these IEG studies add further support to the view that the perirhinal cortex is not the standard route for sensory information required by the hippocampus for spatial processing. Likely routes revealed by IEG studies, anatomical tract tracing, and electrophysiological studies include pathways through the entorhinal cortex, the postrhinal cortex, and the retrosplenial cortex. While lesions in all three areas can disrupt spatial memory tasks in rats, the severity of the deficits cannot match those seen after hippocampal lesions (Aggleton et al., 2000b; Burwell, Bucci, Wiig, Saddoris, & Sanborn, 2002; Burwell, Saddoris, Bucci, & Wiig, 2004). These findings again make it highly likely that parallel routes to the hippocampus can support standard spatial memory tasks, although the IEG data show that of these routes the perirhinal cortex is often the least important in the intact brain. Perhaps, most importantly, these IEG results help to confirm that the perirhinal cortex and the hippocampus have integral roles in quite different forms in memory (Aggleton & Brown, 1999). This information needs to be incorporated in models of how the medial temporal lobe contributes to memory.

REFERENCES

Aggleton, J. P., & Brown, M. W. (1999). Episodic memory, amnesia, and the hippocampal-anterior thalamic axis. *Behavioral Brain Sciences, 22*, 425–489.

Aggleton, J. P., Kyd, R., & Bilkey, D. (2004). When is the perirhinal cortex necessary for the performance of spatial memory tasks? *Neuroscience and Biobehavioural Reviews, 28*, 611–624.

Aggleton, J. P., McMackin, D., Carpenter, K., Hornak, J., Kapur, N., Halpin, S., et al. (2000a). Differential effects of colloid cysts in the third ventricle that spare or compromise the fornix. *Brain, 123*, 800–815.

Aggleton, J. P., & Mishkin, M. (1983). Memory impairments following restricted medial thalamic lesions in monkeys. *Experimental Brain Research, 52*, 199–209.

Aggleton, J. P., Vann, S. D., Oswald, C. J. P., & Good, M. (2000b). Identifying cortical inputs to the rat hippocampus that subserve allocentric spatial processes: A simple question with a complex answer. *Hippocampus, 10*, 466–474.

Bachevalier, J., & Mishkin, M. (1986). Visual recognition impairment follows ventromedial but not dorsolateral prefrontal lesions in monkeys. *Behavioural Brain Research, 20*, 249–261.

Brown, M. W., & Aggleton, J. P. (2001). Recognition memory: What are the roles of the perirhinal cortex and hippocampus? *Nature Reviews Neuroscience, 2*, 51–61.

Burwell, R. D., & Amaral, D. G. (1998). Cortical afferents of the perirhinal, postrhinal and entorhinal cortices of the rat. *Journal of Comparative Neurology, 398*, 179–205.

Burwell, R. D., Bucci, D. J., Wiig, K. A., Saddoris, M. P., & Sanborn, M. R. (2002). Experimental lesions of the parahippocampal region in rats. In M. P. Witter & F. G. Wouterlood (Eds.), *The parahippocampal region* (pp. 217–237). Oxford, UK: Oxford University Press.

Burwell, R. D., Saddoris, M. P., Bucci, D. J., & Wiig, K. A. (2004). Corticohippocampal contributions to spatial and contextual learning. *Journal of Neuroscience, 24*, 3826–3836.

Burwell, R. D., & Witter, M. P. (2002). Basic anatomy of the parahippocampal region in monkeys and rats. In M. P. Witter & F. G. Wouterlood (Eds.), *The parahippocampal region* (pp. 35–59). Oxford, UK: Oxford University Press.

Burwell, R. D., Witter, M. P., & Amaral, D. G. (1995). Perirhinal and postrhinal cortices of the rat: A review of the neuroanatomical literature and comparison with findings from the monkey brain. *Hippocampus, 5*, 390–408.

Bussey, T. J., & Aggleton, J. P. (2002). The 'what' and 'where' of event memory: Independence and interactivity within the medial temporal lobe. In A. Parker, E. Wilding, & T. J. Bussey (Eds.), *The cognitive neuroscience of memory: Encoding and retrieval* (pp. 217–233). Hove, UK: Psychology Press.

Cassaday, H. J., & Rawlins, J. N. P. (1997). The hippocampus, objects, and their contexts. *Behavioral Neuroscience, 111*, 1228–1244.

Eacott, M. J., & Gaffan, E. A. (this issue). The roles of perirhinal cortex, postrhinal cortex, and the fornix in memory for objects, contexts, and events in the rat. *Quarterly Journal of Experimental Psychology, 58B*, 202–217.

Ennaceur, A., Neave, N., & Aggleton, J. P. (1996). Neurotoxic lesions of the perirhinal cortex do not mimic the behavioural effects of fornix transection in the rat. *Behavioural Brain Research, 80*, 9–25.

Fleischmann, A., Hvalby, O., Jensen, V., Strekalova, T., Zacher, C., Layer, L. E., et al. (2003). Impaired long-term memory and NR2A-type NMDA receptor-dependent synaptic plasticity in mice lacking c-Fos in the CNS. *Journal of Neuroscience, 23*, 9116–9122.

Gaffan, D. (1994a). Dissociated effects of perirhinal cortex ablation, fornix transection and amygdalectomy: Evidence for multiple memory systems in the primate temporal lobe. *Experimental Brain Research, 99*, 411–422.

Gaffan, D. (1994b). Scene-specific memory for objects: A model of episodic memory impairment in monkeys with fornix transection. *Journal of Cognitive Neuroscience, 6*, 305–320.

Gaffan, D., & Parker, A. (1996). Interaction of perirhinal cortex with the fornix-fimbria: Memory for objects and 'object-in-place' memory. *Journal of Neuroscience, 16*, 5864–5869.

Guzowski, J. F., Setlow, B., Wagner, E. K., & McGaugh, J. L. (2001). Experience-dependent gene expression in the rat hippocampus after spatial learning: A comparison of the immediate-early genes *Arc*, c-*fos*, and *zif268*. *Journal of Neuroscience, 21*, 5089–5098.

Hall, J., Thomas, K. L., & Everitt, B. J. (2001). Fear communication retrieval induces CREB phosphorylation and Fos expression within the amygdala. *European Journal of Neuroscience, 13*, 1453–1458.

He, J., Yamada, K., & Nabeshima, T. (2002). A role of Fos expression in the CA3 region of the hippocampus in spatial memory formation in rats. *Neuropharmacology, 26*, 259–268.

Herdegen, T. (1996). Jun, Fos, and CREB/ATF transcription factors in the brain: Control of gene expression under normal and pathophysiological conditions. *The Neuroscientist, 2*, 153–161.

Herdegen, T., & Leah, J. D. (1998). Inducible and constitutive transcription factors in the mammalian nervous system: Control of gene expression by Jun, Fos and Krox, and CREB/ATF proteins. *Brain Research Reviews, 28*, 370–490.

Herrera, D. G., & Robertson, H. A. (1996). Activation of c-*fos* in the brain. *Progress in Neurobiology, 50*, 83–107.

Hudon, C., Dore, F. Y., & Goulet, S. (2003). Impaired performance of fornix-transected rats on a distal, but not on a proximal, version of the radial-arm maze cue task. *Behavioral Neuroscience, 117*, 1353–1362.

Jenkins, T. A., Amin, E., Harold, G. T., Pearce, J. M., & Aggleton, J. P. (2003). Different patterns of hippocampal formation activity associated with different spatial tasks: A Fos imaging study in rats. *Experimental Brain Research, 151*, 514–523.

Jenkins, T. A., Amin, E., Pearce, J. M., Brown, M. W., & Aggleton, J. P. (2004). Novel spatial arrangements of familiar stimuli promote activity in the rat hippocampal formation but not the parahippocampal cortices: A c-*fos* expression study. *Neuroscience, 124*, 43–52.

Mileusnic, R., Anokhin, K., & Rose, S. P. R. (1996). Antisense oligodeoxynucleotides to c-fos are amnestic for passive avoidance in the chick. *NeuroReport, 7*, 1269–1272.

Morris, R. G. M., Garrud, P., Rawlins, J. N. P., & O'Keefe, J. (1982). Place navigation impaired in rats with hippocampal lesions. *Nature*, *297*, 681–683.

Mumby, D. G. (2001). Perspective on object-recognition memory following hippocampal damage: Lessons from studies in rats. *Behavioral Brain Research*, *127*, 159–181.

Mumby, D. G., & Pinel, J. P. J. (1994). Rhinal cortex lesions and object recognition in rats. *Behavioural Neuroscience*, *108*, 11–18.

Murray, E. A. (1992). Medial temporal lobe structures contributing to recognition memory: the amygdaloid complex versus the rhinal cortex. In J. P. Aggleton, (Ed.), *The amygdala: Neurobiological aspects of emotion memory and mental dysfunction* (pp. 453–470). New York: Wiley-Liss.

Naber, P. A., Witter, M. P., & da Silva, F. H. L. (1999). Perirhinal cortex input to the hippocampus in the rat: Evidence for parallel pathways, both direct and indirect. A combined physiological and anatomical study. *European Journal of Neuroscience*, *11*, 4119–4133.

O'Keefe, J., & Nadel, L. (1978). *The hippocampus as a cognitive map*. Oxford, UK: Oxford University Press.

Olton, D. S., Becker, J. T., & Handelman, G. E. (1979). Hippocampus, space, and memory. *The Behavioral and Brain Sciences*, *2*, 313–365.

Olton, D. S., & Samuelson, R. J. (1976). Remembrance of places passed: Spatial memory in rats. *Journal of Experimental Psychology: Animal Behaviour Processes*, *2*, 97–116.

Pearce, J. P., Roberts, A. D. L., & Good, M. (1998). Hippocampal lesions disrupt navigation based in cognitive maps but not heading vectors. *Nature*, *396*, 75–77.

Saunders, R. C., Fritz, J. F., & Mishkin, M. (1998). The effects of rhinal cortical lesions on auditory short-term memory (STM) in the rhesus monkey. *Society of Neuroscience, Abstract*, *24*, 1907.

Squire, L. R., & Zola, S. (1996). Structure and function of declarative and nondeclarative memory systems. *Proceedings of the National Academy of Sciences, USA*, *93*, 13515–13522.

Suzuki, W. A. (1996). The anatomy, physiology and functions of the perirhinal cortex. *Current Opinions in Neurobiology*, *6*, 179–186.

Swank, M. W., Ellis, A. E., & Cochran, B. N. (1996). c-Fos antisense blocks acquisition and extinction of conditioned taste aversion in mice. *NeuroReport*, *7*, 1866–1870.

Tischmeyer, W., & Grimm, R. (1999). Activation of immediate early genes and memory formation. *Cellular and Molecular Life Science*, *55*, 564–574.

Touzani, K., Marighetto, A., & Jaffard, R. (2003). Fos imaging reveals ageing-related changes in hippocampal response to radial maze discrimination testing in mice. *European Journal of Neuroscience*, *17*, 628–640.

Vann, S. D., Brown, M., & Aggleton, J. P. (2000a). Fos expression in the rostral thalamic nuclei and associated cortical regions in response to different spatial memory tasks. *Neuroscience*, *101*, 983–991.

Vann, S. D., Brown, M. W., Erichsen, J. T., & Aggleton, J. P. (2000b). Fos imaging reveals differential patterns of hippocampal and parahippocampal subfield activity in response to different spatial memory tasks. *Journal of Neuroscience*, *20*, 2711–2718.

Wan, H., Aggleton, J. P., & Brown, M.W. (1999). Different contributions of the hippocampus and perirhinal cortex to recognition memory. *Journal of Neuroscience*, *19*, 1142–1148.

Wan, H., Warburton, E. C., Kusmiereck, P., Aggleton, J. P., Kowalska, D. M., & Brown, M. W. (2001). Fos imaging reveals differential neuronal activation of areas of rat temporal cortex by novel and familiar sounds. *European Journal of Neuroscience*, *14*, 118–124.

Warburton, E. C., Morgan, A., Baird, A., Muir, J. L., & Aggleton, J. P. (2001). The conjoint importance of the hippocampus and anterior thalamic thalamic nuclei for allocentric spatial learning: Evidence from a disconnection study in the rat. *Journal of Neuroscience*, *21*, 7323–7330.

Winters, B. D., Forwood, S. E., Cowell, R., Saksida, L. M., & Bussey, T. (2004). Double dissociation between the effects of peri-postrhinal cortex and hippocampal lesions on tests of object recognition and spatial memory: Heterogeneity of function within the temporal lobe. *Journal of Neuroscience*, *24*, 5901–5908.

Witter, M. P., Groenewegen, H. J., Lopes da Silva, F. H., & Lohman A. H. M. (1989). Functional organization of the extrinsic and intrinsic circuitry of the parahippocampal region. *Progress in Neurobiology*, *33*, 161–253.

Zengenehpour, S., & Chaudhari, A. (2002). Differential induction and decay curves of *c-fos* and *zif268* revealed through dual activity maps. *Molecular Brain Research*, *109*, 221–225.

Zhu, X. O., McCabe, B. J., Aggleton, J. P., & Brown, M. W. (1996). Mapping recognition memory through the differential expression of the immediate early gene c-*fos* induced by novel or familiar visual stimulation. *NeuroReport*, *7*, 1871–1875.

Zhu, X. O., McCabe, B. J., Aggleton, J. P., & Brown, M. W. (1997). Differential activation of the hippocampus and perirhinal cortex by novel visual stimuli and a novel environment. *Neuroscience Letters, 229,* 141–143.

Zhu, X. O., Brown, M. W., & Aggleton, J. P. (1995a). Neuronal signalling of information important to visual recognition memory in rat rhinal and neighbouring cortices. *European Journal of Neuroscience, 7,* 753–765.

Zhu, X. O., Brown, M. W., McCabe, B. J., & Aggleton, J. P. (1995b). Effects of the novelty or familiarity of visual stimuli on the expression of the intermediate early gene c-*fos* in the rat brain. *Neuroscience, 69,* 821–829.

THE QUARTERLY JOURNAL OF EXPERIMENTAL PSYCHOLOGY
2005, 58B (3/4), 234–245

The perirhinal cortex and long-term familiarity memory

E. T. Rolls, L. Franco, and S. M. Stringer

University of Oxford, UK

To analyse the functions of the perirhinal cortex, the activity of single neurons in the perirhinal cortex was recorded while macaques performed a delayed matching-to-sample task with up to three intervening stimuli. Some neurons had activity related to working memory, in that they responded more to the sample than to the match image within a trial, as shown previously. However, when a novel set of stimuli was introduced, the neuronal responses were on average only 47% of the magnitude of the responses to the set of very familiar stimuli. Moreover, it was shown in three monkeys that the responses of the perirhinal cortex neurons gradually increased over hundreds of presentations (mean = 400 over 7–13 days) of the new set of (initially novel) stimuli to become as large as those to the already familiar stimuli. Thus perirhinal cortex neurons represent the very long-term familiarity of visual stimuli. Part of the impairment in temporal lobe amnesia may be related to the difficulty of building representations of the degree of familiarity of stimuli. A neural network model of how the perirhinal cortex could implement long-term familiarity memory is proposed using Hebbian associative learning.

In this paper, we produce a new model of how the perirhinal cortex implements a long-term form of familiarity memory. In the Introduction, we first describe the neurophysiological data that implicate the perirhinal cortex in long-term familiarity memory and set out what needs to be modelled. Then we compare this type of memory to other types of memory in which the perirhinal cortex is implicated. These other types of memory include, as described below, recognition memory as measured in delayed match–to–sample tasks with short time delays, in which perirhinal cortex neurons typically respond more to the sample than the match stimulus, and delayed match–to–sample tasks with intervening stimuli in which perirhinal cortex neurons may respond more to the match stimulus than the sample; and paired associate learning.

Evidence that the perirhinal cortex is involved in long–term familiarity memory comes from a neuronal recording study in which it was shown that perirhinal cortex neuronal responses in the rhesus macaque gradually increase in magnitude to a set of stimuli as that set is repeated for 400 presentations each 1.3 s long (Hölscher, Rolls, & Xiang, 2003). The

---

Correspondence should be addressed to Professor Edmund T. Rolls, University of Oxford, Department of Experimental Psychology, South Parks Road, Oxford OX1 3UD, UK. Email: Edmund.Rolls@psy.ox.ac.uk

This research was supported by MRC Programme Grant PG9826105 and by the Medical Research Council Interdisciplinary Research Centre for Cognitive Neuroscience.

DOI:10.1080/02724990444000122

single neurons were recorded in the perirhinal cortex in monkeys performing a delayed matching-to-sample task with up to three intervening stimuli, using a set of very familiar visual stimuli used for several weeks. When a novel set of stimuli was introduced, the neuronal responses were on average only 47% of the magnitude of the responses to the familiar set of stimuli. It was shown in eight different replications in three monkeys that the responses of the perirhinal cortex neurons gradually increased over hundreds of presentations of the new set of (initially novel) stimuli to become as large as those to the already familiar stimuli. The mean number of 1.3 s presentations to induce this effect was 400 occurring over 7–13 days. These results show that perirhinal cortex neurons represent the very long-term familiarity of visual stimuli. A representation of the long-term familiarity of visual stimuli may be important for many aspects of social and other behaviour, and part of the impairment in temporal lobe amnesia may be related to the difficulty of building representations of the degree of familiarity of stimuli. It is this type of memory that is modelled in this paper.

The perirhinal cortex is also involved in recognition memory in that damage to the perirhinal cortex produces impairments in recognition memory tasks in which several items intervene between the sample presentation of a stimulus and its presentation again as a match stimulus (Malkova, Bachevalier, Mishkin, & Saunders, 2001; Zola-Morgan, Squire, Amaral, & Suzuki, 1989; Zola-Morgan, Squire, & Ramus, 1994). Indeed, damage to the perirhinal cortex rather than to the hippocampus is believed to underlie the impairment in recognition memory found in amnesia in humans associated with medial temporal lobe damage (Buckley, Booth, Rolls, & Gaffan, 2001; Buckley & Gaffan, 2000). Neurophysiologically, it has been shown that many inferior temporal cortex (a term we use to refer to area TE) neurons (Rolls, 2000; Rolls & Deco, 2002), which provide visual inputs to the perirhinal cortex (Suzuki & Amaral, 1994a, 1994b), respond more to the first than to the second presentation of a stimulus in a running recognition task with trial-unique stimuli (Baylis & Rolls, 1987). In this task, there is typically a presentation of a novel stimulus, and after a delay, which may be in the order of minutes or more and in which other stimuli may be shown, the stimulus is presented again as "familiar", and the monkey can respond to obtain food reward. Most neurons responded more to the "novel" than to the "familiar" presentation of a stimulus, where "familiar" in this task reflects a change produced by seeing the stimulus typically once (or a few times) before. (A small proportion of neurons respond more to the familiar (second) than to the novel (first) presentation of each visual stimulus.) In the inferior temporal cortex this memory spanned up to 1–2 intervening stimuli between the first (novel) and second (familiar) presentations of a given stimulus (Baylis & Rolls, 1987), and as recordings are made more ventrally, towards and within the perirhinal cortex, the memory span increases to several or more intervening stimuli (Brown & Xiang, 1998; Wilson, Riches, & Brown, 1990; Xiang & Brown, 1998).

In a similar task, though, typically performed with non-trial-unique stimuli—a delayed matching-to-sample task with up to several intervening stimuli—some neurons respond more to the match stimulus than to the sample stimulus (Miller, Li, & Desimone, 1998). Many neurons in this task respond more to the sample ("novel") than to the match ("familiar") presentations of the stimuli, and this short-term memory is reset at the start of the next trial (Hölscher & Rolls, 2002). The resetting at the start of each trial shows that the perirhinal cortex is actively involved in the task demands.

A fourth type of memory in which the perirhinal cortex is implicated is paired associate learning (a model of semantic long-term memory), which is represented by a population of neurons in a restricted part of area 36 where the neuronal responses may occur to both members of a pair of pictures used in the paired association task (Miyashita, Kameyama, Hasegawa, & Fukushima, 1998; Miyashita, Okuno, Tokuyama, Ihara, & Nakajima, 1996).

We emphasize that the type of familiarity memory modelled here is a very long-term type of familiarity or recognition memory, which reflects the gradual build-up of neuronal responses over several hundred presentations of a stimulus, and which may thus represent the degree of long-term familiarity of stimuli. We call this long-term familiarity memory in this paper. Figure 1 shows some of the neurophysiological data on long-term familiarity memory from Hölscher et al. (2003) that we wish to model. Figure 1 shows the results of three different replications of the whole investigation in macaque BL. Each replication consisted of starting with a completely new set of "novel" images and using this set for 10 days of testing, in which experiments were performed on many different neurons. The ordinate shows the mean response of a neuron to the set of stimuli in the novel set expressed as a percentage of the response to the set of stimuli in the familiar set. In each replication of the overall investigation, for many neurons early on after the novel set of stimuli was introduced, there were highly significant differences between the mean responses of the neurons to the set of familiar and novel stimuli, as shown by nonparametric (Mann–Whitney U) tests. Indeed, for many cells the difference between the responses to novel and familiar stimuli on the days soon after the novel stimuli were introduced were significant at $p < 10^{-5}$. For the first replication, the degree of variation is indicated by the standard errors of the mean responses of each cell. The slope of the (linear) regression line for each replication (BL1–BL3) was calculated and was highly significant. The intercept of the regression line indicates the average percentage of the neuronal response to novel stimuli compared to very familiar stimuli at the start of testing with novel stimuli. The regression lines show how long it takes neurons to take to respond to the novel set of images as well as to the highly familiar set, shown for hundreds of previous trials so that their maximal response level had been reached. Similar results were found in five further replications in two further monkeys.

Overall, it was found that stimulus selectivity was less than that in the inferior temporal cortex (Rolls, 2000; Rolls & Deco, 2002). This was even confirmed by direct comparison in one of the macaques used in the study, in which when the macaque performed the same task, 15/22 (68%) of inferior temporal cortex (IT) neurons had selective responses to the same set of stimuli. Moreover, the selectivity of the inferior temporal cortex neurons was greater than that of the neurons in the perirhinal cortex (that is, the sparseness of the representation for inferior temporal cortex neurons is lower, see Rolls, 2000; Rolls & Deco 2002). Related perhaps partly to the specificity of the responses of many inferior temporal cortex neurons, in the experiments described here there was no difference apparent in the responses of IT neurons to the novel and the long-term familiar stimuli used. This means at least that long-term familiarity is not made explicit in the responses of inferior temporal cortex neurons, whereas it is made explicit in the responses of neurons in the perirhinal cortex. By "made explicit", we mean that it can be easily read off from the firing rates of one or a small number of neurons, by, for example, dot product decoding (Rolls & Deco, 2002; Rolls & Treves, 1998; Rolls, Treves, & Tovee, 1997a; Rolls, Treves, Tovee, & Panzeri, 1997b).

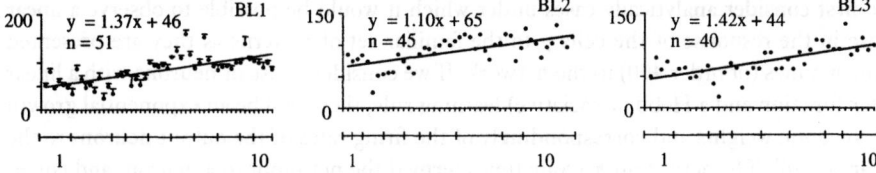

Figure 1. Regressions showing the relative response of each neuron to novel vs. familiar stimuli (expressed as a percentage) as a function of the number of experiments since the novel set of stimuli were investigated (abscissa), for three complete replications (BL1, BL2, and BL3) in macaque BL. Each point on the graph shows the results of one experiment involving 80 trials of the delayed matching-to-sample task on one neuron. The 80 trials included 40 with the novel stimulus set and 40 with the familiar stimulus set, with 5 stimuli on each trial. On some days more than one neuron was analysed in a separate experiment, and the number of days since introduction of the novel set of stimuli is also shown on the abscissa. Results for three separate replications of the whole investigation in one monkey (BL) are shown. Each replication involved starting with a completely novel set of images, and using that novel set on 10 days of testing in which on any day as many experiments as possible were performed, each experiment with a different neuron, and each experiment involving 40 trials with the novel set and 40 trials with the familiar set of images. The first replication (left) involved recordings in 51 experiments from 51 separate neurons over 10 testing days. The slope and intercept of the regression line are shown. The intercept indicates the magnitude of the response to novel stimuli expressed as a percentage of that to familiar stimuli at the start of the replication. During an experiment on each neuron, the set of novel stimuli was shown for approximately 12.5 of the 1.3 s presentations of each novel stimulus during the delayed matching-to-sample task. The results for replication BL1 show the standard error of the mean response of the neuron to the novel relative to the familiar stimuli to give an indication of the degree of accuracy with which this could be estimated. The error bars are omitted from the other replications for clarity.

The model

To analyse how neurons in a receiving brain area such as the perirhinal cortex might build long-term familiarity representations, let us consider the model shown in Figure. 2. This consists of a set of perirhinal cortex neurons y_i receiving synaptic connections w_{ij} from a set of neurons x_j (such as those in the inferior temporal visual cortex), which are tuned to respond to different stimuli and which do not show long-term familiarity-related responses.

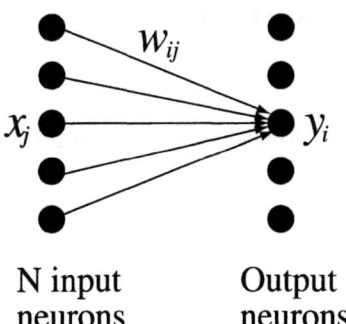

N input
neurons

Output
neurons

Figure 2. The network model for the formation of neurons with responses related to the long-term familiarity of visual stimuli. This consists of a set of perirhinal cortex neurons y_j receiving synaptic connections w_{ij} from a set of neurons x_j (such as those in the inferior temporal visual cortex) which are tuned to respond to different stimuli and which do not show long-term familiarity-related responses.

We first consider analytically cases under which it would be possible to observe a linear increase in the response of the cells y_i to the familiar set of patterns as they are presented very many times (of order 400) to the network. If we consider a case of neurons with a linear transfer function and a Hebb (associative) learning rule, there will be an exponential growth of the synaptic weights and correspondingly of the firing rates of the output neurons to the familiar stimuli. The activation h_i (sometimes termed the net input to a neuron, and corresponding to the depolarization of a neuron produced by its inputs) of neuron i is

$$h_i = \sum_{j=1}^{N} w_{ij} x_j \qquad\qquad 1$$

This can be read as indicating that the depolarization h_i of the neuron is the sum of the activations produced through each synapse, which depend on the presynaptic firing rate x_j and the synaptic strength w_{ij}.

For a linear activation function $y_i = h_i$ relating the firing rate y_i of neuron i to its depolarization or activation h_i, the output, y_i^p, of a neuron i after p presentations of the familiar set of stimuli can be calculated to be

$$y_i^p = \sum_{j=1}^{N} w_{ij}^{p-1} x_j \qquad\qquad 2$$

where w_{ij}^{p-1} are the values of the synaptic weights after the $(p-1)$th presentation, and x_j is the firing rate of input neuron j.

For binary (i.e., 0,1) input patterns with sparseness a (where the sparseness can be thought of as the proportion of the input neurons that respond to any one pattern and thus takes the value 0.2 if 20% of the neurons are active for a stimulus, and 80% are not firing, see Rolls & Deco, 2002; Rolls & Treves, 1998), and for the case of N input neurons we obtain

$$y_i^p = aN w_{ij}^{p-1} \qquad\qquad 3$$

If the weights w_{ij} are updated according to a Hebbian associative learning rule in which the change of weight depends on the product of the presynaptic firing rate x_j and the postsynaptic firing rate y_i shown below in Equation 4,

$$\delta w_{ij} = k y_i x_j \qquad\qquad 4$$

where k is the learning rate, the change in weights at the pth presentation can be calculated to be

$$\delta w_{ij}^p = k(aN w_{ij}^{p-1}) \qquad\qquad 5$$

for active input neuron x_j (using substitution of Equation 3 into Equation 4).

It can then be shown that

$$w_{ij}^p = (1 + kaN)^p w_{ij}^0 \qquad\qquad 6$$

and since $(1 + kaN) > 1$, the weights grow exponentially in this case of using a linear transfer function and an associative learning rule; w_{ij}^0 is the initial small synaptic weight, w_{ij}^p is the weight after the pth presentation, and $(1 + kaN)$ to the power p arises because on every one of the p learning trials, as the synaptic weight increases, it contributes to its own further increase on every trial by influencing the output firing rate on that trial.

Simulation results

We ran simulations using a network comprising 1,000 input neurons projecting to a set of output neurons as shown in Figure 2. The input to the network consisted of random binary patterns that were divided into two sets: the novel and the familiar stimulus sets. In different simulations we tested sparsenesses of the (novel and familiar) input patterns of 0.1 and 0.01.

The growth of the weights summarized in Equation 6 when the network was trained with a Hebb rule and a linear activation function was found in the simulations, and it produced an exponential increase in the firing rate to the familiar patterns, as illustrated in Figure 3. This scenario, therefore, does not account for the approximately linear increase in the firing rate found as a function of the number of training trials, which is illustrated for the real neuronal firing in Figure 1.

Two possible solutions are to use a nonlinear saturating activation function, such as a sigmoid function (which would limit the maximum firing of the output neuron), or to use an associative synaptic modification rule with a decrementing term that captures effects of long-term synaptic depression (LTD). We investigated both.

The results for the case of a sigmoid activation function relating the firing rate y_i of neuron i to its depolarization or activation h_i (shown in Equation 7 where alpha is the threshold, and beta is the slope) are shown in Figure 4. The cell firing rate grows approximately linearly up to its maximum value after approximately 400 repeated presentations.

$$y = \sigma(h) = \frac{1}{1 + e^{-(\beta h + \alpha)}} \qquad\qquad 7$$

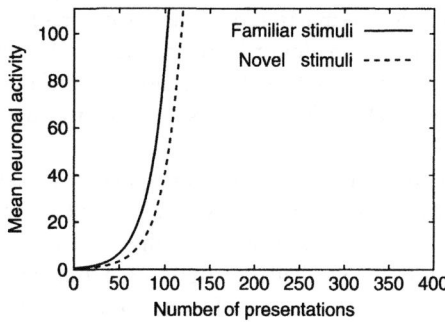

Figure 3. An exponential increase in the firing rate to the familiar patterns as a function of the number of training trials was produced when the network was trained with a Hebb rule and a linear activation function. The familiar set of stimuli was initially novel, and the solid curve shows how the response to this set of stimuli gradually increases as the set becomes familiar over approximately 400 training (i.e., weight modification) presentations. For comparison, the responses to novel stimuli, which were not trained and remained novel, are also shown (dotted curve).

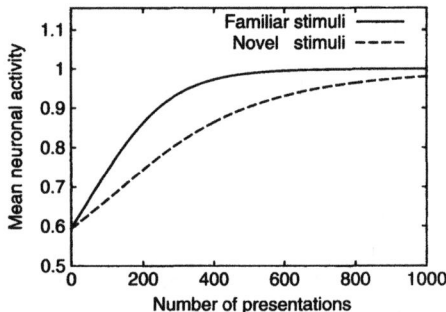

Figure 4. An approximately linear increase in the firing rate to the familiar patterns as a function of the number of training trials was produced when the network was trained with a Hebb rule and a sigmoid activation function (in which beta = 15 and alpha = 0.7). The firing rate saturated at a high value after approximately 500 training trials, and this is a function of the learning rate parameter k in Equation 4. The familiar set of stimuli was initially novel, and the solid curve shows how the response to this set of stimuli gradually increases as the set becomes familiar over approximately 400 training (i.e., weight modification) presentations. For comparison, the responses to novel stimuli, which were not trained and remained novel, are also shown (dotted curve).

We also explored the case of using a modified Hebb rule of the type useful for competitive networks (see Rolls & Deco, 2002; and Rolls & Treves, 1998) in which the synaptic modification, describing heterosynaptic long-term depression in which the synaptic weight can decrease (in proportion to the postsynaptic neuron firing y_i) if the presynaptic firing x_j is lower than the value of the synaptic weight w_{ij}, is:

$$\delta w_{ij} = k(x_j - w_{ij})y_i \qquad\qquad 8$$

This allows active output neurons to increase their synaptic weight if the input x_j is above the existing weight and to decrease the synaptic weight if the input is below the existing weight. (The fact that the weight is subtracted from the presynaptic firing rate in Equation 8 captures the fact that LTD is easier to obtain after long-term potentiation [LTP] has been produced— see Rolls & Deco, 2002.) The weights are clipped to be non-negative. The simulation results using this learning rule and the linear activation function are presented in Figure 5a. After an initial period of learning, there is an almost linear increase in the firing as a function of the number of training trials in the range 0–400, with saturation of the mean neuronal activity to familiar stimuli occurring after that. The response to the novel stimuli remains low. The sparseness was 0.01.

In Figure 5b we show that the model still works well at a higher loading of 0.25. In Figure 5c we show that the model works well with a sigmoid activation function, in which beta = 15 and alpha = 0.7, so that the success of the model is especially related to the modified learning rule, not to the particular activation function chosen.

We then analysed the capacity of the network as the number of patterns stored in the network grows. The capacity in which we are interested is the capacity to respond differently to familiar stimuli after many training trials than to novel stimuli. The results are presented in Figure 6. The training rule shown in Equation 8 was used with a linear activation function. With a loading of 1.0, N patterns are trained when there are N synapses per neuron. (A loading of 1.0 corresponds to the case of training the network with 1,000 random familiar patterns

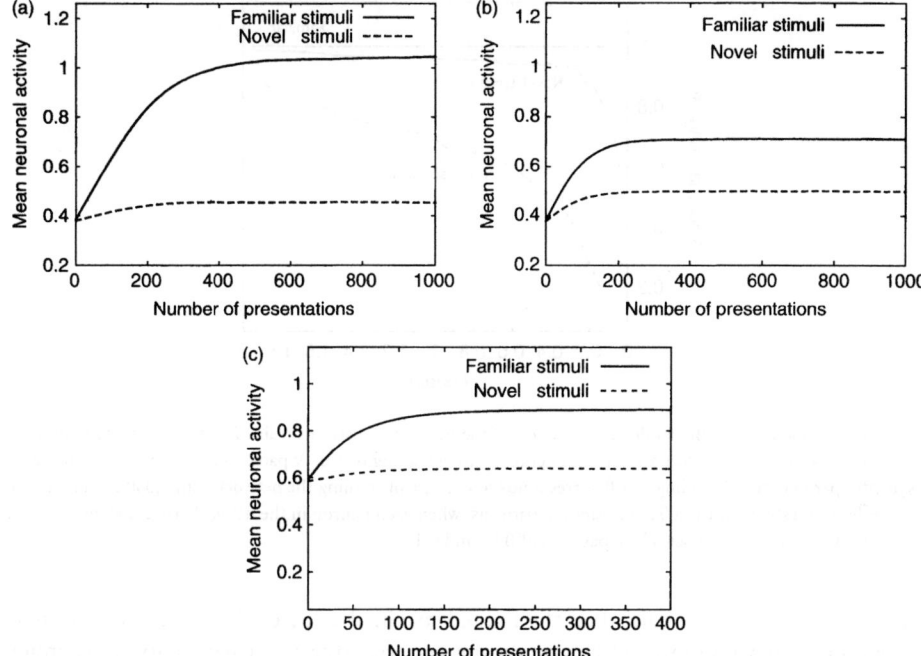

Figure 5. a. (Top left panel) An approximately linear increase in the firing rate to the familiar patterns as a function of the number of training trials was produced when the network was trained with the learning rule shown in Equation 8 and a linear activation function. The firing rate saturated at a high value after approximately 500 training trials, and when this occurs is a function of the learning rate parameter k in Equation 8. The loading was 0.1, and the sparseness a was 0.01. The familiar set of stimuli was initially novel, and the solid curve shows how the response to this set of stimuli gradually increases as the set become familiar over approximately 400 training (i.e., weight modification) presentations. For comparison, the responses to novel stimuli, which were not trained and remained novel, are also shown (dotted curve).
b. (Top right panel) The same simulation as that in Figure 5a, but with a higher loading of 0.25.
c. (Bottom panel) The same simulation as that in Figure 5a, but with a sigmoid activation function, in which beta = 15 and alpha = 0.7.

and testing with 1,000 novel random patterns, when each neuron in the network has 1,000 inputs.) The results are shown in Figure 6 for sparsenesses values of 0.01 and 0.1. (The random initial synaptic weights were initialized to the mean asymptotic value.) Given that the actual neurons in the perirhinal cortex respond to novel stimuli with approximately 50% of the firing rate response to familiar stimuli (see Figure 1 and Hölscher et al., 2003), the model operates within approximately the correct area for a sparseness of the patterns of 0.01 up to a loading of approximately 1. With a sparseness of 0.1 of the binary patterns, the network can be loaded less, up to a loading value of approximately 0.1.

Discussion

The analysis and model show that neurons that gradually increase their response approximately linearly over 400 or more presentations of a stimulus cannot be accounted for by a Hebb rule with a linear activation function. If a sigmoid activation function is used with the

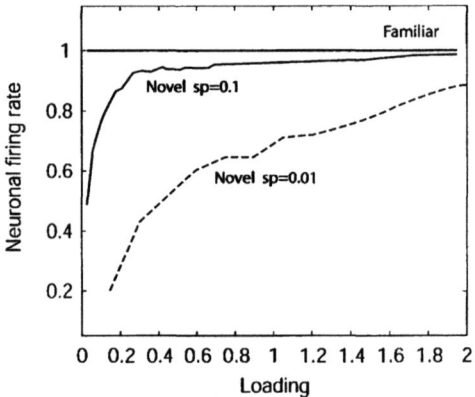

Figure 6. The operation of the model as a function of the number of patterns trained. The training rule shown in Equation 8 was used with a linear activation function. With a loading of 1.0, N patterns are trained when there are N synapses per neuron. (A loading of 1.0 corresponds to the case of training the network with 1,000 random familiar patterns and testing with 1,000 novel random patterns, when each neuron in the network has 1,000 inputs.) The results are shown in sparsenesses of the patterns of 0.01 and 0.1.

same associative learning rule, then the limited maximal value of the firing rate results in only a limited maximum weight change on any one trial, due to y_i having a fixed maximum value in Equation 4. With a relatively low learning rate, gradually more and more neurons have synaptic weight increases that result in the neurons reaching their maximal firing rate, as shown after 400 trials in Figure 4a. After that point in the learning, although the firing rates are at a maximal value to the familiar stimuli, the synaptic weights will continue to increase without bound as further learning trials of the familiar stimuli occur. Unbounded synaptic weights are a well-known problem with a simple associative Hebb learning rule. In this situation, to the extent that the random novel patterns overlap with the familiar patterns, the neuronal responses to the novel patterns will continue to increase after Trial 400.

A more attractive model, because the synaptic weights are self-limiting, is therefore a model with heterosynaptic long-term depression of the type shown in Equation 8 in the learning rule. In this case, as illustrated in Figure 4b, the firing rates to the familiar stimuli gradually saturate to a fixed high value at which the synaptic weights to the familiar stimuli are bounded to within a range set by the presynaptic firing rate, as shown in Equation 8 (the term $x_j - w_{ij}$). At the same time, the firing rates to the novel stimuli remain relatively low, because the values of the synapses that are activated by the novel stimuli tend to remain low as a result of the heterosynaptic long-term depression. In this case, the activation function can be linear, threshold linear, or a sigmoid activation function that approximates a threshold linear activation function. Because the modified Hebb rule shown in Equation 8 is self-limiting, and simulations with it can approximate the neuronal data showing long-term familiarity effects in the perirhinal cortex, this simple model is what we propose could implement the neurophysiological results.

With respect to the model, it is also the case that sparse input representations (in each of which few neurons have high firing rates) will allow larger numbers of stimuli to be in the familiar and novel sets, as shown in Figure 6. Indeed, the loading effects shown in Figure 6 show that the number of patterns in the novel and familiar sets that can be discriminated is in the order of the number of synapses per neuron for sparse representations, as is typical for associative networks (Rolls & Treves, 1990, 1998). Further, we note that the capacities for the network scale with the number of synapses per neuron. Further, if there is sparse connectivity from the set of input neurons x_j to the output neurons y_i, it will be possible to load the whole system to higher capacities, because in effect some patterns will tend to be learned more by some neurons, and other patterns more by other neurons. Indeed, this effect could account for why many neurons, rather than just one or very few neurons, are present in the perirhinal cortex. (These neurons are the y neurons in the model.)

We do not know of other models of how long-term familiarity-related neuronal responses could develop in the perirhinal cortex. Bogacz, Brown, and Giraud-Carrier (2001) modelled how, in what is effectively a short-term memory test, the responses to a novel visual stimulus can be large the first time that the stimulus is shown, but small the second time that the same stimulus is shown. That model thus is not of familiarity-related neurons, in that the neurons in that model respond more to novel stimuli. Also, that model is of a short-term memory process that occurs in one trial. (We note incidentally that that model cannot by itself account for the active resetting of perirhinal cortex neurons in a delayed match-to-sample short-term memory task, in that the responses to the sample stimulus are larger than those to the following match stimulus, even if the sample has been seen very recently of the immediately preceding trial, Hölscher & Rolls, 2002.) In contrast, the model described here is of how perirhinal cortex neurons can gradually increase their responses to stimuli as they become familiar over several hundred presentations. McLaren and Mackintosh (2002) discuss how a number of phenomena in associative learning might arise in systems with distributed representations. We note that the modified Hebb rule we use in Equation 8 (used by Willshaw & von der Malsburg, 1976, and developed by Oja, 1982) has self-limiting properties similar to those of an associative learning rule used by Vogel, Brandon, and Wagner (2003).

With respect to the neurophysiological discovery of perirhinal cortex neurons with activity related to the long-term familiarity of stimuli, with the familiarity building over hundreds of trials (Hölscher et al., 2003), we note that such a time scale has not been investigated previously, since most studies did not record neuronal responses for longer than 24 h and did not allow a slow emergence of increased neuronal responses to images related to their long-term familiarity to be observed (Brown & Xiang, 1998; Erickson & Desimone, 1999; Xiang & Brown, 1998). Although some neurons responding more to familiar than to novel images, and others responding more to novel than familiar images, have been found previously in the perirhinal cortex where novelty and familiarity refer to changes that occur over a few stimulus presentations (Brown & Xiang, 1998; Hölscher & Rolls, 2002; Sobotka & Ringo, 1993), the neurons described by Hölscher et al. (2003) had responses related to a different, long-term type of familiarity in which the increased neuronal responses to familiar images can take days or weeks to develop.

The perirhinal cortex is well placed to form such long-term familiarity representations, because it receives stimulus-selective information about what object is being viewed from the inferior temporal visual cortex. Indeed, it has been shown in a tracer study that the anteroventral part of area TE (TEav) projects diffusely over a wide extent of perirhinal cortex (Saleem & Tanaka, 1996), making the perirhinal cortex anatomically suited for making associations between features in an object (Bussey, Saksida, & Murray, this issue) or the objects in a scene, or computing a general property of all inputs being received, such as how familiar they are. The perirhinal cortex can thus use the stimulus-selective input from potentially most parts of area TE of the inferior temporal visual cortex to form new associations between such selective inputs and thereby to form a unique representation of complex stimuli to identify familiar objects or scenes.

Xiang and Brown (1998) found that in a recognition memory task with trial-unique visual stimuli, the perirhinal cortex neurons tended to respond more to the first than to the second presentation of the stimuli. With stimuli that were already familiar, even this difference was not clear. This may be because the neurons respond at their highest rates to stimuli that are highly familiar, allowing less room for a smaller response to a match than to a nonmatch stimulus in a recognition memory task.

What advantages might the representation of the long-term familiarity of images, a model for which is reported for the first time in this paper, confer? The potential functions of computing the long-term familiarity of objects or images in the brain are many-fold, and include recognition of complex object–environment configurations such as members of one's own social and family group, recognition of one's own possessions, recognition of one's own territory, and so on. Further, it is notable that the loss of the feeling of familiarity for objects and events introduced after medial temporal lobe damage is one of the important symptoms of medial temporal lobe amnesia (Squire, Stark, & Clark, 2004), and this too may be related to these operations that we suggest are being performed by the perirhinal cortex (Henson, this issue; Holdstock, this issue). What we propose is that the identity of the object or face would be represented by area TE of the inferior temporal visual cortex (in the way reviewed by Rolls & Deco, 2002) and that the long-term familiarity of the object would be represented by how strongly the perirhinal cortex neurons analysed here are firing. Together, the two types of neuronal activity encode both identity and long-term familiarity, but allow each type of information to be read out from the system (by other brain areas) independently of the other.

REFERENCES

Baylis, G., & Rolls, E. T. (1987). Responses of neurons in the inferior temporal cortex in short term and serial recognition memory tasks. *Experimental Brain Research, 65,* 614–622.

Bogacz, R., Brown, M. W., & Giraud-Carrier, C. (2001). Model of familiarity discrimination in the perirhinal cortex. *Journal of Computational Neuroscience, 10,* 5–23.

Brown, M., & Xiang, J. (1998). Recognition memory: Neuronal substrates of the judgement of prior occurrence. *Progress in Neurobiology, 55,* 149–189.

Buckley, M. J., Booth, M. C. A., Rolls, E. T., & Gaffan, D. (2001). Selective perceptual impairments following perirhinal cortex ablation. *Journal of Neuroscience, 21,* 9824–9836.

Buckley, M. J., & Gaffan, D. (2000). The hippocampus, perirhinal cortex and memory in the monkey. In J. J. Bolhuis (Ed.), *Brain, perception, memory* (pp. 279–298). Oxford, UK: Oxford University Press.

Bussey, T. J., Saksida, L. M., & Murray, E. A., (this issue). The perceptual-mnemonic/feature conjunction model of perirhinal cortex function. *Quarterly Journal of Experimental Psychology, 58B,* 269–282.

Erickson, C., & Desimone, R. (1999). Responses of macaque perirhinal neurons during and after visual stimulus association learning. *Journal of Neuroscience, 19,* 10404–10416.

Henson, R. (this issue). A mini-review of fMRI studies of human medial temporal lobe activity associated with recognition memory. *Quarterly Journal of Experimental Psychology, 58B,* 340–360.

Holdstock, J. S. (this issue). The role of the human medial temporal lobe in object recognition and object discrimination. *Quarterly Journal of Experimental Psychology, 58B*, 326–339.

Hölscher, C., & Rolls, E. T. (2002). Perirhinal cortex neuronal activity is actively related to working memory in the macaque. *Neural Plasticity, 9*, 41–51.

Hölscher, C., Rolls, E. T., & Xiang, J.-Z. (2003). Perirhinal cortex neuronal activity related to long-term familiarity memory in the macaque. *European Journal of Neuroscience, 18*, 2037–2046.

Malkova, L., Bachevalier, J., Mishkin, M., & Saunders, R. C. (2001). Neurotoxic lesions of perirhinal cortex impair visual recognition memory in rhesus monkeys. *Neuroreport, 12*, 1913–1917.

McLaren, I. P., & Mackintosh, N. J. (2002). Associative learning and elemental representation: II. Generalization and discrimination. *Animal Learning & Behavior, 30*,177–200.

Miller, E.K., Li, L., & Desimone, R. (1998). Activity of neurons in anterior inferior temporal cortex during a short-term memory task. *Journal of Neuroscience, 13*, 1460–1478.

Miyashita, Y., Kameyama, M., Hasegawa, I., & Fukushima, T. (1998). Consolidation of visual associative long-term memory in the temporal cortex of primates. *Neurobiology of Learning and Memory, 1*, 197–211.

Miyashita, Y., Okuno, H., Tokuyama, W., Ihara, T., & Nakajima, K. (1996). Feedback signal from medial temporal lobe mediates visual associative mnemonic codes of inferotemporal neurons. *Cognitive Brain Research, 5*, 81–86.

Oja, E. (1982). A simplified neuron model as a principal component analyzer. *Journal of Mathematical Biology, 15*, 267–73.

Rolls, E. T. (2000). Functions of the primate temporal lobe cortical visual areas in invariant visual object and face recognition. *Neuron, 27*, 205–218.

Rolls, E. T., & Deco, G. (2002). *Computational neuroscience of vision.* Oxford, UK: Oxford University Press.

Rolls, E. T., & Treves, A. (1990). The relative advantages of sparse versus distributed encoding for associative neuronal networks in the brain. *Network, 1*, 407–421.

Rolls, E. T., & Treves, A. (1998). *Neural networks and brain function.* Oxford, UK: Oxford University Press.

Rolls, E. T., Treves, A., & Tovee, M. J. (1997a). The representational capacity of the distributed encoding of information provided by populations of neurons in the primate temporal visual cortex. *Experimental Brain Research, 114*, 149–162.

Rolls, E. T., Treves, A., Tovee, M., & Panzeri, S. (1997b). Information in the neuronal representation of individual stimuli in the primate temporal visual cortex. *Journal Computational Neuroscience, 4*, 309–333.

Saleem, K. S., & Tanaka, K. (1996). Divergent projections from the anterior inferotemporal area TE to the perirhinal and entorhinal cortices in the macaque monkey. *Journal of Neuroscience, 16*, 4757–4775.

Sobotka, S., & Ringo, J. (1993). Investigation of long term recognition and association memory in unit responses from inferotemporal cortex. *Experimental Brain Research, 96*, 28–38.

Squire, L. R., Stark, C. E., & Clark, R. E. (2004). The medial temporal lobe. *Annual Review of Neuroscience, 27*, 279–306.

Suzuki, W., & Amaral, D. (1994a). Perirhinal and parahippocampal cortices of the macaque monkey: Cortical afferents. *Journal of Comparative Neurology, 350*, 497–533.

Suzuki, W., & Amaral, D. (1994b). Topographic organization of the reciprocal connections between the monkey entorhinal cortex and the perirhinal and parahippocampal cortices. *Journal of Neuroscience, 14*, 1856–1877.

Vogel, E. H., Brandon, S. E., & Wagner, A. R. (2003). Stimulus representation in SOP: II. An application to inhibition of delay. *Behavioral Processes, 62*, 27–48.

Willshaw, D. J., & von der Malsburg C. (1976). How patterned neural connections can be set up by self-organization. *Proceedings of the Royal Society of London. Biological Sciences, 194*, 431–45.

Wilson, F. A., Riches, I. P., & Brown, M. W. (1990). Hippocampus and medial temporal cortex: Neuronal activity related to behavioural responses during the performance of memory tasks by primates. *Behavioral Brain Research, 40*, 7–28.

Xiang, J., & Brown, M. (1998). Differential neuronal encoding of novelty, familiarity and recency in regions of the anterior temporal lobe. *Neuropharmacology, 37*, 657–676.

Zola-Morgan, S., Squire, L. R., Amaral, D. G., & Suzuki, W. A. (1989). Lesions of perirhinal and parahippocampal cortex that spare the amygdala and hippocampal formation produce severe memory impairment. *Journal of Neuroscience, 9*, 4355–4370.

Zola-Morgan, S., Squire, L. R., & Ramus, S. J. (1994). Severity of memory impairment in monkeys as a function of locus and extent of damage within the medial temporal lobe memory system. *Hippocampus, 4*, 483–494.

THE QUARTERLY JOURNAL OF EXPERIMENTAL PSYCHOLOGY,
2005, 58B (3/4), 246–268

The role of the perirhinal cortex and hippocampus in learning, memory, and perception

Mark J. Buckley

Oxford University, UK

One traditional and long-held view of medial temporal lobe (MTL) function is that it contains a system of structures that are exclusively involved in memory, and that the extent of memory loss following MTL damage is simply related to the amount of MTL damage sustained. Indeed, human patients with extensive MTL damage are typically profoundly amnesic whereas patients with less extensive brain lesions centred upon the hippocampus typically exhibit only moderately severe anterograde amnesia. Accordingly, the latter observations have elevated the hippocampus to a particularly prominent position within the purported MTL memory system. This article reviews recent lesion studies in macaque monkeys in which the behavioural effects of more highly circumscribed lesions (than those observed to occur in human patients with MTL lesions) to different subregions of the MTL have been examined. These studies have reported new findings that contradict this concept of a MTL memory system. First, the MTL is not exclusively involved in mnemonic processes; some MTL structures, most notably the perirhinal cortex, also contribute to perception. Second, there are some forms of memory, including recognition memory, that are not always affected by selective hippocampal lesions. Third, the data support the idea that regional functional specializations exist within the MTL. For example, the macaque perirhinal cortex appears to be specialized for processing object identity whereas the hippocampus may be specialized for processing spatial and temporal relationships.

What is so interesting about the perirhinal cortex?

When one considers the cerebral cortex and the cognitive functions it mediates, it would not be that surprising if the significance of a small cortical region located in the anterior medial part of the temporal lobe, known as the perirhinal cortex, was overlooked as this region accounts for only approximately 0.5% of the total human cortical volume. Nevertheless, despite its relatively small size it has become increasingly apparent over the last decade that

Correspondence should be addressed to Dr Mark J. Buckley, Department of Experimental Psychology, University of Oxford, South Parks Road, Oxford OX1 3UD, UK. Email: mark.buckley@psy.ox.ac.uk

DOI:10.1080/02724990444000186

the perirhinal cortex is likely to be the most crucial cortical region that supports object recognition memory (a role previously attributed to the hippocampus). However, whilst its contribution to recognition memory is now widely accepted, its contribution to other forms of memory as well as perception remains a controversial issue. One influential view of medial temporal lobe (MTL) functional organization considers the MTL (composed of the hippocampus, amygdala, and perirhinal, entorhinal, and parahippocampal cortices) to constitute a functional homogenous memory system, which plays no part whatsoever in perception (Buffalo, Ramus, Squire, & Zola, 2000; Buffalo, Reber, & Squire, 1998a; Buffalo, Stefanacci, Squire, & Zola, 1998b; Squire & Zola-Morgan, 1991; Zola-Morgan, Squire, & Ramus, 1994). However, a growing body of recent research appears to indicate, to the contrary, that not only are there functional subdivisions within the MTL but that the perirhinal cortex does indeed contribute to perception (reviewed in Buckley & Gaffan, 2000; Murray & Bussey, 1999; and Murray & Richmond, 2001). Our understanding of the functional architecture of the MTL and neural mechanisms supporting memory and perception depends crucially upon resolving these controversies, and so this article examines the conflicting data in some detail to assess whether the opposing data-sets can yet be reconciled. Beyond this theoretical debate, understanding the role of this region of the brain is important for other reasons. For example, a more precise characterization of the role of the perirhinal cortex may pave the way to the development of more sensitive clinical diagnostic tests that could potentially provide an earlier diagnosis for Alzheimer's disease, in which early neuropathological changes are seen to occur in the rhinal cortical region (Arnold, Hyman, & Van Hoesen, 1994; Braak & Braak, 1985; Van Hoesen et al., 1991). This research may also help guide research into treatments for this disease, and other conditions, which affect the normal function of the MTL.

Where is the perirhinal cortex located, and what are its connections?

The perirhinal cortex (which corresponds to Brodmann's areas 35 and 36) is located in the anterior and medial part of the ventral aspect of the temporal lobe (see Figure 1). It is important to appreciate the pattern of anatomical connectivity of the perirhinal cortex in order to better appreciate its role. These are reviewed briefly here (for a more extensive review, see Suzuki, 1996a). The most prominent input into the perirhinal cortex arises from laterally adjacent inferotemporal cortex (unimodal visual areas TE and TEO). Substantial input also arises from polymodal parahippocampal cortex (areas TH and TF) and to a lesser extent from a wide range of other polymodal areas including dorsal superior temporal sulcus, orbitofrontal cortex, and cingulate cortex. Unimodal inputs also arise from somatosensory insular cortex and auditory superior temporal gyrus. In short, the perirhinal cortex receives convergent sensory information from a range of unimodal and polymodal cortical areas and is therefore well placed to integrate complex sensory information suitable for mediating one of its proposed roles—namely, representing objects. The perirhinal cortex, also has prominent interconnections with the entorhinal cortex, to which it is robustly and mutually innervated. The entorhinal cortex in turn serves as the source of the perforant pathway into the hippocampal system and is in receipt of return projections from the hippocampus suggesting that the influence is bidirectional. Visuo-spatial information from the dorsal visual

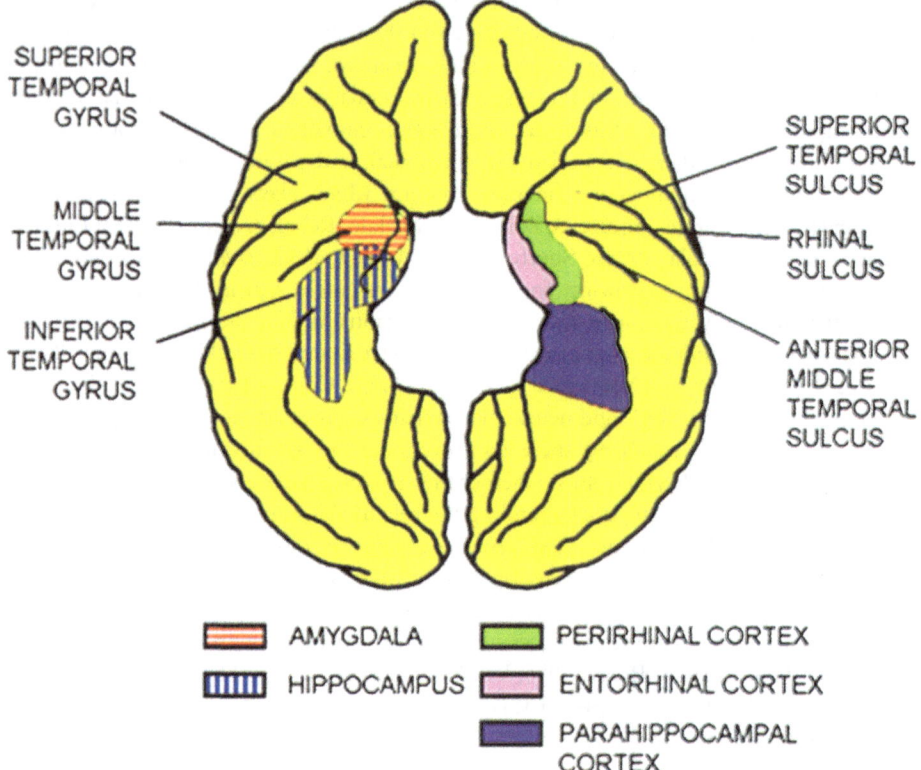

SUPERIOR
TEMPORAL
GYRUS

MIDDLE
TEMPORAL
GYRUS

INFERIOR
TEMPORAL
GYRUS

SUPERIOR
TEMPORAL
SULCUS

RHINAL
SULCUS

ANTERIOR
MIDDLE
TEMPORAL
SULCUS

AMYGDALA   PERIRHINAL CORTEX

HIPPOCAMPUS   ENTORHINAL CORTEX

PARAHIPPOCAMPAL
CORTEX

Figure 1.   Schematic drawing of a ventral view of a macaque brain; the gyri and sulci, which are visible in this view of the MTL, are labelled on the left and right side of the figure, respectively. In addition, the relative location and extent of the hippocampus and amygdala within the MTL are also depicted on the left side of the figure, and the location and extent of the underlying cortical fields—namely, the perirhinal, entorhinal, and parahippocampal cortex—are shown on the right of the figure. The extent of the perirhinal cortex shown on this figure is conservative and matches the extent of the perirhinal cortex targeted in our own lesion studies; however, it has been suggested, on the basis of connectional grounds, that the lateral extent of the perirhinal cortex may even extend further laterally, to the anterior middle temporal sulcus (Amaral, Insausti, & Cowan, 1987; Insausti, Amaral, & Cowan, 1987; Suzuki, Zola-Morgan, Squire, & Amaral, 1993).

stream regions may project into the MTL via connections between the posterior parietal cortex and the parahippocampal cortex, which in turn innervates the rhinal cortex (Lavenex, Suzuki, & Amaral, 2004). Thus, spatial information (including any that may be processed in the hippocampus itself) has the means to interact with highly processed information in the perirhinal cortex regarding the identity of stimuli; the association of such information across multiple fixations may contribute to the representation of entire scenes. The fornix provides another major input and output pathway of the hippocampus. Prominent outputs include projections to other regions implicated in memory such as the mammillary bodies and anterior nucleus of the thalamus. The fornix also contains other fibres including cholinergic fibres from the basal forebrain, some of which do not synapse in the hippocampus itself but project to the entorhinal cortex directly. It has been hypothesized that these cholinergic fibres may provide part of a reinforcement signal for learning of information processed in the MTL cortex. Consistent with this, interruption of all of the cholinergic projections from the basal forebrain into the MTL leads to dense amnesia in macaques (Easton, Ridley, Baker, & Gaffan, 2002; Gaffan, Parker, & Easton, 2001), and fornix transection itself impairs learning but not recall of spatial information (Buckley, Charles, Browning, & Gaffan, 2004a). Finally, the perirhinal cortex also has strong interconnections with amygdala, which may be important for learning about the reward value of complex stimuli such as food items and other important objects, including objects and faces that have already been processed and identified to a high degree by the perirhinal cortex (Malkova, Gaffan, & Murray, 1997).

Human MTL amnesia

In the 1950s a number of patients suffering from severe epilepsy received neurosurgical MTL excisions as a treatment for their condition (Milner, 1958; Scoville & Milner, 1957); whilst this relieved the severity of their epilepsy, it left the patients profoundly amnesic. The most famous of these patients is known by his initials H.M., whose memory deficits and spared cognitive abilities have been probed extensively ever since by neuropsychologists (see Corkin, 2002, for a recent review). Although H.M.'s perceptual and intellectual abilities were unchanged, H.M. was no longer able to acquire new long-term memories about consciously experienced stimuli and events. Similarly, while H.M. was also able to learn to perform various motor tasks and improve upon his performance with practice, he was unable to recollect having done the tasks before. In short, the pattern of spared and intact abilities in H.M. following his bilateral MTL excision suggested that memory may be a distinct cerebral function dissociable from other perceptual and cognitive processes, and that memory itself may be subdivided into various different mnemonic processes.

The development of animal models of human amnesia

Animal models of human amnesia have continued to evolve ever since H.M. became amnesic. To pursue the evolutionary analogy further, just as speciation itself appears to follow a course of punctuated equilibrium, changes in textbook accounts of memory (quite aptly analogous to the fossil-record) also exhibit considerable stasis prior to the emergence of new accepted forms. There are at least three plausible reasons to account for this state of affairs. First, it is natural that the development of new investigative techniques that overcome some of the limitations of older techniques will lead to conceptual advances. For example, the development of neurotoxic lesion techniques allowed cell bodies in a region of interest to be targeted without also damaging nonsynapsing fibres-of-passage (axons) that may course through a region (which are damaged with aspiration or electro-cautery lesion techniques). Second, if the tools for the job are not optimum then progress will be slow; thus another reason why this article argues that animal models of human amnesia require updating is that the choice of behavioural tasks settled upon as the benchmarks for assessing amnesia in developing these animal models in the past may not have been the most appropriate. Two of the tasks employed most extensively to assess the extent of memory loss following MTL lesions in animals are recognition memory (either delayed matching-to-sample, DMS, or delayed nonmatching-to-sample, DNMS). DMS and DNMS tasks proceed by requiring the subject to view a sample stimulus (or a list of sample stimuli), and then after a delay in which no stimuli are presented, two choice stimuli are provided between which the subject must choose to express its recognition of familiarity. In concurrent discrimination learning on the other hand, there are generally two stimuli in each trial, both of which are equally familiar, and the subject is rewarded by choosing the one stimulus out of that pair that is associated with reward (the object–reward associations are learnt by trial and error with progressive runs through the list of problems in the concurrent learning set). Despite discrimination learning being the more mnemonic of the two tasks (as it requires associative memory in addition to stimulus recognition) recognition memory became the preeminent task for assessing amnesia in nonhuman primates. Third, new theories generally only gain popular acceptance after previous hypotheses are no longer tenable in the face of mounting contradictory evidence. Perhaps understandably then, the more popular and influential a theory, the stronger its resistance to change.

Early animal models

Early attempts to model the memory deficits in patients like H.M. proved on the whole to be unsuccessful in that monkeys that received similarly extensive MTL lesions were found not to be greatly impaired in stimulus memory (Correll & Scoville, 1965a, 1965b, 1967; Orbach, Milner, & Rasmussen, 1960). However, subsequent modifications to the design of the behavioural tasks, including the adoption of trial–unique stimuli and the introduction of performance tests to further tax memory, led to considerable progress in the development of the models (Gaffan, 1974; Mishkin & Delacour, 1975). One particularly influential early study (Mishkin, 1978) assessed recognition memory over a range of delay lengths and list lengths of sample objects and found that monkeys with either amygdala or hippocampal aspiration lesions alone could perform the task almost as well as control monkeys (see Figure 2, panel A). However, significant performance deficits emerged in monkeys with combined amygdala and hippocampal lesions at longer delay lengths (see Figure 2, panel A). It was concluded that conjoint amygdala and hippocampal damage led to recognition memory impairment in monkeys, and accordingly that H.M., having extensive damage to both structures, was also amnesic for the same reason.

Around the same time an alternative account for H.M.'s memory deficit was proposed by Horel (1978) who suggested that damage to the white-matter fibres in the temporal lobe (known as the temporal stem) was the primary cause of H.M.'s amnesia. The temporal stem covers the dorsal and lateral aspect of the inferior horn of the lateral ventricle and amygdala, and sectioning these fibres in monkeys, using a frontal surgical approach similar to that used by Scoville during H.M.'s surgery, had been found to impair the learning and retention of visual discriminations (Horel & Misantone, 1976). However, Zola-Morgan, Squire, and Mishkin (1982) rejected this challenge to the amygdala-hippocampus account of amnesia as their data showed that while macaques with temporal stem section performed normally on DNMS but were impaired upon the visual discrimination learning task, amygdala-hippocampal lesioned animals showed the reversed pattern of deficits. Subsequent confirmation that monkeys with large aspiration lesions of the amygdala and hippocampus could

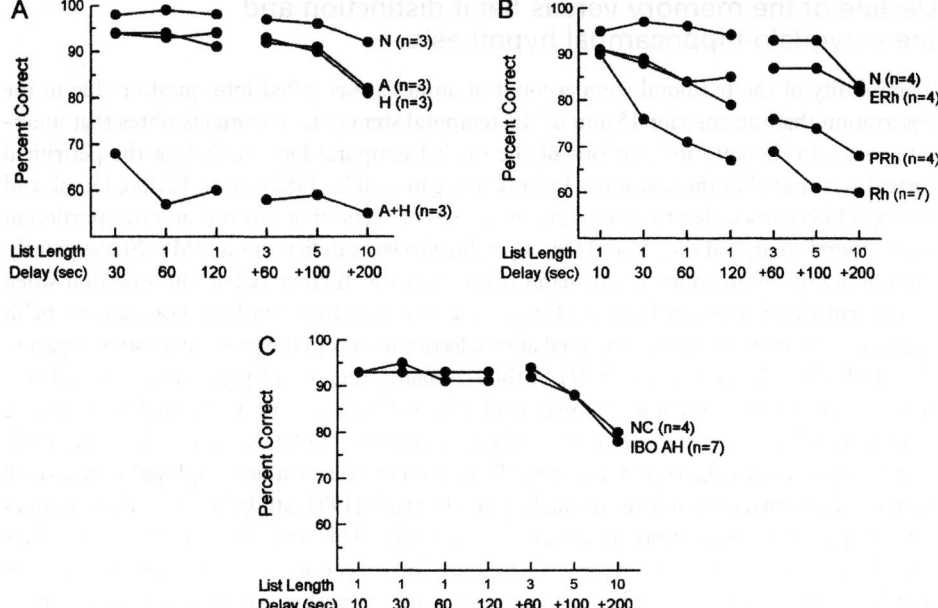

Figure 2.   Panel A replots data originally published in Mishkin (1978) regarding the performance of four groups of macaques on DNMS; the groups indicated are normal controls (N), amygdala aspiration lesions (A), hippocampal aspiration lesions (H), and combined amygdalo–hippocampal aspiration lesions (A+H). Panel B replots data originally published in Meunier et al. (1993) regarding the performance of another four groups of macaques on DNMS; the groups indicated are normal controls (N), entorhinal cortex aspiration lesions (ERh), perirhinal cortex aspiration lesions (PRh), and combined amygdalo–hippocampal aspiration lesions (Rh). Panel C replots data originally published in Murray and Mishkin (1998) regarding the performance of two further two groups on DNMS; the groups indicated are normal controls (NC) and ibotenic acid neurotoxic lesions of the amygdala and hippocampus (IBO AH). For each groups in each study, the number of animals in the group are indicated by the value of *n* in parentheses.

learn without impairment a list of 20 concurrent discrimination problems with 24-hr inter-trial intervals (Malamut, Saunders, & Mishkin, 1984) gave credence to the theory that recognition memory and discrimination learning assessed distinctly different processes—namely, memory and habit learning, respectively (Mishkin, Malamut, & Bachevalier, 1984; Mishkin & Petri, 1984). According to this view, memories (i.e., stored associations between representations of events) were considered to rely upon an intact MTL memory system whereas habit learning (i.e., behavioural changes mediated by acquired direct associations between a stimulus and a response) was believed to be supported by interaction of the neocortex with the striatum. Thus the popular idea that memories and habits relied upon different brain systems was born, and the idea of a MTL-based memory system became firmly established (Squire & Zola-Morgan, 1991), and, accordingly, recognition memory became the task of choice for assessing amnesia.

Decline of the memory versus habit distinction and the amygdalo-hippocampal hypothesis

The validity of the temporal stem account of amnesia was called into question due to the observation that the anterior 15 mm of the temporal stem (which contains fibres that inner-vate cortex in the anterior portions of the medial temporal lobe, including the perirhinal cortex) was spared in the macaques in Zola-Morgan et al.'s (1982) study. Cirillo, Horel, and George (1989) proceeded to investigate the effects of sectioning just this anterior portion of the temporal stem, and they found that it produced a large deficit upon DMS. Subsequently, the validity of the memory versus habit distinction was further called into question when concurrent discrimination learning (i.e., a task conceptualized to be a nonmemory habit learning task) was shown to be impaired after selective lesions to the perirhinal cortex, a part of the MTL (Buckley & Gaffan, 1997). Although many studies had previously reported that recognition memory, but not concurrent discrimination learning, was impaired following lesions to MTL structures (Eacott, Gaffan, & Murray, 1994; Gaffan & Murray, 1992; Zola-Morgan, Squire, Clower, & Rempel, 1993), each of these studies employed only a small number of concurrent problems. In Buckley and Gaffan's (1997) study, however, the macaques were required to learn more problems concurrently. This manipulation makes the task considerably more taxing in terms of stimulus identification as discriminating between members of a larger stimulus set requires each stimulus to be represented more precisely (for example, by requiring particular configurations of elemental stimulus features to be encoded). Around the same time it was shown that neo-cortical interaction with the striatum (i.e., the pathway that was believed to support habit learning) could also support memory (Gaffan, 1996), indicating that there was no longer any convincing reason to distinguish between the neural substrates of memory and habit learning systems on the basis of differential impairments on recognition memory and concurrent discrimination learning tasks.

Furthermore, the theory that conjoint amygdala and hippocampal damage was necessary to impair recognition memory was also facing challenges from increasing numbers of experiments that reported impairments in recognition memory in lesioned animals with either the amygdala or hippocampus left intact (Mahut, Zola-Morgan, & Moss, 1982; Murray & Mishkin, 1986; Zola-Morgan & Squire, 1986). It was clear that a closer anatomical analysis of previous lesions made to the MTL region was required to resolve the matter.

The rhinal cortex hypothesis

Figure 1 illustrates that whereas the anterior part of the rhinal cortex underlies the amyg-dala, the posterior part of the rhinal cortex underlies the hippocampus. During aspiration lesions of the amygdala, while some anterior rhinal cortex is damaged the posterior rhinal cortex is left intact; conversely, during hippocampal aspiration lesions, while some posterior rhinal and parahippocampal cortex is damaged the anterior rhinal cortex is left intact. Combined amygdala and hippocampus lesions, however, would lead to extensive rhinal and some parahippocampal cortical damage together with disruption of efferents from the region. Thus the possibility remained then that damage to the underlying cortex itself, rather than damage to the amygdala and hippocampus, was responsible for the severe recognition memory impairments that followed combined amygdala and hippocampus ablation (Mishkin, 1978). Conclusive evidence that damage to the underlying cortical areas was necessary and sufficient to produce recognition memory impairments was finally provided by Meunier, Bachevalier, Mishkin, and Murray (1993) who showed that combined perirhinal and entorhinal ablation, which left the amygdala and hippocampus intact, led to deficits almost as severe as those that followed Mishkin's combined amygdala and hippocampal lesions (compare panels A and B in Figure 2). Of these underlying cortical areas the perirhinal cortex appeared to be the most crucial; indeed, lesions to the entorhinal cortex alone result in only mild or transient effects (Leonard, Amaral, Squire, & Zola-Morgan, 1995; Meunier et al., 1993). Further confirmation was provided by Murray and Mishkin (1998) who showed that ibotenic acid lesions of the amygdala and hippocampus alone (without damage to underlying cortical regions) produced absolutely no impairment at all on recognition memory even when lists of sample objects and long delays up to 40 minutes were given (Figure 2, panel C).

How distinct is the role of the perirhinal cortex?

The pattern of impairments after perirhinal cortex lesions can be doubly dissociated from those following ablation of the middle temporal gyrus, which corresponds to the dorsal part of visual area TE. Buckley, Gaffan, and Murray, (1997) found that whereas lesions of perirhinal cortex impaired object recognition memory but not colour discrimination, middle temporal gyrus lesions showed the reversed pattern of relative deficits. Furthermore, the impairments following perirhinal lesions have also been doubly dissociated from those that follow fornix transection. Gaffan (1994a) showed that transecting the fornix led to marked impairment on simple spatial discrimination learning but only mild effects upon recognition memory for complex naturalistic scenes (in this case the scenes were individual still frames from a movie that were each judged to be distinctly different), whereas perirhinal cortex lesions, in contrast, resulted in severe impairments in the scene recognition task but spared simple spatial discrimination learning. These double dissociations show that the perirhinal cortex is functionally distinct from the structures to which it is connected both in the ventral visual processing stream and in the MTL (see also Aggleton & Brown, this issue; Eacott & Gaffan, this issue). The nature of the particular functional specialization of the perirhinal cortex, however, remained an issue of contention.

The memory versus perception debate

According to some authors, the macaque perirhinal cortex is exclusively involved in memory and does not contribute to stimulus perception (Buffalo et al., 1998b, 1999, 2000). Stimulus perception, according to this view, is dependent upon cortical areas earlier in the ventral processing stream, particularly TE and TEO. Therefore, to begin, the data presented in support of this view are examined in detail in this section.

For instance, it has been argued by Buffalo and colleagues (1998b) that that the extent of inadvertent damage to area TE (in monkeys that had received MTL lesions in their laboratory) correlated more positively than did the extent of perirhinal cortex damage with the number of errors made upon learning concurrent discriminations. However, this pattern of impairments is simply consistent with the idea that the particular concurrent discrimination learning task employed (with only eight pairs of objects) was just not sensitive enough to perirhinal cortex damage as it placed too little demand upon object identification. Indeed, one may reason that the entire set of 16 objects employed may be discriminated solely on the basis of simple feature differences of the kind known to be represented by TE neurons (Tanaka, 1996). As discussed earlier, increasing the set size of concurrent discrimination learning tasks has been found to reveal impairments after perirhinal cortex lesions (Buckley & Gaffan, 1997). Thus, Buffalo et al.'s (1998b) data can be more parsimoniously explained as the expected result of performance upon a TE-sensitive task being correlated with the extent of TE damage.

Buffalo et al. (1999) likewise concluded that the perirhinal cortex is important for memory formation whereas TE is important for visual perceptual processing on account of three observations: (a) that TE lesions impaired visual but not tactile recognition memory whereas perirhinal lesions impair both visual and tactile recognition memory, (b) that a visual paired comparison task was impaired at both long and short delays after TE lesions but only impaired at long delays after perirhinal lesions, and (c) that perirhinal cortex but not TE lesions impaired the learning of individual discrimination problems, whereas concurrent discrimination learning was impaired by TE lesions but not perirhinal lesions. However, none of these observations, either individually or together, is persuasive.

The first observation is simply attributable to perirhinal cortex, unlike TE, being a polymodal cortical area (Suzuki, 1996a, 1996b) imbuing it with the ability to represent polymodal feature conjunctions without necessarily indicating any greater mnemonic role for the perirhinal cortex. To respond to the second observation, the nature of the visual comparison task needs to be briefly explained. In this task, macaques are habituated to a visual stimulus, and then, after a delay, that stimulus and a novel stimulus item are presented concurrently, and the time spent looking at each stimulus is recorded. If the animal spends longer on average looking at the novel stimulus then this is taken to indicate that the monkey can recognize the familiar stimulus. The authors argue that with very short delays the importance of stimulus memory is lessened so that the differential looking found at short delays in the TE group indicates that TE but not the perirhinal cortex contributes to stimulus perception. However, even at the shortest delay (1 s) used in this task, this task does not provide an adequate test of perception as stimulus memory is still required to bridge this delay. In contrast, simultaneous matching-to-sample has no delay over which memory is required to bridge, and on this task rhinal cortex lesioned macaques are impaired (Eacott et al., 1994), providing positive support to the alternative idea that the rhinal cortex does contribute to stimulus perception. Furthermore, given that the aim of Buffalo et al.'s (1999) visual paired comparison task was to investigate lesioned monkeys' abilities to perceive differences between objects then very similar (i.e., hard to distinguish) stimulus objects should have been employed and not the high-contrast, black-and-white, two-dimensional shapes that were used. In effect, this version of the visual paired comparison task assesses the ability to discriminate simple features (of the kind represented in TE, not the perirhinal cortex), and Buffalo et al.'s data from this task are more parsimoniously explained as the entirely expected outcome of asking TE lesioned monkeys to discriminate stimuli of the kind that are represented in TE.

With respect to the third observation, the impairment on the (eight-pair) concurrent discrimination learning task after TE lesions (but not after perirhinal lesions) is attributable to the fact that TE (not the perirhinal cortex) represents the simple features that performance on this task depends upon as already discussed. The lack of impairment on single problem learning after TE lesions may be due to the stimuli in each problem being able to be distinguished on the basis of a simple feature (e.g., colour) of the kind that are represented even earlier in the ventral visual processing stream, in TEO (Tanaka, 1996), and which would remain available to guide discriminations in TE lesioned animals. Nevertheless, it is the impairments on single-problem discrimination learning after perirhinal lesions that are potentially most problematic for our hypothesis that perirhinal lesion effects depend upon the demands placed upon object identification. However, there are at least three circumstances in which such impairments would be expected to follow perirhinal cortex lesions (all of which may be contributing factors in Buffalo et al., 1999). First, proactive interference between stimuli will be expected in cases where numerous stimuli have already been experienced previously, irrespective of whether subsequent stimuli are presented singly or concurrently. Second, if the task has not been trained preoperatively then impairments in single-pair learning may be attributable to a failure to learn the rules of the task particularly if the monkeys have been trained on other different tasks previously. Third, if the stimuli are very similar so that there is extensive feature overlap making the stimuli hard to distinguish, then impairments after perirhinal cortex lesions will be expected even in discriminating single problems (Bussey, Saksida, & Murray, 2002; Bussey, Saksida, & Murray, this issue).

A more recent study by Buffalo and colleagues (2000) similarly concluded that the perirhinal cortex is important for memory and not perception on the basis that DNMS is impaired at short 0.5-s delays after TE but not after perirhinal lesions, whereas the perirhinal group, unlike the control group, were at chance at long 10-min delays. As discussed above, short delay conditions (0.5 s) do not increase the perceptual demands of the task, and the finding that TE lesions are more disruptive is merely consistent with TE having a role in processing the 'simple features' that distinguish the objects in the DNMS task. Furthermore, their data were not particularly consistent with the perirhinal cortex having a robust mnemonic role either, as no significant delay-dependent deficits were found after perirhinal lesions. Indeed, the performance of perirhinal lesioned monkeys in this study was not statistically different from that of the controls at any delay length tested, including the long 10-min delay; indeed, the best animal at the longest delay length tested was actually from the perirhinal lesioned group.

Thus, as can be seen, the data from macaque lesion studies that have been presented as evidence that the perirhinal cortex contributes exclusively to memory can be reinterpreted within the hypothesis that TE contributes to perception and memory of simple features and that the perirhinal cortex contributes to perception and memory of complex features or feature conjunctions. In the next section, the experimental data supporting the latter hypothesis are similarly presented in some detail.

Why the perirhinal cortex is now thought to contribute to both memory and perception

One of the earliest indications that the perirhinal cortex might not be exclusively involved in mnemonic processing was provided by Eacott et al. (1994). In this study, macaques with rhinal cortex lesions were found to be impaired at simultaneous matching-to-sample as well as in DMS. In addition, evidence from associative learning tasks also converges upon the view that the perirhinal cortex contributes to stimulus perception as well as to stimulus memory. Concurrent discrimination learning is found to be impaired after perirhinal lesions when the demands that the task places upon object identification (and the representation of feature conjunctions) are increased, for example by increasing the number of distracting stimuli or by increasing the number of problems (Buckley & Gaffan, 1997). Buckley and Gaffan (1998a) also showed that perirhinal lesioned macaques were impaired even in a 10-pair concurrent discrimination learning task if both stimuli in each trial were presented in one of three different views (chosen independently for each stimulus) in different trials of the same problem. This manipulation places higher demands upon object identification in a different way by requiring the subjects to recognize objects across different views in order to solve the individual problems.

In another experiment, Buckley and Gaffan (1998b) showed that the same macaques were subsequently impaired at recognizing these objects in views that, although similar, had not been seen before. This latter task was designed with the specific intention of increasing the demands upon stimulus identification (by requiring new views of the familiar objects to be recognized) without increasing the demands upon memory (the solutions to each problem being unchanged). Hampton and Murray (2002; Hampton, this issue) recently challenged the impaired stimulus identification interpretation of our data by suggesting that our use of a trials-to-criterion measure may have confounded generalization to new views (reflected in performance on the first exposure to a new view) and relearning (reflected in the rate at which accuracy increases towards the criterion level over sessions). Whether our impairment may be attributable wholly or partly to a deficit in relearning rather than a deficit entirely in stimulus identification may easily be evaluated by looking at what happens on the very first trial of each problem when the problems are presented with new views. This new analysis of the data from Buckley and Gaffan (1998b) is illustrated in Figure 3, which clearly shows that on the first block of trials, in which each of the 40 problems were presented for the very first time in new views, animals in the control group do not make any more errors than they made on preceding blocks with the old familiar views. Yet, in contrast, those in the perirhinal lesioned group make a significantly greater number of errors on the first block and then pro-ceed to improve (as shown by the regression line). Therefore the data from Buckley and Gaffan (1998b) can be interpreted as evidence for an impairment in stimulus identification or perception as we originally stated and is not attributable to an impairment in relearning as suggested by Hampton and Murray (2002).

Animals with perirhinal lesions were also tested on their ability to learn two more tasks: configural learning (a task also referred to as biconditional discrimination learning task) and paired-associate learning (Buckley & Gaffan, 1998c). In configural learning the macaques were required to learn eight problems concurrently. Each problem consisted of two compound stimuli, one on the left of the screen and one on the right (the compound stimuli were each comprised of two coloured alpha–numeric elemental stimuli abutted together. These problems were arranged so that they could only be solved by learning the association between particular compounds of elemental stimuli and reward as the problems were counter-balanced so that each elemental stimulus appeared equally in rewarded and unrewarded compounds. The perirhinal lesioned macaques were significantly impaired relative to controls on this task, making on average twice as many errors to learn the task to a criterion (Buckley

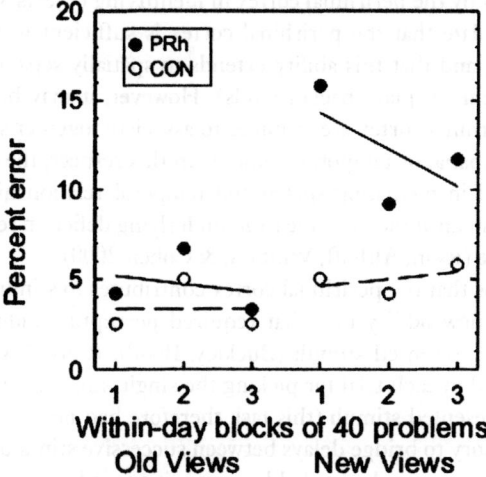

**Concurrent discrimination learning
(transfer to new views of objects)**

Figure 3. New analysis of data originally published in Buckley and Gaffan (1998b) showing that when macaques with perirhinal cortex lesions are tested on familiar concurrent discrimination problems with the stimulus objects presented in previously unseen views, that they are impaired, relative to controls, even on the very first trials with stimuli in new views. The filled circles represent the mean percentage of error of the perirhinal lesioned group and the unfilled circles represent the mean percentage of error of the perirhinal lesioned group on each of six consecutive blocks of 40 problems over two days (i.e., the three blocks within the last preoperative session and the three blocks within the first postoperative session). The solid and dashed lines depict the regression lines for each group respectively (for the three pre-and postoperative blocks).

& Gaffan, 1998c). In the paired-associate learning task, the macaques were presented with three coloured alpha-numeric (elemental) stimuli in each trial; one was positioned on the left, one on the right, and one in the middle of the touch-screen. There were eight different stimuli in total divided into four pairs. On each trial one member of a particular pair was presented in the centre, and the other member of that pair was presented on one of the two sides (decided pseudorandomly); the remaining position was occupied by one member of a different stimulus pair. The macaques had to choose the lateral stimulus that was paired with the central stimulus to obtain reward (a correct choice was followed by an abutting together of the correct pairing of stimuli on the screen to visually emphasize the compound). Both controls and perirhinal lesioned groups found this task harder than the configural learning task, which may be attributable to the spatial separation between the elemental stimuli in this task and those in the configural learning task. Nevertheless, the macaques with bilateral perirhinal cortex lesions were also significantly impaired on the paired-associate learning task relative to controls, again amassing on average twice as many errors to learn the task to a criterion (Buckley & Gaffan, 1998c). Both tasks involve memory for the association together of elemental stimuli that make up compound stimuli, which may account for why both tasks were impaired to a similar extent. The demands to learn about the conjunctions of elemental stimuli are also likely to be evident in the process of learning to discriminate objects, which are themselves composed of particular combinations of features. Thus the impairments in learning about the configuration of elemental features in these tasks may be related to the more general role played by the perirhinal cortex in identifying objects. On the basis of these findings, one may speculate that the perirhinal cortex is sufficient to encode the relations between object features, and that this ability extends to spatially separable features (at least when these features occur on plain backgrounds). However, it may be the case that brain regions beyond the perirhinal cortex are required to associate together stimuli that are separable by more complex spatial or temporal relations. In this respect, the macaque hippocampus has been implicated in mediating spatial and temporal relationships (see later in this article). The idea that human amnesia is due to an underlying deficit in relational memory has been proposed elsewhere (Ryan, Althoff, Whitlow, & Cohen 2000).

To test the hypothesis that the perirhinal cortex contributes to stimulus perception more directly we designed a new oddity task that required perceptual judgements to be made between simultaneously presented stimuli (Buckley, Booth, Rolls, & Gaffan, 1998, 2001). Macaques were rewarded in each trial for picking the single stimulus that differed from the other simultaneously presented stimuli (this task therefore had no requirement for associative or recognition memory to bridge delays between successive stimulus presentation ranging from seconds to hours as are demanded by recognition and associative memory tasks.). There were two preoperative conditions, which varied in the demands that they played upon object discrimination. In the first condition (same-view problems; see panel A1 of Figure 4), five identical greyscale views of one object were presented with one view of a different object. We hypothesized that perirhinal lesioned macaques would remain unimpaired on this task, as it can be easily be solved on the basis of discrimination between any number of simple features. In the second condition (different-view problems; see panel B of Figure 4), we presented five different views of the same object together with one view of a different object. In trials of this kind, without any obvious simple features to consistently distinguish the two objects, we hypothesized that this task would be perirhinal dependent, because discriminating the odd object would require the ability to discriminate stimuli at an object rather than at a simple-feature level. Each of the six different views of each of the 10 stimulus objects were equally likely to appear as a rewarded or unrewarded stimulus in any given trial, which ensured that there were no consistent stimulus–reward associations that could be learnt to guide performance. Thus, successful performance on each trial depends upon perceptual comparisons between simultaneously presented stimuli. Just as predicted, the macaques with perirhinal lesions were unimpaired relative to controls at performing the same-view problems, but were impaired at performing the different-view problems, particularly on more difficult problems (see panel A3 of Figure 4).

To rule out that the impairment was not just one of making difficult discriminations, we also conducted a series of control tasks, which required perceptual comparisons between stimuli that varied only in simple features, and in which perceptual difficulty could be manipulated between trials. For example, panels B1 and C1 of Figure 4 illustrate easy colour and shape discrimination problems, respectively, whereas panels B2 and C2 of Figure 4 depict more difficult colour and shape discrimination problems. We hypothesized that the perirhinal lesioned group would be unimpaired at all of these problems irrespective of difficulty, as each could be solved by distinguishing between simple features that are understood to be represented upstream of the perirhinal cortex in the ventral visual processing

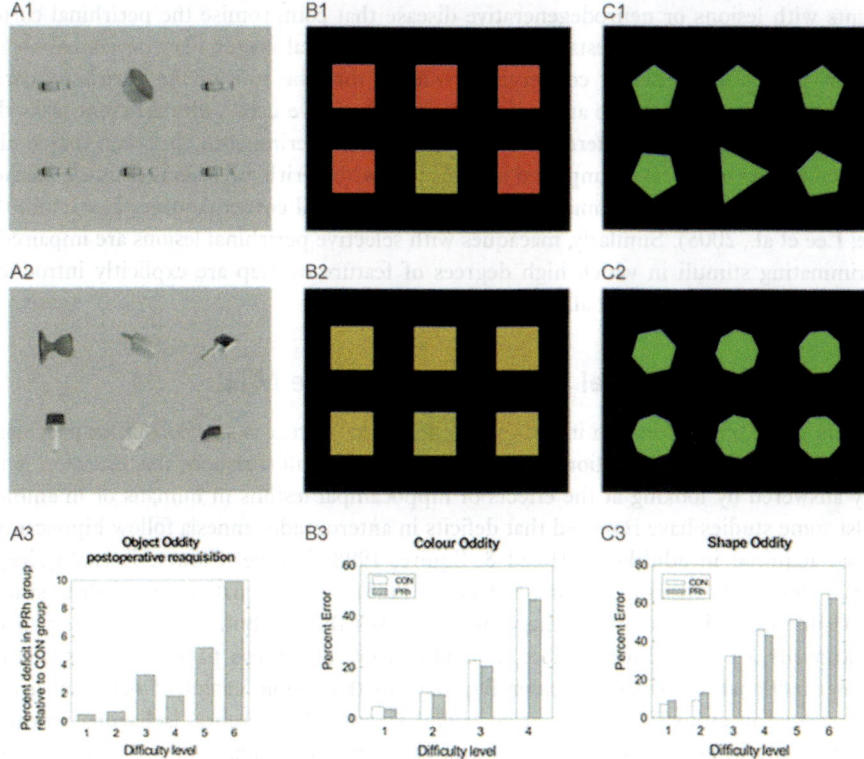

Figure 4. Panel A1 shows an example of a single trial from the same-view object oddity task, and panel A2 shows an example of a single trial from the different-view object oddity task. Panels B1 and B2 illustrate an example of an easy and a difficult trial, respectively, from the colour oddity task, and panels C1 and C2 illustrate an easier and a harder trial from the shape oddity task, respectively. Panels A3, B3, and C3 show data originally published in Buckley et al. (2001) on each of these three tasks; panel A3 shows that macaques with perirhinal lesions were impaired post-operatively, relative to controls, at difficult different-view object-oddity problems whereas panels B3 and C3 show that these two simple-feature oddity discrimination tasks are unimpaired even at very hard difficulty levels.

stream. As panels B3 and C3 of Figure 4 show, the data supported our hypothesis for simple feature discriminations (likewise, there was no impairment in the perirhinal cortex group in making very difficult size discriminations too). In contrast, when we tested the macaques on oddity problems that could not be solved on the basis of simple features they were consistently impaired. Thus, same-view human face and same-view monkey face problems were unimpaired but different-view faces (both human and monkey) were impaired. The pattern of results in every stimulus domain investigated was entirely in accord with the perceptual hypothesis of perirhinal cortex function—namely, that simple-feature discrimination does not depend upon the perirhinal cortex but discriminations at an object level (including faces) do depend upon the perirhinal cortex. This experiment therefore provides convincing evidence in favour of the idea that the perirhinal cortex contributes to object perception and does not have a role limited to memory. The paper by Lee and colleagues (Lee, Barense, & Graham, this issue) describes how these oddity tasks have since been used in testing human patients with lesions or neurodegenerative disease that compromise the perirhinal cortex, and the pattern of data suggests that the human perirhinal cortex likewise contributes to stimulus perception. Further converging evidence that the role of the perirhinal cortex extends to perception has also arisen from studies that have used quite different tasks that tax object identification in different ways. For example, discriminating between very similar morphed images of objects is impaired in macaques with perirhinal lesions (Bussey, Saksida, & Murray, 2003) as it is in human patients after perirhinal cortex damage (Lee et al., this issue; Lee et al., 2005). Similarly, macaques with selective perirhinal lesions are impaired at discriminating stimuli in which high degrees of feature overlap are explicitly introduced (Bussey et al., 2002; Bussey et al., this issue).

The nature of functional specializations in the MTL

If, as the preceding discussion indicates, the perirhinal cortex is specialized for processing objects, what is the specialization of the hippocampus? Unfortunately, this question is not easily answered by looking at the effects of hippocampal lesions in humans or in animals. Whilst some studies have reported that deficits in anterograde amnesia follow hippocampal damage acquired in adulthood, (Reed & Squire, 1998; Rempel-Clower, Zola, Squire, & Amaral, 1996; Zola-Morgan, Squire, & Amaral, 1986), all of the patients from these studies have visible extrahippocampal damage, which makes the attribution of the deficit to the hippocampus less safe. Furthermore, it is also likely that these patients have structural and/or functional abnormalities extending beyond the visible extent of neuronal loss as demonstrated in animals; Mumby et al. (1996) showed that cerebral ischaemia produced experimentally in animals can, as in clinical cases, result in neuronal loss limited to the hippocampus, but the memory impairments are not explicable by the hippocampal damage observed, since they are more severe than those after surgical removal of the hippocampus.

Interpreting the effects of hippocampal lesions in animals is also problematic. In fact, hippocampal aspiration lesions are rarely performed in macaque monkeys these days because they are notoriously hard to interpret due to the unavoidable damage sustained to other cortical regions and neuronal pathways in the MTL (as well as the likely interruption to the blood flow in branches of the posterior cerebral artery that cross the parahippocampal gyrus en route to area TE (Gaffan & Lim, 1991) during the course of making the lesion). Therefore, the role of the hippocampal system in macaques has typically been probed by two alternative lesion techniques. First, the fornix, a major input and output pathway of the hippocampus, can be sectioned. This technique, whilst providing only an indirect means of assessing the role of the hippocampal system, has the distinct advantage that it is highly replicable as the fornix can be reliably cut in its entirety, bilaterally, and without damaging other structures. An alternative technique is to make neurotoxic lesions of the hippocampus (Jarrard, 2002), which may (in the absence of neurotoxin leakage) be targeted towards the hippocampus itself, albeit with considerable variability in the extent of cell loss achievable between hemispheres and between subjects (e.g., see Table 1 in Murray, Baxter, & Gaffan, 1998). The findings from the relatively few neurotoxic hippocampal lesion studies so far carried out can be summarized here. One of the most striking findings, referred to earlier, is that neurotoxic hippocampal lesions fail to impair object recognition memory in macaques that were trained on the task preoperatively (Murray & Mishkin, 1998). However, two other studies did find deficits on this recognition memory task (Beason-Held, Rosene, Killiany, & Moss, 1999; Zola et al., 2000), and it is not yet established which of a number of methodological differences might account for the disparate findings, although both of the studies that found impairments employed only postoperative training of the task consistent with acquisition of the task rules rather than pure 'performance' being affected. Monkeys with neurotoxic hippocampal lesions have also been found to be impaired at the visual paired-comparison' task at longer delays (Zola et al., 2000), 'delayed recognition span performance (Beason-Held et al., 1999), and learning about objects embedded in unique contexts (Dore, Thornton, White, & Murray, 1998). However, in all three cases the tasks were again only trained postoperatively, so likewise one cannot rule out that these impairments might be attributed to deficits in learning the rules of the task rather than to pure performance measures. Murray and Mishkin (1998) also showed that that delayed nonmatching-to-position (spatial recognition memory), like object recognition memory, was also unimpaired following neurotoxic hippocampal lesions. It seems safe to conclude that recognition memory is not a crucial function of the hippocampus. In another study, Murray et al. (1998) showed that neurotoxic hippocampal lesions also failed to impair place reversal learning; a mild impairment in object reversal learning was observed although this, as the authors themselves advise, should be interpreted with caution as previous studies found no effects on object reversal learning even with large aspirative lesions of the hippocampus (Mahut, 1971). Like the effects of fornix transection (Gaffan, 1994b), neurotoxic hippocampal lesions also impair place-in-scene learning (Murray et al., 1998) although preoperative training on this task was again omitted. Despite the attractiveness of the neurotoxic lesion technique, it is clear from the prominent negative results, as well as the problems in interpreting these deficits that have been observed, that the precise role of the macaque hippocampus has not yet been determined on the basis of neurotoxic hippocampal lesion studies.

Whilst the conclusion that the hippocampus is not crucial for object recognition memory sits comfortably with the idea that the perirhinal cortex mediates such a role, the absence of any spatial recognition memory or spatial reversal learning impairments after neurotoxic hippocampal lesions (Murray & Mishkin, 1998; Murray et al., 1998) seems at odds with considerable evidence from rodent studies indicating that the hippocampus contributes to spatial memory (e.g., Aggleton & Brown, this issue; Eacott & Gaffan, this issue; Jarrard, 1995; Morris, Garrud, Rawlins, & O'Keefe, 1982; O'Keefe & Nadel, 1978). Fornix transection, while by no means functionally equivalent to the effects of hippocampal lesions, does lend support to the idea that there are functional subdivisions, according to stimulus domain, within the MTL. In addition to the double dissociation between the effects of fornix transection and perirhinal cortex ablation (Gaffan, 1994a), a more recent study in our laboratory has shown that fornix transection produces impairments in spatial concurrent discrimination learning that mirror, in some ways, the impairments in object concurrent discrimination learning following perirhinal cortex ablation. Buckley et al. (2001) devised a spatial version of the concurrent discrimination learning task in which macaques had to remember which one of two stimuli in each trial was rewarded; the stimuli that we employed varied in position, angle of orientation, and length—three spatial features the discrimination of which is known not to be impaired by TE lesions (Gaffan, Harrison, & Gaffan, 1986; Holmes & Gross, 1984). These particular stimuli were also chosen for use in this task because we had previously found that patients with unilateral fornix transection were impaired at recognizing these stimuli when they were presented to the hemifield contralateral to the patients' fornix transection (Buckley et al., 2004b). Relearning of a set of 40 preoperatively acquired spatial concurrent discrimination problems was unaffected by fornix transection; however, concurrent learning of new sets of only 10 problems was significantly impaired. Notably, the most marked deficit in this spatial concurrent discrimination learning task was found in new learning of a small number of concurrent problems when increased numbers of foils were incorporated into the task. This parallels, in the spatial domain, our earlier finding that postoperative new learning of small sets of object concurrent discrimination learning problems is impaired after perirhinal cortex lesions when the number of incorrect object foils is increased (Buckley & Gaffan, 1997). Furthermore, in the same study Buckley et al. (2004b) also showed that spatial configural learning is impaired after fornix transection, which mirrors the impairment found in object configural learning after perirhinal lesions discussed earlier (Buckley & Gaffan, 1998c).

Should we conclude from all of this that the macaque hippocampal system is specialized for spatial memory? It is certainly not as simple as this, and the evidence is again somewhat difficult to interpret. For example, impairments in 'spatial' tasks such as spatial delayed response tasks and spatial reversal learning are inconsistently observed after hippocampal system disruption (see Buckley & Gaffan, 2004a, for a review). The effects of fornix transection and hippocampal lesions upon scene memory certainly seem more robust (Gaffan, 1992, 1994a, 1994b; Murray et al., 1998) and have also been noted in human patients with fornix damage (Aggleton et al., 2000); however, scene learning tasks do not provide a pure test of spatial memory as the contribution of object memory to these tasks cannot be ruled out (as scenes contain objects, and even 'object-less' locations within a scene contain background features that could be processed as object-like configurations of visual features). Unambiguous demonstrations of spatial memory deficits after fornix transection in macaques were until recently limited to paradigms (such as the T-maze or a Wisconsin General Test Apparatus) in which the animal learns about the location in which it finds food reward (Gaffan, 1994a; Gaffan & Harrison, 1989; Murray, Davidson, Gaffan, Olton, & Suomi, 1989;). The impairments in spatial discrimination learning shown by Buckley et al. (2004a) add to this by showing that spatial impairments after fornix transection also extend to the context of the touchscreen apparatus in which the animal receives food in a constant location.

One should note that the impairments that follow fornix transection in the macaque are not limited to the spatial domain. For example, Brasted, Bussey, Murray, and Wise (2003) recently demonstrated that fornix transection led to impairments in the ability of macaques to learn associations between visual stimuli and their own temporally (but nonspatially) differentiated responses. In our own laboratory we have also examined the effects of fornix transection upon a different kind of nonspatial task requiring judgements of relative recency (Charles, Gaffan, & Buckley, 2004). Our experiment contained three types of trial. The first trial types were called within-session-recency (WSR) trials; in these trials the macaques were presented with a series of five successive patterned clipart stimuli followed by a choice trial in which two of the stimuli in the sequence were presented together on the touchscreen, upon the occurrence of which the macaque was rewarded for choosing the stimulus that had appeared most recently in the sequence. The second trial types were absolute-novelty (AN) trials; in these trials a sequence of five stimuli was presented, but at the choice stage only one stimulus from the sequence was paired with a novel stimulus never previously seen by the macaque. In AN trials the macaque was rewarded for picking the familiar stimulus. The third trial types were between-session-recency (BSR) trials; these trials were similar to AN trials in that only one of the two choice items occurred in the sequence, but they differed in that both choice stimuli had been experienced previously by the macaques in earlier sessions. The fornix-transected macaques were found to be generally impaired at making recency judgements in the WSR trials, but remained unimpaired upon recognition memory (AN trials). However, the BSR task was also impaired; in other words, impairments on the object recognition memory task are dependent upon the novelty status of the stimuli employed. Whereas object recognition memory with trial-unique stimuli is unimpaired by fornix transection, if the same task is administered with familiar stimuli then impairments are found, presumably due to the inability of the fornix transacted macaques to judge how recently they saw those stimuli. These data provide further convincing evidence that the role of the fornix extends beyond the spatial domain to include temporal processing.

Overview and conclusions

In this article, considerable evidence has been presented to argue that the perirhinal cortex has a role that extends beyond memory to include the perception of objects. This has led to the conclusion that the perirhinal cortex is specialized for processing of stimuli at an object level, and that the MTL is not a functionally homogeneous unit. We have also seen that recent research has established that fornix transection results in deficits in spatial as well as temporal memory processes, and therefore an obvious question arises; if these deficits are attributable to hippocampal dysfunction then might the hippocampal system also contribute to spatial and temporal perception as well as to memory? There are some early reports (Graham et al., 2004; and see Lee et al., this issue) of intriguing evidence from human patients with hippocampal lesions being selectively impaired at a spatial version of the oddity task described in this paper. If the hippocampus, like the perirhinal cortex, also contributes to perception as well as memory then it will be increasingly apparent that there really is no such thing as a MTL memory system, but rather a network of MTL structures with specializations largely along stimulus rather than functional domains.

That is not to say that the hippocampus, perirhinal cortex, and MTL structures in general do not contribute directly to memory and that any mnemonic deficits that are observed to follow damage to these structures are merely secondary to perceptual deficits. Rather, the impairments following perirhinal cortex lesions are consistent with the idea that the perirhinal cortex is involved in the binding together of the features that constitute objects, an ability that has both mnemonic and perceptual applications. For example, representing objects may be considered to be mnemonic, in that it depends upon maintaining associative linkages between particular combinations of component features that together define a particular object and distinguish it from other objects. Thus a failure to bind together object features after perirhinal cortex lesions may account for the range of deficits observed in mnemonic tasks such as configural and paired-associate learning (Buckley & Gaffan, 1998c), as well as deficits in recognition memory (Eacott et al., 1994; Meunier et al., 1993) and concurrent discrimination learning tasks (Buckley & Gaffan, 1997) when large stimulus sets are employed in either case. The same underlying deficit would also account for impairments on object oddity tasks in which it can be argued that configurations of features have to be processed in order to solve the task (Buckley et al.,

2001). While it is recognized that the oddity tasks may require the ability to compare stimuli over multiple fixations, the absence of any impairment in any of the simple-feature oddity tasks rules out a general short-term memory deficit explanation of the object-oddity deficits. In a different context, the ability to bind together stimulus features may be considered to play a perceptual role too, for example in allowing one to perceptually group together foreground features that make up coherent objects, thereby allowing foreground objects to be distinguished from visual features that make up the background. Experimental data certainly support this account; Buckley and Gaffan (1998b) showed that performance on a set of well learnt concurrent object discrimination problems was impaired when the very same problems were required to be solved in the context of an unfamiliar scene (the stimuli used in this task were digitized photographs of the pairs of familiar objects standing in unfamiliar scenes), and Buckley et al. (2001) showed that scene oddity was itself impaired after perirhinal cortex ablation. In this variant of the macaque oddity task, three identical views of one scene were presented with one view of a different scene (although foreground objects were different between all of the scenes used, some overlap between background features did exist in those trials that paired similar scenes). Note that the impairment upon this task rules out the argument that deficits on oddity task after perirhinal lesions only occur in conditions in which objects have to be compared across different views; if this were not the case, then one could argue that impairments upon oddity tasks involved some form of mental rotation or manipulation of a mnemonic representation, rather than perceptual comparison as we are arguing here.

Given the mounting evidence that the hippocampus plays a more selective role in spatial and temporal processing, which may extend to perceptual as well as mnemonic processes, then one may speculate that the hippocampus might also contribute to the binding together of stimuli, specifically the binding together of spatial and temporal configurations of stimuli that constitute events or episodes. By this account, hippocampal damage would lead, not only to deficits in episodic memory, but also to deficits in other tasks that place sufficient demands upon binding together either the spatial or the temporal relationships between stimuli. Consistent with this hypothesis, concurrent learning of multiple object-in-place or place-in-scene associations are impaired after fornix or hippocampal system disruption in macaques (Gaffan, 1994b; Murray et al., 1998). In contrast, the learning of only a limited amount of spatial information is typically unimpaired after hippocampal lesions. For example, Angeli, Murray, and Mishkin (1993) showed that hippocampal-lesioned macaques could remember one place but not two after short delays, and Beason-Held et al. (1999) showed that increasing memory span revealed general impairments after hippocampal lesions. Furthermore, our preliminary data on scene-oddity in human patients with hippocampal lesions (Graham et al., 2004) show that perceptual matching in the spatial domain is only impaired in the different-view condition which involves processing complex spatial information, as opposed to spatial matching in the same-view condition, which can be solved solely on the basis of matching any simple spatial feature (see Lee et al., this issue). Similarly, with respect to the temporal domain, we can see how the binding hypothesis would account for the fact that simply remembering whether a stimulus is familiar (without any spatial or temporal context) remains unimpaired after fornix transection whereas remembering the relative recency of occurrence of items (i.e., with a temporal context introduced) within or between sessions is impaired (Charles et al., 2004).

To conclude, traditional ideas regarding MTL function have come under increasing challenge from recent data, both from animal models themselves and from related human neuropsychological tests that have developed out of these models. Far from being a homogeneous memory system as previously conceived, the MTL instead exhibits multiple specializations with MTL structures contributing to both memory and perception. The goal of future research will be to evaluate and challenge these new theories of MTL function to see whether they can stand the test of time.

REFERENCES

Aggleton, J. P., & Brown, M. W. (this issue). Contrasting hippocampal and perirhinal cortex function using immediate early gene imaging. *Quarterly Journal of Experimental Psychology, 58B*, 218–233.

Aggleton, J. P., McMackin, D., Carpenter, K., Hornak, J., Kapur, N., Halpin, S., et al. (2000). Differential cognitive effects of colloid cysts in the third ventricle that spare or compromise the fornix. *Brain, 123*, 800–815.

Amaral, D. G., Insausti, R., & Cowan, W. M. (1987). The entorhinal cortex of the monkey. I. Cytoarchitectonic organization. *The Journal of Comparative Neurology, 230*, 465–496.

Angeli, S. J., Murray, E. A., & Mishkin, M. (1993). Hippocampectomized monkeys can remember one place but not two. *Neuropsychologia, 31*, 1021–1030.

Arnold, S. E., Hyman, B. T., & Van Hoesen, G. W. (1994). Neuropathologic changes of the temporal pole in Alzheimer's disease and Pick's disease. *Archives of Neurology, 51*, 145–150.

Beason-Held, L. L., Rosene, D. L., Killiany, R. J., & Moss, M. B. (1999). Hippocampal formation lesions produce memory impairment in the rhesus monkey. *Hippocampus, 9*, 562–574.

Braak, H., & Braak, E. (1985). On areas of transition between entorhinal allocortex and temporal ioscortex in the human brain: Normal morphology and lamina-specific pathology in Alzheimer's disease. *Acta Neuropathologica, 68*, 325–332.

Brasted, P. J., Bussey, T. J., Murray, E. A., & Wise, S. P. (2003). Role of the hippocampal system in associative learning beyond the spatial domain. *Brain, 126*, 1202–1223.

Buckley, M. J., & Gaffan, D. (1997). Impairment of visual object-discrimination learning after perirhinal cortex ablation. *Behavioral Neuroscience, 111*, 467–475.

Buckley, M. J., & Gaffan, D. (1998a). Learning and transfer of object–reward associations and the role of the perirhinal cortex. *Behavioral Neuroscience, 112*, 15–23.

Buckley, M. J., & Gaffan, D. (1998c). Perirhinal cortex ablation impairs configural learning and paired-associate learning equally. *Neuropsychologia, 36*, 535–546.

Buckley, M. J., & Gaffan, D. (1998b). Perirhinal cortex ablation impairs visual object identification. *Journal of Neuroscience, 18*, 2268–2275.

Buckley, M. J., & Gaffan, D. (2000). The hippocampus, perirhinal cortex and memory in the monkey. In J. J. Bolhuis (Ed.), *Brain, perception, memory: Advances in cognitive neuroscience* (pp. 279–298). New York: Oxford University Press.

Buckley, M. J., Booth, M. C. A., Rolls, E. T., & Gaffan, D. (1998). Selective visual-perceptual deficits following perirhinal cortex ablation in the macaque. *Society for Neuroscience Abstracts, 24 (14.6)*, 18.

Buckley, M. J., Booth, M. C. A., Rolls, E. T., & Gaffan, D. (2001). Selective perceptual impairments after perirhinal cortex ablation. *Journal of Neuroscience, 21*, 9824–9836.

Buckley, M. J., Charles, D. P., Browning, P. G. F., & Gaffan, D. (2004a). Learning and retrieval of concurrently presented spatial discrimination tasks: Role of the fornix. *Behavioral Neuroscience, 118*, 138–149.

Buckley, M. J., Gaffan, D., & Murray, E. A. (1997). Functional double dissociation between two inferior temporal cortical areas: Perirhinal cortex versus middle temporal gyrus. *Journal of Neurophysiology, 77*, 587–598.

Buckley, M. J., Williams, H., & Hornak, J. (2004b). Material-specific hemispheric asymmetries in visual recognition memory observed in normal controls are absent in epilepsy patients with unilateral medial temporal lobe excissions: Evidence for functional reorganization. *Society for Neuroscience Abstracts* (200.10).

Buffalo, E. A., Ramus, S. J., Clark, R. E., Teng, E., Squire, L. R., & Zola, S. M. (1999). Dissociation between the effects of damage to perirhinal cortex and area TE. *Learning and Memory, 6*, 572–599.

Buffalo, E. A., Ramus, S. J., Squire, L. R., & Zola, S. M. (2000). Perception and recognition memory in monkeys following lesions of area TE and perirhinal cortex. *Learning and Memory*, *7*, 375–382.

Buffalo, E. A., Reber, P. J., & Squire, L. R. (1998a). The human perirhinal cortex and recognition memory. *Hippocampus*, *8*, 330–339.

Buffalo, E. A., Stefanacci, L., Squire, L. R., & Zola, S. M. (1998b). A reexamination of the concurrent discrimination learning task: The importance of anterior inferotemporal cortex, area TE. *Behavioral Neuroscience*, *112*, 3–14.

Bussey, T. J., Saksida, L. M., & Murray, E. A. (2002). Perirhinal cortex resolves feature ambiguity in complex visual discriminations. *European Journal of Neuroscience*, *15*, 365–374.

Bussey, T. J., Saksida, L. M., & Murray, E. A. (2003). Impairments in visual discrimination after perirhinal cortex lesions: Testing 'declarative' vs. 'perceptual-mnemonic' views of perirhinal cortex function. *European Journal of Neuroscience*, *17*, 649–660.

Bussey, T. J., Saksida, L. M., & Murray, E. A. (this issue). The perceptual-mnemonic/feature conjuction model of perirhinal cortex function. *Quarterly Journal of Experimental Psychology*, *58B*, 269–282.

Charles, D. P., Gaffan, D., & Buckley, M. J. (2004). Impaired recency judgements and intact novelty judgements following fornix transection in monkeys. *Journal of Neuroscience*, *28*, 2037–2044.

Cirillo, R. A., Horel, J. A., & George, P. J. (1989). Lesions of the anterior temporal stem and the performance of delayed match-to-sample and visual discriminations in monkeys. *Behavioural Brain Research*, *34*, 55–69.

Corkin, S. (2002). What's new with the amnesic patient H.M.? *Nature Reviews Neuroscience*, *3*, 153–160.

Correll, R. E., & Scoville, W. B. (1965a). Effects of medial temporal lesions on visual discrimination performance. *Journal of Comparative and Physiological Psychology*, *60*, 175–181.

Correll, R. E., & Scoville, W. B. (1965b). Performance of delayed match following lesions of medial temporal lobe structures. *Journal of Comparative and Physiological Psychology*, *60*, 360–367.

Correll, R. E., & Scoville, W. B. (1967). Significance of delay in the performance of monkeys with medial temporal lobe resections. *Experimental Brain Research*, *4*, 86–96.

Dore, F. Y., Thornton, J. A., White, N. M., & Murray, E. A. (1998). Selective hippocampal lesions yield nonspatial memory impairments in rhesus monkeys. *Hippocampus*, *8*, 323–329.

Eacott, M. J., & Gaffan, E. A. (this issue). The roles of the perirhinal cortex, postrhinal cortex and the fornix in memory for objects, contexts, and events in the fat. *Quarterly Journal of Experimental Psychology*, *58B*, 202–217.

Eacott, M. J., Gaffan, D., & Murray, E. A. (1994). Preserved recognition memory for small sets, and impaired stimulus identification for large sets, following rhinal cortex ablations in monkeys. *European Journal of Neuroscience*, *6*, 1466–1478.

Easton, A., Ridley, R. M., Baker, H. F., & Gaffan, D. (2002). Unilateral lesions of the cholinergic basal forebrain and fornix in one hemisphere and inferior temporal cortex in the opposite hemisphere produce severe learning impairments in rhesus monkeys. *Cerebral Cortex*, *12*, 729–736.

Gaffan, D. (1974). Recognition impaired and association intact in the memory of monkeys after transection of the fornix. *Journal of Comparative and Physiological Psychology*, *86*, 1100–1109.

Gaffan, D. (1992). Amnesia for complex naturalistic scenes and for objects following fornix transection in the rhesus-monkey. *European Journal of Neuroscience*, *4*, 381–388.

Gaffan, D. (1994a). Dissociated effects of perirhinal cortex ablation, fornix transection and amygdalectomy: Evidence for multiple memory systems in the primate temporal lobe. *Experimental Brain Research*, *99*, 411–422.

Gaffan, D. (1994b). Scene-specific memory for objects: A model of episodic memory impairment in monkeys with fornix transection. *Journal of Cognitive Neuroscience*, *6*, 305–320.

Gaffan, D. (1996). Memory, action and the corpus striatum: Current developments in the memory-habit distinction. *Seminars in the Neurosciences*, *8*, 33–38.

Gaffan, D., & Harrison, S. (1989). Place memory and scene memory: Effects of fornix transection in the monkey. *Experimental Brain Research*, *74*, 202–212.

Gaffan, D., Harrison, S., & Gaffan, E. A. (1986). Visual identification following inferotemporal ablation in the monkey. *Quarterly Journal of Experimental Psychology*, *38B*, 5–30.

Gaffan, D., & Lim, C. (1991). Hippocampus and the blood supply to TE: Parahippocampal pial section impairs visual discrimination learning in monkeys. *Experimental Brain Research*, *87*, 227–231.

Gaffan, D., & Murray, E. A. (1992). Monkeys (Macaca-Fascicularis) with rhinal cortex ablations succeed in object discrimination-learning despite 24-hr intertrial intervals and fail at matching to sample despite double sample presentations. *Behavioral Neuroscience*, *106*, 30–38.

Gaffan, D., Parker A., & Easton, A. (2001). Dense amnesia in the monkey after transection of fornix, amygdala and anterior temporal stem. *Neuropsychologia, 39*, 51–70.

Graham, K. S., Lee, A. C. H., Buckley, M. J., Spiers, H. J., Scahill, V., Gaffan, D., et al. (2004). Object and spatial perception deficits following medial temporal lobe lesions. *Society for Neuroscience Abstracts* (201.9).

Hampton, R. R. (this issue). Monkey perirhinal cortex is critical for visual memory, but not for visual perception: Re-examination of the behavioural evidence from monkeys. *Quarterly Journal of Experimental Psychology, 58B*, 283–299.

Hampton, R. R., & Murray, E. A. (2002). Learning of discriminations is impaired, but generalization to altered views is intact, in monkeys (Macaca mulatta) with perirhinal cortex removal. *Behavioral Neuroscience, 116*, 363–377.

Holmes, E. J., & Gross, C. G. (1984). Effects of inferior temporal lesions on discrimination of stimuli differing in orientation. *Journal of Neuroscience, 4*, 3063–3068.

Horel, J. A. (1978). The neuroanatomy of amnesia: A critique of the hippocampal memory hypothesis. *Brain, 101*, 403–445.

Horel, J. A., & Misantone, L. J. (1976). Visual discrimination impaired by cutting temporal lobe connections. *Science, 193*, 336–338.

Insausti, R., Amaral, D. G., & Cowan, W. M. (1987). The entorhinal cortex of the monkey. II. Cortical afferents. *The Journal of Comparative Neurology, 264*, 356–395.

Jarrard, L. E. (1995). What does the hippocampus really do? *Behavioural Brain Research, 71*, 1–10.

Jarrard, L. E. (2002). Use of excitotoxins to lesion the hippocampus: Update. *Hippocampus, 12*, 405–414.

Lavenex, P., Suzuki, W. A., & Amaral, D. G. (2004). Perirhinal and parahippocampal cortices of the macaque monkey: Intrinsic projections and interconnections. *Journal of Comparative Neurology, 472*, 371–394.

Lee, A. C. H., Barense, M. D., & Graham, K. S. (this issue). The contribution of the human medial temporal to perception: Bridging the gap between animal and human studies. *Quarterly Journal of Experimental Psychology, 58B*, 300–325.

Lee, A. C. H., Bussey, T. J., Murray, E. A., Saksida, L. M., Epstein, R. A., Kapur, N., et al. (2005). Perceptual deficits in amnesia: Challenging the medial temporal lobe 'mnemonic' view. *Neuropsychologia, 43*, 1–11.

Leonard, B. W., Amaral, D. G., Squire, L. R., & Zola-Morgan, S. (1995). Transient memory impairment in monkeys with bilateral lesions of the entorhinal cortex. *Journal of Neuroscience, 15*, 5637–5659.

Mahut, H. (1971). Spatial and object reversal learning in monkeys with partial temporal lobe ablations. *Neuropsychologia, 9*, 409–424.

Mahut, H., Zola-Morgan, S., & Moss, M. (1982). Hippocampal resections impair associative learning and recognition memory in the monkey. *Journal of Neuroscience, 2*, 1214–1229.

Malamut, B. L., Saunders, R., & Mishkin, M. (1984). Monkeys with combined amygdalo-hippocampal lesions suceed in object discrimination learning despite 24-hour intertrial intervals. *Behavioral Neuroscience, 98*, 759–769.

Malkova, L., Gaffan, D., & Murray, E. A. (1997). Excitotoxic lesions of the amygdala fail to produce impairment in visual learning for auditory secondary reinforcement but interfere with reinforcer devaluation effects in rhesus monkeys. *Journal of Neuroscience, 17*, 6011–6020.

Meunier, M., Bachevalier, J., Mishkin, M., & Murray, E. A. (1993). Effects on visual recognition of combined and separate ablations of the entorhinal and perirhinal cortex in rhesus monkeys. *Journal of Neuroscience, 13*, 5418–5432.

Milner, B. (1958). Psychological defects produced by temporal lobe excision. In H. C. Solomon, S. Cobb, & W. Penfield (Eds.), *The brain and human behavior.* [*Research publications of the Association for Research in Nervous and Mental Disease*] (pp. 244–257). Baltimore: The Williams & Wilkins Company.

Mishkin, M. (1978). Memory in monkeys severely impaired by combined but not by separate removal of amygdala and hippocampus. *Nature, 273*, 297–298.

Mishkin, M., & Delacour, J. (1975). An analysis of short-term visual memory in the monkey. *Journal of Experimental Psychology: Animal Behavior Processes, 1*, 326–334.

Mishkin, M., & Petri, H. L. (1984). Memories and habits: Some implications for the analysis of learning and retention. In L. Squire & N. Butters (Eds.), *Neuropsychology of memory* (1st ed., pp. 287–296). New York: Guilford Press.

Mishkin, M., Malamut, B. L., & Bachevalier, J. (1984). Memories and habits: Two neural systems. In G. Lynch, J. L. McGaugh, & N. M. Weinberger (Eds.), *Neurobiology of learning and memory*. New York: Guilford Press.

Morris, R. G. M., Garrud, P., Rawlins, J. N. P., & O'Keefe, J. (1982). Place navigation impaired in rats with hippocampal-lesions. *Nature, 297*, 681–683.

Mumby, D. G., Wood, E. R., Duva, C. A., Kornecook, T. J., Pinel, J. P. J., & Phillips, A. G. (1996). Ischemia-induced object-recognition deficits in rats are attenuated by hippocampal ablation before or soon after ischemia. *Behavioral Neuroscience, 110,* 266–281.

Murray, E. A., Baxter, M. G., & Gaffan, D. (1998). Monkeys with rhinal cortex damage or neurotoxic hippocampal lesions are impaired on spatial scene learning and object reversals. *Behavioral Neuroscience, 112,* 1291–1303.

Murray, E. A., & Bussey, T. J. (1999). Perceptual-mnemonic functions of the perirhinal cortex. *Trends in Cognitive Sciences, 3,* 142–151.

Murray, E. A., Davidson, M., Gaffan, D., Olton, D. S., & Suomi, S. (1989). Effects of fornix transection and cingulate cortical ablation on spatial memory in rhesus monkeys. *Experimental Brain Research, 74,* 173–186.

Murray, E. A., & Mishkin, M. (1986). Visual recognition in monkeys following rhinal cortical ablations combined with either amygdalectomy or hippocampectomy. *Journal of Neuroscience, 6,* 1991–2003.

Murray, E. A., & Mishkin, M. (1998). Object recognition and location memory in monkeys with excitotoxic lesions of the amygdala and hippocampus. *Journal of Neuroscience, 18,* 6568–6582.

Murray, E. A., & Richmond, B. J. (2001). Role of perirhinal cortex in object perception, memory, and associations. *Current Opinion in Neurobiology, 11,* 188–193.

O'Keefe, J., & Nadel, L. (1978). *The hippocampus as a cognitive map.* London: Oxford University Press.

Orbach, J., Milner, B., & Rasmussen, T. (1960). Learning and retention in monkeys after amygdala-hippocampal resections. *Archives of Neurology, 3,* 1214–1229.

Reed, J. M., & Squire, L. R. (1998). Retrograde amnesia for facts and events: Findings from four new cases. *Journal of Neuroscience, 18,* 3943–3954.

Rempel-Clower, N. L., Zola, S. M., Squire, L. R., & Amaral, D. G. (1996). Three cases of enduring memory impairment after bilateral damage limited to the hippocampal formation. *Journal of Neuroscience, 16,* 5233–5255.

Ryan, J. D., Althoff, R. R., Whitlow, S., & Cohen, N. J. (2000). Amnesia is a deficit in relational memory. *Psychological Science, 11,* 454–461.

Scoville, W. B., & Milner, B. (1957). Loss of recent memory after bilateral hippocampal lesions. *Journal of Neurology, Neurosurgery and Psychiatry, 20,* 11–21.

Squire, L. R., & Zola-Morgan, S. (1991). The medial temporal lobe memory system. *Science, 253,* 1380–1386.

Suzuki, W. A. (1996a). Neuroanatomy of the monkey entorhinal, perirhinal and parahippocampal cortices: Organization of cortical inputs and interconnections with amygdala and striatum. *Seminars in the Neurosciences, 8,* 3–11.

Suzuki, W. A. (1996b). The anatomy, physiology and functions of the perirhinal cortex. *Current Opinion in Neurobiology, 6,* 179–186.

Suzuki, W., Zola-Morgan, S., Squire, L., & Amaral, D. (1993). Lesions of the perirhinal and parahippocampal cortices in the monkey produce long-lasting memory impairment in the visual and tactual modalities. *Journal of Neuroscience, 13,* 2430–2451.

Tanaka, K. (1996). Inferotemporal cortex and object vision. *Annual Review of Neuroscience, 19,* 109–139.

Van Hoesen, G. W., Hyman, B. T., & Damasio, A. R. (1991). Entorhinal cortex pathology, in Alzheimer's disease. *Hippocampus, 1,* 1–8.

Zola, S. M., Squire, L. R., Teng, E., Stefanacci, L., Buffalo, E. A., & Clark, R. E. (2000). Impaired recognition memory in monkeys after damage limited to the hippocampal region. *Journal of Neuroscience, 20,* 451–463.

Zola-Morgan, S., & Squire, L. R. (1986). Memory impairment in monkeys following lesions limited to the hippocampus. *Behavioral Neuroscience, 100,* 155–160.

Zola-Morgan, S., Squire, L., & Amaral, D. G. (1986). Human amnesia and the medial temporal region—enduring memory impairment following a bilateral lesion limited to field Ca1 of the hippocampus. *Journal of Neuroscience, 6,* 2950–2967.

Zola-Morgan, S., Squire, L., Clower, R., & Rempel, N. (1993). Damage to the perirhinal cortex exacerbates memory impairment following lesions to the hippocampal formation. *Journal of Neuroscience, 13,* 251–265.

Zola-Morgan, S., Squire, L. R., & Mishkin, M. (1982). The neuroanatomy of amnesia—amygdala-hippocampus versus temporal stem. *Science, 218,* 1337–1339.

Zola-Morgan, S., Squire, L. R., & Ramus, S. J. (1994). Severity of memory impairment in monkeys as a function of locus and extent of damage within the medial temporal-lobe memory system. *Hippocampus, 4,* 483–495.

THE QUARTERLY JOURNAL OF EXPERIMENTAL PSYCHOLOGY
2005, 58B (3/4), 269–282

The perceptual-mnemonic/feature conjunction model of perirhinal cortex function

Timothy J. Bussey and Lisa M. Saksida

University of Cambridge, UK

Elisabeth A. Murray

National Institute of Mental Health, Bethesda, MD, USA

The perirhinal cortex was once thought to be "silent cortex", virtually ignored by researchers interested in the neurobiology of learning and memory. Following studies of brain damage associated with cases of amnesia, perirhinal cortex is now widely regarded as part of a "medial temporal lobe (MTL) memory system". This system is thought to be more or less functionally homogeneous, having a special role in declarative memory, and making little or no contribution to other functions such as perception. In the present article, we summarize an alternative view. First, we propose that components of the putative MTL system such as the hippocampus and perirhinal cortex have distinct and dissociable functions. Second, we provide evidence that the perirhinal cortex has a role in visual discrimination. In addition, we propose a specific role for perirhinal cortex in visual discrimination: the contribution of complex conjunctive representations to the solution of visual discrimination problems with a high degree of "feature ambiguity". These proposals constitute a new view of perirhinal cortex function, one that does not assume strict modularity of function in the occipito-temporal visual stream, but replaces this idea with the notion of a hierarchical representational continuum.

There was a time when the perirhinal cortex—a thin strip of tissue on the anterior ventromedial aspect of the temporal lobe—was referred to as "silent cortex", of little interest to researchers in the field of learning and memory (Penfield & Jasper, 1954). Today, this thin strip of cortex commands an entire issue of a major journal. What happened?

In our opinion, two major things have happened. First, it has turned out that the perirhinal cortex is critical for important types of memory traditionally assumed to be the domain of the hippocampus. Specifically, it is now clear that object recognition memory, thought to provide a relatively pure measure of declarative memory (Manns, Stark, & Squire, 2000),

Correspondence should be addressed to Timothy J. Bussey, Department of Experimental Psychology, University of Cambridge, Cambridge, CB2 3EB, UK. Email: t.bussey@psychol.cam.ac.uk

© 2005 The Experimental Psychology Society
DOI:10.1080/02724990544000004

depends much more on the perirhinal cortex than the hippocampus (Baxter & Murray, 2001b; Forwood, Winters, & Bussey, in press; Holdstock, this issue; Meunier, Bachevalier, Mishkin, & Murray, 1993; Murray & Mishkin, 1998; Winters, Forwood, Cowell, Saksida, & Bussey, 2004). Second, perirhinal cortex seems to have a role not only in declarative memory, but in perception, thus seemingly putting to rest the standard view of perirhinal cortex as part of a "medial temporal lobe memory system" that has no role in perception (Squire & Zola-Morgan, 1991).

Both of these ideas remain controversial, many authors maintaining that the hippocampus is critical for object recognition memory, and that perirhinal cortex has no role in perception (Squire, Stark, & Clark, 2004). Regarding the first, the reports of absolutely no effect of complete hippocampal removal on object recognition memory, in both monkeys and rats, even after long retention intervals (40 minutes and 48 hours in monkeys and rats, respectively), taken together with impairments on the same task following perirhinal cortex lesions, provide quite convincing evidence that the perirhinal cortex is the more important structure for this type of memory (Forwood et al., in press; Meunier et al., 1993; Murray & Mishkin, 1998; Winters et al., 2004). Even when impairments after hippocampal damage are found, they are relatively mild and may be due to spatial or temporal contextual components of the task (Charles, Gaffan, & Buckley, 2004; Forwood et al., in press; Nadel, 1995; Nemanic, Alvarado, & Bachevalier, 2004; Winters et al., 2004). Indeed, one analysis has reported an *inverse* correlation between the amount of hippocampal damage and the magnitude of impairment in object recognition memory (Baxter & Murray, 2001b). Furthermore, in rats, double dissociations of function have been reported between perirhinal cortex and hippocampus on tests of object recognition and spatial memory (Winters et al., 2004). As for the second idea regarding the role of perirhinal cortex in perception, our argument is put forward in the following pages. Briefly, there are now several studies indicating a role for perirhinal cortex in perception. After a discussion of the historical context, we summarize our work in this area, beginning with a hypothesis that we call the perceptual-mnemonic/feature conjunction (PMFC) model, which ascribes a specific role to the perirhinal cortex in learning, memory, and perception. This is followed by a summary of studies designed to test the model.

Perirhinal cortex and "object identification"

The first hint that perirhinal cortex might have a role in perception came from a study by Eacott, Gaffan, and Murray (1994), who tested macaque monkeys on the delayed match-to-sample (MTS) test of object recognition memory. Because the monkeys were tested in an automated apparatus, versions of simultaneous MTS and MTS with short delays could be run. Eacott et al. reported that, relative to intact controls, monkeys with removal of perirhinal and entorhinal cortex showed a delay-dependent deficit. That is, the operated monkeys showed severe impairments at long delays, with no impairment at the shortest delays. This latter aspect of the delay-dependent deficit is usually taken to mean that perception is intact (Buffalo et al., 1999; Buffalo, Reber, & Squire, 1998; Holdstock, Shaw, & Aggleton, 1995; Squire et al., 2004; Stark & Squire, 2000; but see discussion below). However, when these authors increased the perceptual load in the task, impairments emerged even in the simultaneous and shortest delay conditions, at which there would be minimal or no memory load.

Following the publication of Eacott et al. (1994), Buckley and Gaffan began examining the role of perirhinal cortex, not in object recognition, but in visual discrimination learning. Based on the results of several experiments (Buckley, this issue; Buckley & Gaffan, 1997, 1998a) these authors argued for a role for perirhinal cortex beyond object recognition memory. Likewise Goulet and Murray (2001), who studied tactual-to-visual associations derived from individual objects, found evidence for perirhinal cortex contributions beyond recognition. Together, these studies argued for a more general role for perirhinal cortex in what has been called "object identification" (Bright, Moss, Stamatakis, & Tyler, this issue; Buckley & Gaffan, 1998b; Eacott & E. Gaffan, this issue; Murray, 2000; Murray & Bussey, 1999; Murray, Malkova, & Goulet, 1998).

The PMFC model

Several studies focusing on the discrimination of pairs of visual stimuli, including those mentioned above, indicated that although damage to perirhinal cortex can lead to impairments in such discriminations, it does so only under certain circumstances. It was initially unclear exactly what these conditions were. In an attempt to make sense of these findings, we (Bussey & Saksida, 2002; Murray & Bussey, 1999) began to think about the perirhinal cortex, not only as part of a mnemonic system in which perirhinal cortex works in concert with the hippocampus, but also as part of the ventral visual stream (VVS) or "what" pathway (Ungerleider & Mishkin, 1982). Electrophysiological evidence indicates that neurons in rostral regions of the VVS code more complex visual representations than do neurons in more caudal regions (Desimone & Ungerleider, 1989). We proposed that effects of perirhinal cortex lesions on visual discriminations may be explained by assuming that perirhinal cortex has visual information processing properties similar to those of other regions within the VVS, and that perirhinal cortex may be the final station in this pathway.

We have instantiated this idea in a neural network model referred to as the perceptual-mnemonic/feature-conjunction (PMFC) model. We proposed that the apparently inconsistent effects of lesions within this region can be understood by considering the hierarchical organization of representations in the VVS. As shown in Figure 1, caudal regions in the VVS are held to represent simple features, whereas more rostral regions in the VVS are held to represent more complex conjunctions of features. If this is true, an animal with damage to perirhinal cortex will not have access to highly complex visual representations, which consist of the conjunction of large numbers of features. Without such representations, the animal must rely on representations of simple features in earlier regions of the VVS to attempt to solve visual discriminations. Specifically, the model suggests that impairments in visual discrimination learning are caused by perirhinal cortex lesions because animals with such damage have impoverished conjunctive representations of complex stimuli, and the remaining representations of simple features alone are insufficient for discrimination of certain types of object.

Thus, the PMFC model claims that the effects of perirhinal cortex lesions on visual discrimination are due to compromised representations, rather than impairment of a specific type of learning or memory. Further, the model predicts that it is the nature of the stimulus material used in a task that determines whether performance impairments are seen following perirhinal cortex lesions. To test the model, lesions were made in the component of the network corresponding to perirhinal cortex, and the resulting effects were compared with previously

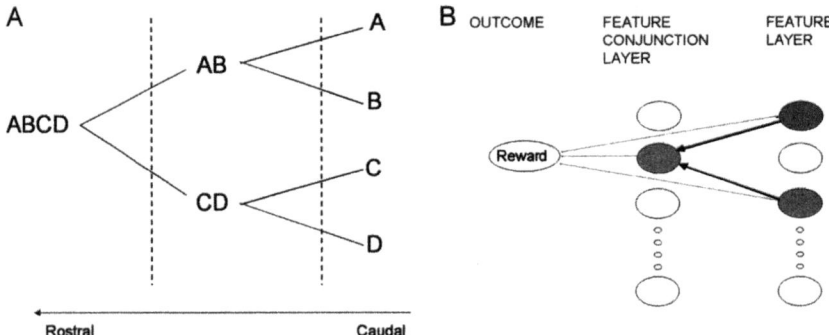

Figure 1.   (A) The proposed organization of object feature representations in the ventral visual stream. A, B, C, and D represent simple visual features encoded in caudal regions of the ventral visual stream. More complex representations of the conjunctions of these features are stored in more rostral regions. These conjunctive representations may reach maximum complexity in perirhinal cortex. (B) Diagram of the connectionist model. The network consists of two layers of units—the feature layer and the feature conjunction layer—as well as an outcome node representing a consequent event (e.g., reward). Each node in the feature layer represents a single element of a stimulus. The feature layer is connected to the feature conjunction layer via a set of fixed weights. Thus the structure of the model is similar to that of Pearce (1994), with the important exception that "elemental" representations in the feature layer are connected to the outcome node, allowing learning to proceed via both the feature and the feature conjunction layers. Active units are shown in grey. Note that, for simplicity, only two feature layer units are shown to be active, whereas in the simulations 10 units per stimulus are active. Both the feature layer and the feature conjunction layer are fully connected to the reward node. These weights are adjustable via an associative mechanism. The feature conjunction layer represents perirhinal cortex, and the feature layer represents more caudal regions of the ventral visual stream.

reported effects of lesions in perirhinal cortex in monkeys (Bussey & Saksida, 2002). The model was able to simulate with remarkable accuracy the effects of lesions of perirhinal cortex.

Although we first simulated extant data, the real value of a neural network model is seen when it makes predictions that can be tested in subsequent experiments. To this end, we designed a task for monkeys that would require them to use representations of complex conjunctions of features. We did this by using a visual discrimination problem in which *combinations* of features, rather than any individual feature, predict the correct item. In the context of discrimination learning, this can be arranged by having individual features appear as part of both the correct (S+) and the incorrect (S−) objects in a pair. We have termed this property of discrimination problems "feature ambiguity"; that is, feature ambiguity occurs when a feature is rewarded when it is part of one object but is not rewarded when it is part of another object. In discrimination problems with this property, an individual feature is insufficient to indicate whether a particular object is rewarded; only a combination of features will serve to guide correct responses. Here we review three different experiments designed to test directly the PMFC model. In the first two experiments, we examined the effects of systematically varying feature ambiguity on acquisition of visual discriminations. Experiment 1 used specially constructed stimuli, and Experiment 2 used morphed (blended) stimuli to modulate feature ambiguity. Experiment 3, like Experiment 2, used morphed images to achieve a high degree of feature ambiguity; this time, however, we examined monkeys' choices when feature ambiguity was abruptly introduced to already-learned discriminations.

Experiment 1: Feature ambiguity versus set size

Buckley and Gaffan (1997), in their series of studies on the contributions of perirhinal cortex to discrimination learning, reported that monkeys with perirhinal cortex lesions were impaired relative to unoperated control monkeys when learning a large but not a small number of concurrent pair-wise visual discriminations. According to the PMFC model, however, it is not the number of objects that is the critical factor, but rather the degree of feature ambiguity. According to this view, perirhinal cortex lesions affect concurrent discrimination of large but not small numbers of object pairs because the former possess greater feature ambiguity: As the number of object pairs to be discriminated becomes larger, the probability increases that a given object feature will be rewarded when part of one object, but unrewarded when part of another.

To test directly the model's prediction that animals with perirhinal cortex lesions have difficulty acquiring visual discriminations under conditions of high feature ambiguity, we assessed performance of control monkeys and monkeys with lesions of perirhinal cortex on a series of concurrent discriminations in which the number of items to be discriminated was held constant, but the degree of feature ambiguity was varied (Bussey, Saksida, & Murray, 2002). Pairs of "objects", each consisting of two "features", were presented on each trial, and monkeys were tested in three conditions: maximum feature ambiguity, in which all four features were explicitly ambiguous (AB+, CD+, BC−, AD−; the biconditional problem; Saavedra, 1975); minimum feature ambiguity, in which no features were explicitly ambiguous (AB+, CD+, EF−, GH−); and intermediate feature ambiguity, in which two features were explicitly ambiguous (AB+, CD+, CE−, AF−; see Figure 2). As shown in Figure 3, the pattern of results closely matched that predicted by simulations using the connectionist network: Monkeys with perirhinal cortex lesions were unimpaired in the minimum feature ambiguity condition, mildly impaired in the intermediate feature ambiguity condition, and severely impaired in the maximum feature ambiguity condition. Because only feature ambiguity was varied, while holding the set size constant, this finding showed that it is feature ambiguity, and not the number of items to be discriminated per se, that is the critical factor (see also Lee, Barense, & Graham, this issue). This account also explains why Buckley and Gaffan (1997, 1998a), using a relatively small number of stimulus pairs, observed a substantial impairment in monkeys with perirhinal cortex lesions on a "configural" discrimination task, whereas a much larger stimulus set was required to reveal an impairment in concurrent discrimination learning. This is because in configural discrimination problems, feature ambiguity is explicitly arranged to be at a maximum. By contrast, in concurrent discrimination problems, feature ambiguity increases with increasing set size because as the number of to-be-discriminated objects increases, the probability increases that a given feature will be rewarded when part of one object, but not when part of another; features become ambiguous by chance. In this sense concurrent discrimination of large numbers of object pairs can be thought of as "partial" configural discriminations, similar to the intermediate condition of Experiment 1. This is perhaps why experiments investigating the effects of lesions including perirhinal cortex on acquisition of single-pair or concurrent visual discriminations have yielded conflicting results, with some yielding significant impairments (Baxter & Murray, 2001a; Buckley & Gaffan, 1997; Buffalo et al., 1999) and others not (Buckley & Gaffan, 1997; Buffalo et al., 1999; Eacott et al., 1994;

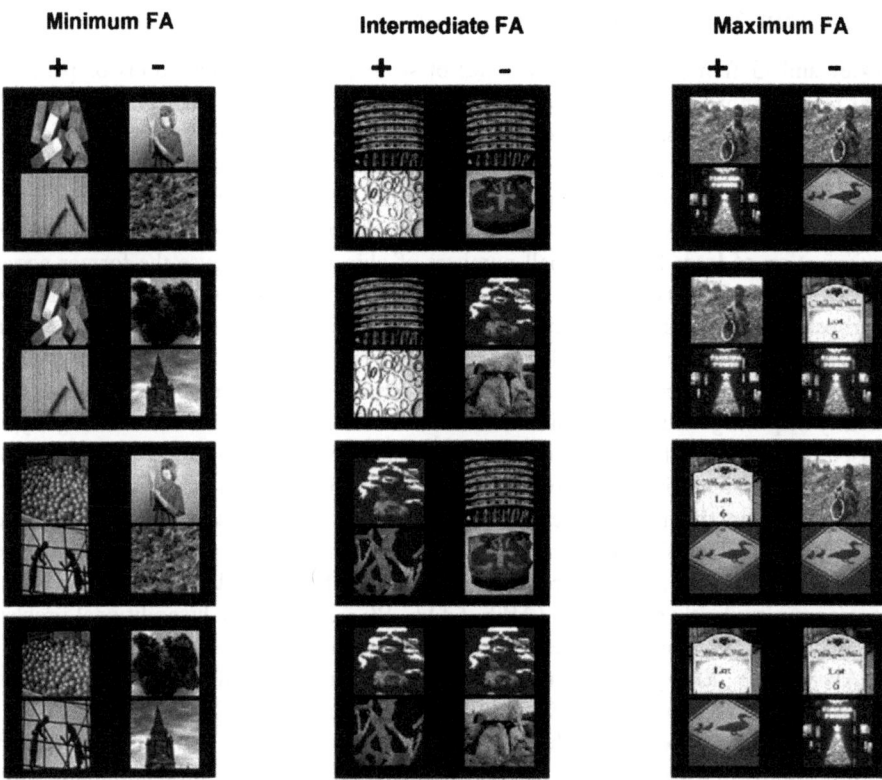

Figure 2.   Example of stimulus pairs in the minimum, intermediate, and maximum feature ambiguity (feature ambiguity) conditions of Experiment 1. In the minimum condition, no features were explicitly ambiguous (i.e., each feature was consistently either rewarded or nonrewarded). In the intermediate condition, two features were explicitly ambiguous. In the maximum condition, all four features were explicitly ambiguous. The "features" comprising each two-feature stimulus were randomly selected greyscale photographs obtained from a commercially available clip-art collection. The + and − indicate that the stimuli on the left were the correct (rewarded) images in the pair, whereas those on the right were incorrect (unrewarded). For pairs of stimuli presented in the test sessions, the location of the S + (left or right) followed a pseudorandom order.

Gaffan & Murray, 1992; Thornton, Rothblat, & Murray, 1997). Because the stimulus material and object set sizes varied considerably across these studies, the amount of feature ambiguity probably varied considerably as well, with the result that perirhinal cortex lesions impaired acquisition in some cases and not others.

Experiment 2: Acquisition of single-pair visual discriminations

A second experiment tested the role of perirhinal cortex in perception more directly. Although acquisition of single-pair visual discriminations typically is unaffected by perirhinal cortex lesions, we postulated that if feature ambiguity is important in revealing an impairment, then

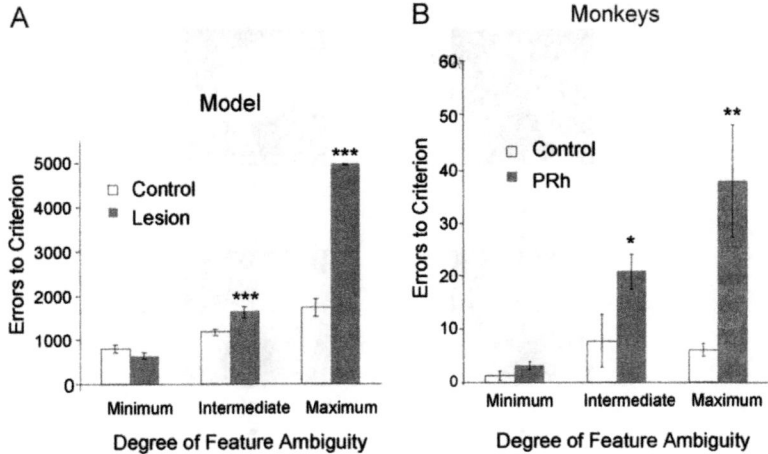

Figure 3. (A) Simulation data generated by the connectionist network. Bars indicate the number of errors committed by intact networks and networks with "lesions" of the feature conjunction layer during acquisition of four-pair concurrent discriminations in each of the minimum, intermediate, and maximum feature ambiguity conditions. Error bars indicate $+/-$ standard error of the mean. Asterisks indicate significant differences between the groups; $p < .00001$. Control-intact networks ($N = 4$); Lesion-networks with the feature conjunction layer removed ($N = 4$). (B) Acquisition by control monkeys and monkeys with lesions of the perirhinal cortex of four-pair concurrent discriminations in each of the minimum, intermediate, and maximum feature ambiguity conditions. Scores are the group mean errors to criterion for four sets of problems in each condition. Error bars indicate $+/-$ standard error of the mean. Asterisks indicate significant differences between the groups; * $p < .05$; ** $p = .01$. Control-unoperated control monkeys ($N = 4$). PRh-monkeys with bilateral lesions of the perirhinal cortex ($N = 4$). Data from Bussey, Saksida, and Murray (2002).

we should be able to induce a deficit even with acquisition of a single-pair discrimination. In this set of experiments (Bussey, Saksida, & Murray, 2003), we increased feature ambiguity in greyscale picture discriminations by blending together the stimuli using commercially available "morphing" software (see Figure 4). As the pictures are blended together, they contain more features in common. As in the case of the constructed objects used in Experiment 1, we reasoned that successful discrimination would require representations of feature conjunctions. Thus it was predicted that monkeys with perirhinal cortex lesions should be disproportionately impaired in acquisition of discrimination problems with high feature ambiguity, relative to those with low feature ambiguity.

Monkeys were tested on a series of "low feature ambiguity" (Figure 4, Images 1 and 40) and "high feature ambiguity" (Figure 4, Images 17 and 24) single-pair discrimination problems. In the high feature ambiguity pair, the stimuli share many common features, which are rewarded when part of the S+ but not when part of the S−. As shown in Figure 5, results from the monkey study confirmed the predictions of the model: Monkeys with perirhinal cortex lesions were severely impaired in acquiring the high feature ambiguity discriminations, whereas there were no significant differences between the groups in the low feature ambiguity condition. The same monkeys were not impaired on difficult size or colour discriminations (which did not involve ambiguous features), indicating that impairments in discrimination learning were not related simply to task difficulty. According to the PMFC

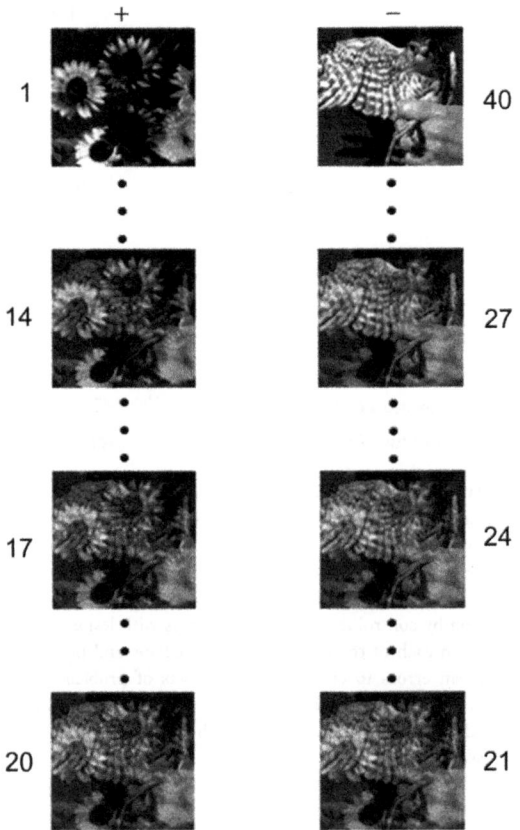

Figure 4. Example of greyscale picture stimuli used in Experiments 2 and 3. To create each pair of stimuli, two pictures were "morphed" (blended) together to create a series of 40 images, the first images in this series consisting mostly of features from Picture 1, the latter images consisting mostly of features from Picture 40. The numbers next to an image indicate the position of the image in the series. In Experiment 2 the original pictures, images 1 and 40, were used as the low feature ambiguity pair, and the 14th and 27th Images were used as the high feature ambiguity test pair. In Experiment 3, the original images were used as the trained pair, the 14th and 27th images were used as Morph Pair 1, and the 17th and 24th images were used as Morph Pair 2. In both of these experiments, monkeys were tested on four such stimulus sets, and the data were averaged across sets. The + and − indicate that the images on the left, in this example those most similar to the photograph of the sunflowers, were the correct (rewarded) images in the pair, whereas those on the right were incorrect (unrewarded).

model, perirhinal cortex lesions yield impairment in the high feature ambiguity condition because complex conjunctive representations in this region serve to disambiguate the discriminanda. That is, conjunctive representations in perirhinal cortex are thought to resolve feature ambiguity by providing the additional information that the features present in the stimulus "belong together": "The whole is greater than the sum of the parts". Interestingly, recent electrophysiological evidence supports the existence of neurons in inferotemporal cortex that respond selectively to stimulus wholes and respond negligibly to the features that comprise those wholes (Baker, Behrmann, & Olson, 2002).

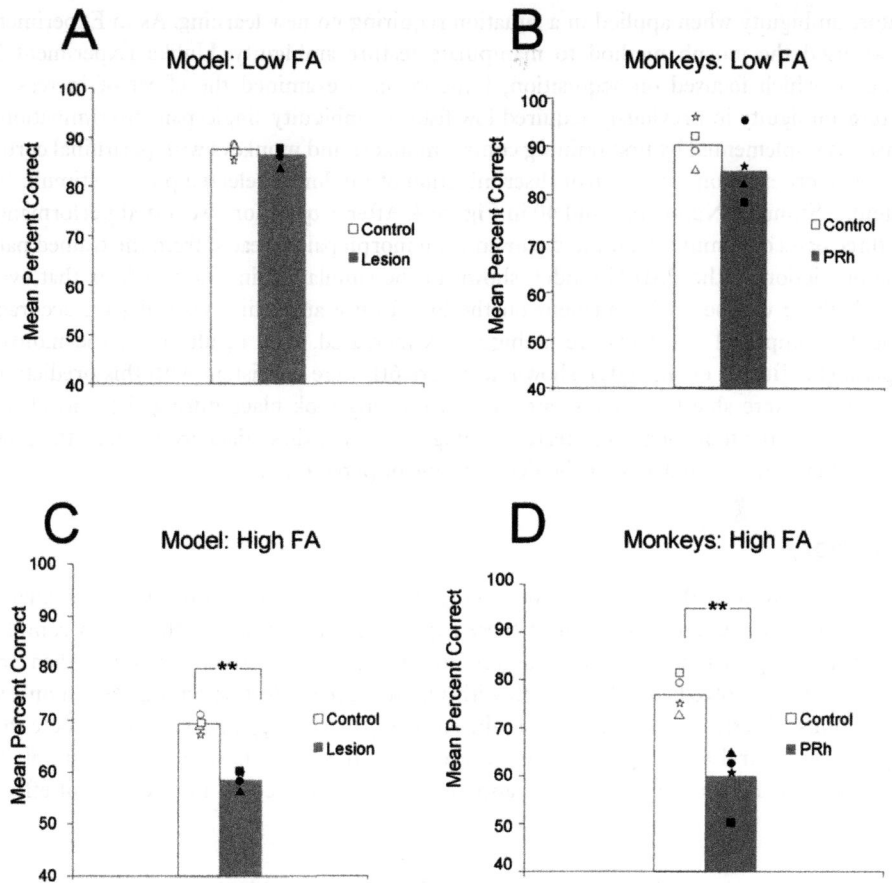

Figure 5. (A) Simulation of the acquisition of discriminations of stimuli with low feature ambiguity, using the connectionist network. Data are percentage of correct scores averaged across training trials. (B) Acquisition of discriminations of stimuli with low feature ambiguity, by control monkeys and monkeys with lesions of perirhinal cortex. (C) Simulation of acquisition of discriminations of stimuli with high feature ambiguity. (D) Acquisition of discriminations of stimuli with high feature ambiguity, by control monkeys and monkeys with lesions of perirhinal cortex. Asterisks indicate significant differences between the groups; **$p < .01$. Other conventions as in Figure 3. Data are from Bussey, Saksida, and Murray (2003).

Experiment 3: Performance of single-pair visual discriminations

Although Experiments 1 and 2 supported the PMFC model, they could be criticized on the grounds that they have a memory component. That is, even though we designed the experiments so that the memory demands were equal across conditions, and even though we were able to rule out the possibility that the impairment is related to difficulty, per se, there might be an interaction of perception and memory that could influence the outcome of those experiments (see Hampton, this issue). Accordingly, we decided to measure the effects of feature ambiguity when applied in a situation requiring no new learning. As in Experiment 2, we used the morph method to manipulate feature ambiguity. Unlike Experiment 2, however, which focused on acquisition, Experiment 3 examined the effect of increasing feature ambiguity in previously acquired low feature ambiguity single-pair discriminations. This was implemented by first training control monkeys and monkeys with perirhinal cortex lesions to criterion on a single-pair discrimination of randomly selected picture stimuli: for example, Stimulus Numbers 1 and 40 in Figure 4. After acquisition, we tested performance on three pairs of stimuli: the trained pair and two morph pairs created from the trained pair. The prediction of the PMFC model, shown in the simulation in Figure 6A, is that even though there will be no impairment on the low feature ambiguity trained pair, accuracy should be impaired when feature ambiguity is increased. The results from the monkey experiment (Bussey et al., 2003), shown in Figure 6B, were consistent with this prediction. Because we were able to demonstrate that no learning took place during the critical test sessions with the morphed (high feature ambiguity) pairs, these data are perhaps the most compelling evidence to date that the deficit is one of perception.

Summary

The studies outlined above provide converging evidence in support of our PMFC model of perirhinal cortex function. In each of these experiments it was shown that when complex conjunctive representations were required for normal performance—that is, when the discrimination involved complex stimuli with a high degree of feature ambiguity—monkeys with lesions of perirhinal cortex were significantly impaired. It is important also to note that appropriate control tasks were included in each case; these effects were not due simply to task difficulty. Thus the PMFC model appears to be able to account for a variety of effects

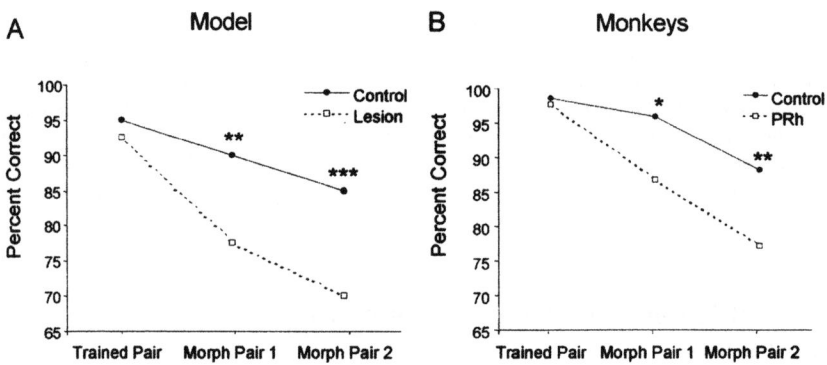

Figure 6.   (A) Discrimination performance of intact networks, and networks with the feature conjunction layer removed, on the low feature ambiguity *trained pair*, and on the high feature ambiguity *Morph Pair 1* and *Morph Pair 2* (Experiment 3). Asterisks indicate significant differences between the groups; **$p < .01$; ***$p = .001$. (B) Discrimination performance of control monkeys and monkeys with lesions of perirhinal cortex (PRh) on the low feature ambiguity *trained pair*, and on the high feature ambiguity *Morph Pair 1* and *Morph Pair 2* (Experiment 3). Asterisks indicate significant differences between the groups; *$p < .05$; **$p < .01$. Data from Bussey, Saksida, and Murray (2003).

of perirhinal cortex lesions on visual discriminations. How such a model fits into a more comprehensive theory of perirhinal function, encompassing, for example, object recognition and stimulus–stimulus associative learning, is a target for future research.

The foregoing view is very different from the view that perirhinal cortex is important for memory—specifically declarative memory—and has little to do with visual information processing. In line with our conception of perirhinal cortex as part of the ventral visual stream, the impairments outlined above could be regarded as perceptual: Monkeys with perirhinal cortex lesions do not exhibit a general impairment in the ability to associate stimuli with rewards, but they do exhibit impairments in certain perceptually challenging discrimination tasks. Thus it appears that perirhinal cortex is important for both normal memory and normal perception. The effects of perirhinal cortex lesions on visual discrimination appear to be due to compromised representations, rather than impairment of a specific type of learning or memory (although see Hampton, this issue).

Concluding comments

The model and results outlined above suggest a new way of understanding the effects of brain damage in the occipito-temporal processing stream. This new view has, predictably, been challenged. As indicated earlier, some investigators have argued that perirhinal cortex cannot be involved in perception because perirhinal cortex lesions do not impair the short-delay condition of object recognition tasks (Buffalo et al., 1998, 1999; Holdstock et al., 1995). There is a problem, however, with this interpretation. First, these studies do not distinguish between perceptual and mnemonic function because although memory load is increased by extending the delay until an impairment emerges, nothing is done to challenge perception. The conclusion that perception is normal is therefore unjustified. Second, when studies have increased perceptual load under the short delay condition to see if an impairment might emerge, impairment was indeed observed (Eacott et al., 1994). Because Eacott et al. used an automated apparatus, they were able to evaluate performance on both simultaneous and 0-s delay conditions; it was these conditions—those with little or no memory requirement—on which monkeys with lesions involving perirhinal cortex were impaired.

In another critique (Squire et al., 2004), much emphasis is placed on a single study (Hampton & Murray, 2002) that found no perceptual impairments in monkeys with perirhinal cortex lesions, even when perception was challenged. These authors trained monkeys on a large number of visual discrimination problems and tested for transfer under several conditions including those in which stimuli were rotated, made smaller or larger, or degraded with a mask. In each case, the manipulation reduced accuracy, indicating that the conditions increased difficulty, but there was no interaction of condition and lesion. However, as Hampton and Murray point out, their findings are in no way incompatible with the PMFC model. The model predicts specifically that a critical factor in yielding such impairments is feature ambiguity, not difficulty per se; because Hampton and Murray did not manipulate feature ambiguity, there is no conflict.

Finally, some authors have taken the angle that the findings in monkeys may not be applicable to humans (Stark & Squire, 2000). This radical pronouncement appears to be premature. Lee, Graham, and colleagues have applied some of the tests used in monkeys to humans with focal medial temporal lobe lesions. These results indicate that humans with perirhinal cortex damage, like monkeys with such damage, have difficulty discriminating objects (and faces) in conditions of high feature ambiguity (Lee et al., this issue; Lee et al., 2005).

In favour of the PMFC model are several additional findings, some of which are reviewed in this issue. For example, Buckley and colleagues have examined visual object processing in monkeys with perirhinal cortex lesions, using methods different from those used here, and conclude that perirhinal cortex lesions disrupt perception (Buckley, this issue). In addition, Eacott and Gaffan, in experiments in rats, have likewise found that perirhinal cortex is essential for learning visual discriminations with high feature ambiguity (Eacott & E. Gaffan, this issue). These authors were also able to show, using a "constant negative" discrimination designed to minimize memory requirements, that rats with perirhinal cortex damage have difficulty representing aspects of object information, but not spatial information. Furthermore, the recent finding of impairments following inactivation of perirhinal cortex selectively during the encoding phase of an object recognition task is consistent with our perceptual-mnemonic view (Winters & Bussey, 2005, in press). Still, more work is needed to elucidate the conditions under which perirhinal cortex is essential for visual perception and object identification, and what neural mechanisms underlie these functions. We will no doubt be hearing much more from the so-called "silent" cortex....

REFERENCES

Baker, C. I., Behrmann, M., & Olson, C. R. (2002). Impact of learning on representation of parts and wholes in monkey inferotemporal cortex. *Nature Neuroscience*, *5*, 1210–1216.

Baxter, M. G., & Murray, E. A. (2001a). Impairments in visual discrimination learning and recognition memory produced by neurotoxic lesions of rhinal cortex in rhesus monkeys. *European Journal of Neuroscience*, *13*, 1228–1238.

Baxter, M. G., & Murray, E. A. (2001b). Opposite relationship of hippocampal and rhinal cortex damage to delayed nonmatching-to-sample deficits in monkeys. *Hippocampus*, *11*, 61–71.

Bright, P., Moss, H. E., Stamatakis, E. A., & Tyler, L. K. (this issue). The anatomy of object processing: The role of anteromedial temporal cortex. *Quarterly Journal of Experimental Psychology*, *58B*, 361–377.

Buckley, M. J. (this issue). The role of the perirhinal cortex and hippocampus in learning, memory, and perception. *Quarterly Journal of Experimental Psychology*, *58B*, 246–268.

Buckley, M. J., & Gaffan, D. (1997). Impairment of visual object-discrimination learning after perirhinal cortex ablation. *Behavioral Neuroscience*, *111*, 467–475.

Buckley, M. J., & Gaffan, D. (1998a). Perirhinal cortex ablation impairs configural learning and paired-associate learning equally. *Neuropsychologia*, *36*, 535–546.

Buckley, M. J., & Gaffan, D. (1998b). Perirhinal cortex ablation impairs visual object identification. *Journal of Neuroscience*, *18*, 2268–2275.

Buffalo, E. A., Ramus, S. J., Clark, R. E., Teng, E., Squire, L. R., & Zola, S. M. (1999). Dissociation between the effects of damage to perirhinal cortex and area TE. *Learning and Memory*, *6*, 572–599.

Buffalo, E. A., Reber, P. J., & Squire, L. R. (1998). The human perirhinal cortex and recognition memory. *Hippocampus*, *8*, 330–339.

Bussey, T. J., & Saksida, L. M. (2002). The organization of visual object representations: A connectionist model of effects of lesions in perirhinal cortex. *European Journal of Neuroscience*, *15*, 355–364.

Bussey, T. J., Saksida, L. M., & Murray, E. A. (2002). Perirhinal cortex resolves feature ambiguity in complex visual discriminations. *European Journal of Neuroscience*, *15*, 365–374.

Bussey, T. J., Saksida, L. M., & Murray, E. A. (2003). Impairments in visual discrimination after perirhinal cortex lesions: Testing declarative versus perceptual-mnemonic views of perirhinal cortex function. *European Journal of Neuroscience*, *17*, 649–660.

Charles, D. P., Gaffan, D., & Buckley, M. J. (2004). Impaired recency judgments and intact novelty judgments after fornix transection in monkeys. *Journal of Neuroscience*, *24*, 2037–2044.

Desimone, R., & Ungerleider, L. G. (1989). Neural mechanisms of visual processing in monkeys. In F. Boller & J. Grafman (Eds.), *Handbook of neuropsychology* (Vol. 2, pp. 267–299). New York: Elsevier Science.

Eacott, M. J., & Gaffan, E. A. (this issue). The roles of the perirhinal cortex, postrhinal cortex, and the fornix in memory for objects, contexts, and events in the rat. *Quarterly Journal of Experimental Psychology, 58B*, 202–217.

Eacott, M. J., Gaffan, D., & Murray, E. A. (1994). Preserved recognition memory for small sets, and impaired stimulus identification for large sets, following rhinal cortex ablations in monkeys. *European Journal of Neuroscience, 6*, 1466–1478.

Forwood, S. E., Winters, B. D., & Bussey, T. J. (in press). Hippocampal lesions that abolish spatial maze performance spare object recognition memory at delays of up to 48 hours. *Hippocampus, 15*.

Gaffan, D., & Murray, E. A. (1992). Monkeys (*Macaca fascicularis*) with rhinal cortex ablations succeed in object discrimination learning despite 24-hr intertrial intervals and fail at matching to sample despite double sample presentations. *Behavioral Neuroscience, 106*, 30–38.

Goulet, S., & Murray, E. A. (2001). Neural substrates of crossmodal association memory in monkeys: The amygdala versus the anterior rhinal cortex. *Behavioral Neuroscience, 115*, 271–284.

Hampton, R. R. (this issue). Monkey perirhinal cortex is critical for visual memory, but not for visual perception: Reexamination of the behavioural evidence from monkeys. *Quarterly Journal of Experimental Psychology, 58B*, 283–299.

Hampton, R. R., & Murray, E. A. (2002). Stimulus representations in rhesus monkeys with perirhinal cortex lesions. *Behavioral Neuroscience, 116*, 363–377.

Holdstock, J. S. (this issue). The role of the human medial temporal lobe in object recognition and object discrimination. *Quarterly Journal of Experimental Psychology, 58B*, 326–339.

Holdstock, J. S., Shaw, C., & Aggleton, J. P. (1995). The performance of amnesic subjects on tests of delayed matching-to-sample and delayed matching-to-position. *Neuropsychologia, 33*, 1583–1596.

Lee, A. C. H., Barense, M. D., & Graham, K. S. (this issue). The contribution of the human medial temporal lobe to perception: Bridging the gap between animal and human studies. *Quarterly Journal of Experimental Psychology, 58B*, 300–325.

Lee, A. C., Bussey, T. J., Murray, E. A., Saksida, L. M., Epstein, R. A., Kapur, N., et al. (2005). Perceptual deficits in amnesia: Challenging the medial temporal lobe 'mnemonic' view. *Neuropsychologia, 43*, 1–11.

Manns, J. R., Stark, C. E., & Squire, L. R. (2000). The visual paired-comparison task as a measure of declarative memory. *Proceedings of the National Academy of Science USA, 97*, 12375–12379.

Meunier, M., Bachevalier, J., Mishkin, M., & Murray, E. A. (1993). Effects on visual recognition of combined and separate ablations of the entorhinal and perirhinal cortex in rhesus monkeys. *Journal of Neuroscience, 13*, 5418–5432.

Murray, E. A. (2000). Memory for objects in nonhuman primates. In M. S. Gazzaniga (Ed.), *The new cognitive neurosciences*. London: The MIT Press.

Murray, E. A., & Bussey, T. J. (1999). Perceptual-mnemonic functions of perirhinal cortex. *Trends in Cognitive Sciences, 3*, 142–151.

Murray, E. A., Malkova, L., & Goulet, S. (1998). Crossmodal associations, intramodal associations, and object identification in macaque monkeys. In A. D. Milner (Ed.), *Comparative neuropsychology* (pp. 51–67). Oxford, UK: Oxford University Press.

Murray, E. A., & Mishkin, M. (1998). Object recognition and location memory in monkeys with excitotoxic lesions of the amygdala and hippocampus. *Journal of Neuroscience, 18*, 6568–6582.

Nadel, L. (1995). The role of the hippocampus in declarative memory: A comment on Zola-Morgan, Squire, and Ramus (1994). *Hippocampus, 5*, 232–239.

Nemanic, S., Alvarado, M. C., & Bachevalier, J. (2004). The hippocampal/parahippocampal regions and recognition memory: Insights from visual paired comparison versus object-delayed nonmatching in monkeys. *Journal of Neuroscience, 24*, 2013–2026.

Pearce, J. M. (1994). Similarity and discrimination: A selective review and a connectionist model. *Psychological Review, 101*, 587–607.

Penfield, W., & Jasper, H. (1954). *Epilepsy and the functional anatomy of the human brain*. Boston: Little, Brown.

Saavedra, M. A. (1975). Pavlovian compound conditioning in the rabbit. *Learning and Motivation, 6*, 314–326.

Squire, L. R., Stark, C. E., & Clark, R. E. (2004). The medial temporal lobe. *Annual Reviews Neuroscience, 27*, 279–306.

Squire, L. R., & Zola-Morgan, S. (1991). The medial temporal lobe memory system. *Science, 253*, 1380–1386.

Stark, C. E., & Squire, L. R. (2000). Intact visual perceptual discrimination in humans in the absence of perirhinal cortex. *Learning and Memory, 7*, 273–278.

Thornton, J. A., Rothblat, L. A., & Murray, E. A. (1997). Rhinal cortex removal produces amnesia for preoperatively learned discrimination problems but fails to disrupt postoperative acquisition and retention in rhesus monkeys. *Journal of Neuroscience, 17*, 8536–8549.

Ungerleider, L. G., & Mishkin, M. (1982). Two cortical visual systems. In D. J. Ingle, M. A. Goodale, & R. J. W. Mansfield (Eds.), *Analysis of visual behavior* (pp. 549–586). Cambridge, MA: MIT Press.

Winters, B. D., & Bussey, T. J. (2005). Transient inactivation of perirhinal cortex disrupts encoding, retrieval, and consolidation of object recognition memory. *Journal of Neuroscience, 25*, 52–61.

Winters, B. D., & Bussey, T. J. (in press). Glutamate recceptors in perirhinal cortex mediate encoding, retrieval, and consolidation of object recognition memory. *Journal of Neuroscience.*

Winters, B. D., Forwood, S. E., Cowell, R. A., Saksida, L. M., & Bussey, T. J. (2004). Double dissociation between the effects of peri-postrhinal cortex and hippocampal lesions on tests of object recognition and spatial memory: Heterogeneity of function within the temporal lobe. *Journal of Neuroscience, 24*, 5901–5908.

THE QUARTERLY JOURNAL OF EXPERIMENTAL PSYCHOLOGY
2005, 58B (3/4), 283–299

Monkey perirhinal cortex is critical for visual memory, but not for visual perception: Reexamination of the behavioural evidence from monkeys

Robert R. Hampton

Emory University, Atlanta, GA, USA

Overdependence on discrimination learning paradigms to assess the function of perirhinal cortex has complicated understanding of the cognitive role of this structure. Impairments in discrimination learning can result from at least two distinct causes: (a) failure to accurately apprehend and represent the relevant stimuli, or (b) failure to form and remember associations between stimulus representations and reward. Thus, the results of discrimination learning experiments do not readily differentiate deficits in perception from deficits in learning and memory. Here I describe studies that do dissociate learning and memory from perception and show that perirhinal cortex damage impairs learning and/or memory, but not perception. Reanalysis and reconsideration of other published data call into further question the hypothesis that the monkey perirhinal cortex plays a critical role in visual perception.

The perirhinal cortex appears well placed in the brain to participate in both memory and perception. It is located at the downstream end of the ventral visual processing stream where it could contribute to the final stages of visual perceptual processing (Murray & Bussey, 1999). Alternatively, this access to highly processed visual input, combined with strong connections to the hippocampus via entorhinal cortex (Suzuki, 1996), could support a critical role in memory. Neurons in perirhinal cortex also have properties consistent with either role. They have large receptive fields (Jagadeesh, Chelazzi, Mishkin, & Desimone, 2001) and respond selectively to complex visual stimuli (Logothetis, 1998). Behavioural studies of monkeys with perirhinal cortex removed have been used to support both mnemonic and perceptual roles for perirhinal cortex. Delay-dependent deficits in recognition memory have

Correspondence should be addressed to Robert R. Hampton, Department of Psychology and Yerkes National Primate Research Center, Emory University, 532 Kilgo Circle, Atlanta, GA 30322, USA. Email: robert.hampton@emory.edu

The experimental work conducted by the author was supported by the National Institute of Mental Health Intramural Research Program. Preparation of the manuscript was supported by Yerkes National Primate Research Center Base Grant RR00165. Heather Kirby provided assistance with the figures and provided comments on earlier drafts.

© 2005 The Experimental Psychology Society
DOI:10.1080/02724990444000195

been observed following removal of perirhinal cortex (e.g., Buffalo et al., 1999; Buffalo, Ramus, Squire, & Zola, 2000; Eacott, Gaffan, & Murray, 1994; Meunier, Bachevalier, Mishkin, & Murray, 1993; Zola-Morgan, Squire, Amaral, & Suzuki, 1989) and have generally been interpreted as reflecting amnesia. Other deficits following perirhinal cortex removal, although not inconsistent with a mnemonic role, have suggested contributions to perception. For example, the formation and maintenance of stimulus–stimulus associations between individual objects depend on the perirhinal cortex, and such associations may contribute to the ability to perceive whole objects (Buckley & Gaffan, 1998a; Eacott & Gaffan, this issue; Murray, 2000; Murray, Gaffan, & Mishkin, 1993). Based on the inference of interference between visual representations in discrimination learning, Buckley and Gaffan (1997, 1998b) argued that perirhinal cortex lesions cause impairments in perception or representation of stimuli. Thus, recent reports have posited a role for perirhinal cortex either exclusively in memory (Buffalo et al., 1999; Hampton & Murray, 2002) or in memory as well as in perception (Buckley, this issue; Buckley & Gaffan, 1997, 1998b; Bussey, Saksida, & Murray, this issue; Eacott & Heywood, 1995; Lee, Barense, & Graham, this issue; Murray, 2000; Murray & Bussey, 1999; Murray, Malkova, & Goulet, 1998).

In this paper I argue that there is strong evidence indicating that the perirhinal cortex is critical for learning and memory in monkeys, whereas support for the hypothesis that this structure contributes significantly to perceptual processes is comparatively weak. In making this argument I adopt specific definitions of perception and memory. While there is no a priori reason to presume that perception and memory do not share neural substrates, potentially including the perirhinal cortex (e.g., Murray & Bussey, 1999), the processes of perception and memory can be distinguished behaviourally. *Perception* involves the immediate apprehension and initial representation of stimuli currently impinging on a sensory surface. Complementing this role of perception in initially representing stimuli, *memory* involves the retention and retrieval of representations when stimuli are absent and therefore are not stimulating a sensory surface. For example, an otherwise normal subject that cannot detect the difference in colour between simultaneously presented squares is said to fail to perceive the colour difference. In contrast, a memory failure is inferred if the subject can correctly indicate that two squares differ in colour, but cannot indicate which one was recently seen and then withdrawn from view. Use of these definitions makes it straight forward to implement experiments that test the two processes independently. However, adopting these definitions does not commit one to the position that the neural substrates of memory and perception are distinct. Deficits in both perception and memory could be observed following selective brain damage to a given structure. Finally, it should be recognized that memory is not confined to retention of isolated representations, but can entail encoding of relationships such as lightening predicts thunder, and cardinals are birds that are red.

I apply these definitions of perception and memory to a review of the evidence regarding the function of the monkey perirhinal cortex, focusing exclusively on work with monkeys. In the first section I present a critique of some of the reports often cited in support of the hypothesis that perirhinal cortex is critical for visual perception and argue that reanalysis of the results indicates memory deficits rather than perceptual deficits. In the second section I summarize a series of experiments that demonstrate intact perception but impaired memory in monkeys lacking perirhinal cortex. In the third section, the significance of experiments testing the perceptual-mnemonic model of perirhinal cortex function in monkeys is considered.

What evidence indicates that perirhinal cortex contributes to perception?

The significance of the perirhinal cortex first became apparent through comparison with the function of the hippocampus, a structure widely believed to be critical for some types of memory. Much of the early work on the perirhinal cortex in monkeys was therefore motivated by interest in memory rather than perception. A series of studies conducted in the 1980s and 1990s showed that lesions of the perirhinal cortex produced recognition memory deficits (as measured by matching- and nonmatching-to-sample) nearly as severe as those resulting from large medial temporal lobe aspiration lesions that included the hippocampus (Meunier et al., 1993; Murray & Mishkin, 1986, 1998; Zola-Morgan et al., 1989). However, the observation of a set size effect (Eacott et al., 1994) in one early experiment directed thinking towards perceptual functions of perirhinal cortex. Using an automated matching-to-sample procedure, these authors found that the magnitude of the recognition memory impairment observed following perirhinal cortex removal was a function of the number of discriminanda employed in the memory tests. When stimuli were essentially trial unique, a large impairment was observed, whereas if a small set of highly familiar images was used, the deficit was relatively mild (and the task was very difficult for control and experimental animals alike). Because these results showed that the monkeys lacking perirhinal cortex were impaired on some but not other memory tests, they triggered a search for non mnemonic accounts of perirhinal cortex function.

Possibly stimulated by the set size effects observed in recognition memory by Eacott and colleagues, Buckley and Gaffan (1997) investigated the role of set size in discrimination learning. In discrimination learning, subjects learn over repeated presentations that one of two simultaneously presented images or objects is associated with food reward while the other is not. Successful performance requires both the ability to discriminate the items in each pair from one another and the ability to learn and remember the association of one of the items with food reward. In concurrent object discrimination learning, more than one discrimination problem is presented in each test session. Set size refers to the number of discrimination problems learned simultaneously, or concurrently. Buckley and Gaffan (1997) reported that monkeys with perirhinal cortex lesions were impaired in learning large, but not small, sets of object discrimination problems. This is probably the piece of evidence most often cited in support of the hypothesis that perirhinal cortex is critical for perception. Reanalysis of the original data demonstrates that there is in fact no set size effect.

Buckley and Gaffan (1997) proposed that stimuli in larger sets are more likely to be perceptually confused with one another than are stimuli in smaller sets. Due to the perceptual challenge inherent in learning a large set of discrimination problems simultaneously, they predicted that monkeys with perirhinal cortex removed would be especially impaired in learning large problem sets. The critical comparison in this study was of the number of errors made learning sets of 20, 40, or 80 object discriminations concurrently. While the results initially appear to demonstrate a perceptual impairment following perirhinal cortex removal, further examination shows this not to be the case. The following concrete example helps clarify the issue.

Two groups of monkeys make 5 and 10 errors, respectively, in learning a single discrimination problem. By a simple process of multiplication the groups would be expected to make 50 and 100 errors, respectively, in learning a set of 10 problems of equal difficulty. Note that the magnitude of the difference between the groups is a function of the number of problems learned. The groups differ by 5 errors in learning a single problem, but differ by 50 errors in learning 10 problems (10 times as many problems, 10 times as many errors). However, in both cases the monkeys make 5 and 10 errors, respectively, *per problem*, and the ratio of errors made by the two groups is a constant 2 to 1. Analysis of variance (ANOVA) of the errors made learning 1 problem and learning the set of 10 problems would yield a significant Group × Set Size interaction. Such a result could give the false impression that learning becomes more difficult as the number of problems learned increases. If it were actually more difficult for the slower group to learn discriminations as the set size is increased, a nonlinear increase in the difference in errors between the groups would be observed. That is, the slow group would commit more errors per problem in learning a larger set, and the number of errors made learning each problem by the two groups would change as set size increased. To assess the difficulty of learning each discrimination problem as a function of set size, performance should be measured as errors per problem (a ratio measure), rather than total errors.

In comparing the rate at which normal monkeys and monkeys lacking perirhinal cortex learned discrimination problems in different set sizes, Buckley and Gaffan (1997) failed to take into account the difference between errors per problem and total errors. The importance of this difference is highlighted graphically in Figure 1, which plots data from Buckley and Gaffan (1997; data for individual monkeys were provided in Table 1 of that paper). The left panel of Figure 1, which plots total errors, gives the impression of an interaction between group and set size. In contrast, the right panel indicates no such interaction. Rather, there is a consistent, marginally significant deficit in the monkeys lacking perirhinal cortex (mean errors/problem at each set size for monkeys lacking perirhinal cortex and control monkeys, respectively: Set Size 20, 6.07 vs. 3.34; Set Size 40, 7.26 vs. 3.44; Set Size 80, 7.44 vs. 3.5);

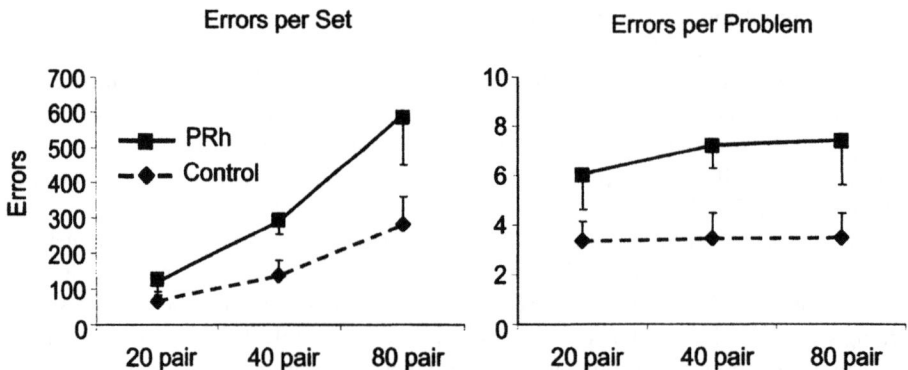

Figure 1.   Graphs of data from Buckley and Gaffan (1997). Both the left and the right panels plot errors made learning three sets of discrimination problems by control monkeys and monkeys lacking perirhinal cortex. Total errors committed while learning the discrimination problem sets is plotted in the left panel. The more appropriate measure of errors per discrimination problem is plotted in the right panel. Note that the left panel gives the impression of an interaction between group and set size, while the right panel does not.

group, $F(1, 6) = 5.17$, $p = .06$; Set Size, $F(2,12) = 0.53$, $p = .60$; Group $\times$ Set Size, $F(2, 12) = 0.66$, $p = .53$; log 10 transformed data. Statistically identical results are obtained with the untransformed data. This consistent difference in errors per problem is exactly the result predicted by a deficit in learning or memory rather than a deficit in perception. In evaluating the significance of this negative evidence, it should be noted that the perceptual account of perirhinal cortex might be interpreted to predict a huge increase in errors when monkeys learn the largest set because they have to learn this set after already learning many other discriminations earlier in the study.

In another set of influential studies, Buckley and Gaffan (1998b) presented monkeys with digitized images of objects taken from six different perspectives. After the discriminations had been learned from three of the six perspectives, the monkeys were tested for their ability to relearn discriminations of the same objects using the photographs taken from the remaining three perspectives. Because generalization to the new views is thought to be perceptually challenging, Buckley and Gaffan (1998b) predicted that monkeys with perirhinal cortex removed would be impaired relative to controls. Indeed, the perirhinal group was impaired in relearning the discriminations. But this result in isolation does not demonstrate that the deficit was due to a perceptual impairment. If tests with the new views taxed perception, and monkeys with perirhinal cortex removed were perceptually impaired, then the impairment in this group should be greater in relearning tests than in the initial learning of the discriminations. Otherwise the newly observed deficit simply recapitulates the deficit observed in initial learning. In initial learning of 40 discrimination problems, Buckley and Gaffan's (1998b) operated monkeys made, on average, nearly twice as many errors as did experimental animals (214 vs. 120 errors; inferential statistics were not reported). On the transfer test, controls made an average of 11 errors, while operated monkeys made an average of 41 errors (a significant difference by the one-tailed test reported). Given that one of the three operated monkeys made more than twice as many errors as did the other two, thus substantially biasing the mean, it seems unlikely that these scores represent a significant change from the 2:1 difference observed in initial learning. Monkeys lacking perirhinal cortex may have been impaired on the transfer test, not because of problems with generalization to the new views per se, but because of a general impairment in visual discrimination learning. This learning impairment could occur because in the experimental design used by Buckley and Gaffan (1998b) objects were photographed from different vantage points. As a result, new visual features appeared on the transfer tests (e.g., the back side of the objects). Because the transfer tests thereby required learning of new associations between visual features and reward, the observed impairment in monkeys lacking perirhinal cortex could be due to a learning or memory impairment (see Buckley, this issue, for further reanalysis of their data).

In a follow-up experiment the same monkeys were required to discriminate a subset of the familiar objects embedded in novel, cluttered scenes (Buckley & Gaffan, 1998b). Embedding the images in cluttered scenes was intended to tax perception. Monkeys with perirhinal cortex lesions were again impaired in relearning the discriminations to criterion, accruing double the errors made by controls (246 vs. 123 errors). Again, however, this manipulation of perceptual difficulty did not increase the magnitude of the deficit observed in operated animals over that found in initial learning (i.e., 2 to 1), so this result, too, is consistent with a general role for perirhinal cortex in visual discrimination learning rather than in perception.

Taken together, the foregoing reexaminations of published results cast doubt upon the idea that perirhinal cortex is critical for visual perception. Reanalysis of the discrimination set size experiments of Buckley and Gaffan (1997) shows that large set size does not exacerbate the deficit in discrimination learning observed in monkeys lacking perirhinal cortex. Rather, there is an invariant deficit in learning that is evident across set sizes. Failure to find such deficits in small sets is probably an issue of statistical power, which will increase with set size. While it is the case that monkeys lacking perirhinal cortex are impaired in learning to discriminate new views of familiar objects and familiar objects in new settings, these results do not demonstrate a role for perirhinal cortex in perception. As in the case of the set size manipulation, it appears that the deficits observed were invariant across variations in perceptual demand, a result that is inconsistent with the view that perirhinal cortex is especially important under conditions of perceptual difficulty.

Exhaustive review of the findings of Buckley and colleagues is beyond the scope of this paper. However, one more set of results deserves attention, as it is put forward as evidence contradictory to the foregoing analyses. Buckley, Booth, Rolls, and Gaffan (2001) presented normal monkeys and monkeys lacking perirhinal cortex with arrays of images in which all but one of the images were the same. The monkeys were required to identify the single image that differed from the others. In one type of problem, images deemed the same were exactly identical (image oddity). In the other type of problem, images deemed the same were views of the same stimulus from different vantage points, while the odd image was a view of a different stimulus (object oddity). Monkeys lacking perirhinal cortex were reported to be impaired in learning object oddity tasks but not impaired in learning image oddity tasks. In assessing these results it is critical to keep two points in mind. First, the dependent measures were again measures of *learning*, rather than perceptual generalization on a single trial, so perceptual and mnemonic processes were confounded. Second, because stimuli deemed to be the same in the object oddity task were different views of the same object, they necessarily contained different visual information, much as the front and back of an object can be visually distinct. It would appear that the only way these differing views could come to be treated as the same by monkeys is through learning. Therefore the impairment in acquiring object oddity problems is at least as likely to be due to a learning or memory impairment as to a perceptual impairment.

On a more technical note, the magnitude of the differences between the groups across the different tasks does not correlate with the statistical significance of the results in a sensible way (Buckley et al., 2001). For example, in the image oddity task employing human faces, monkeys lacking perirhinal cortex made about four times as many errors as did normal monkeys, but the difference was not significant. In the case of object oddity with human faces this ratio of errors is about 1.5, and the difference is significant. A similar inversion of the effect size and the statistical results occurs for these tests with monkey faces. In most of the remaining object oddity–image oddity contrasts, direct comparisons cannot be made because the data were treated differently, and different performance criteria were employed, for the two types of task (Buckley et al., 2001). In summary, these reanalyses and reexaminations of the data from Buckley and colleagues do not call into doubt the fact that monkeys lacking perirhinal cortex are impaired in various types of discrimination learning. However, the source of the impairment seems to be difficulty with learning or memory, not a difficulty in perceptual processing.

Evidence for intact perception and impaired retention

The reanalyses presented in the preceding section call into question the hypothesis that perirhinal cortex is critical for perception. To more directly distinguish between perceptual and mnemonic accounts of perirhinal cortex function, Hampton and Murray (2002) designed a series of experiments that independently assessed perceptual and mnemonic function. The first study in the series also assessed whether images that differed in visual complexity might be differentially difficult for monkeys lacking perirhinal cortex to learn to discriminate. If perirhinal cortex is responsible for the highest level of a hierarchical organization of perceptual processing in the ventral visual processing stream (Bussey et al., this issue; Murray & Bussey, 1999; Saksida & Bussey, 2002), then removal of this area might impair processing of complex images while leaving processing of simpler images intact.

Effect of stimulus complexity on learning

To test whether the complexity of the images to be remembered is a determinant of the involvement of perirhinal cortex in discrimination learning, Hampton and Murray (2002) examined acquisition of discrimination problems using different types of visual stimuli: (a) *Black-and-white* images were filled outline drawings of various objects resembling silhouettes; (b) *colour* images were similar to the black and white images in form, but included one or more colours and some internal details; (c) *detailed* images were perspective renderings approaching photographic quality (see Hampton & Murray, 2002, Figure 1, for examples of the stimuli). Monkeys were trained on each of the three types of image concurrently, receiving one trial with each discrimination problem each day. Monkeys with perirhinal cortex removed were impaired in learning all three types of discrimination, but were no more impaired in learning the more complex detailed images than they were in learning the simple images (Figure 2); Group, $F(1, 8) = 9.34$, $p < .05$; Image type, $F(2, 16) = 10.39$, $p < .01$; Group × Image type, $F(2, 16) = 0.13$. While the discriminations involving the detailed images were learned more rapidly by both groups of monkeys, the operated group made roughly two errors to every one error made by the control group across the three types of problem. These results suggest that the perirhinal cortex is not especially critical for discriminating complex stimuli.

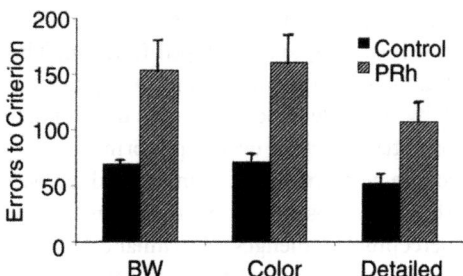

Figure 2. Average errors committed in learning object discrimination problems involving stimuli of three levels of complexity (Hampton & Murray, 2002, Exp. 1). Control monkeys, black bars; Monkeys lacking perirhinal cortex, striped bars. Vertical lines indicate standard errors. Monkeys lacking perirhinal cortex were impaired equally in learning all three types of discrimination problem.

Whereas these results do clearly show that monkeys with perirhinal cortex removed are impaired in learning new visual discriminations, they do not allow differentiation between impairment caused by deficits in visual perception and deficits in learning or memory. This is because the deficit in discrimination learning could be due either to a deficit in perception of the stimuli or to a deficit in learning or retention of the association between food reward and one of the images in each discrimination problem. In a series of subsequent experiments, Hampton and Murray (2002) therefore assessed visual perception free of the requirement for new learning. If the deficit in discrimination learning is due to problems in perception, then monkeys lacking perirhinal cortex should continue to be impaired in tests that isolate perceptual function.

Perceptual tests free from new learning

After both intact monkeys and monkeys lacking perirhinal cortex had learned a large set of discrimination problems to criterion, they were challenged with difficult discriminations consisting of altered versions of the now very familiar discriminanda. The discriminations were made difficult by altering both images in each discrimination problem in the following ways: shrinking, removing colour, rotating, exchanging colours between objects, and masking with an opaque screen (see Hampton & Murray, 2002, for examples of these manipulations). Critically, each altered pair of discriminanda was presented only once or twice, intermixed with normal trials, to preclude the occurrence of new learning. By this method performance on probe trials using the altered stimuli assessed perceptual generalization unconfounded with new learning.

Each of the manipulations of the stimuli had the intended effect of making the discriminations difficult, indicated by the fact that the performance of all monkeys decreased significantly. But monkeys with perirhinal cortex removed had no more difficulty with any of these perceptually difficult discriminations than did normal monkeys. In one example experiment, the discriminations were made difficult by placing a mask over the discriminanda (Hampton & Murray, 2002, Exp. 6). The masks always covered 50% of the area of the test image, and there were six types of mask, differing in the size of the blocks making up the mask. When masks made with the smallest size blocks were used, the effect was essentially to dim the images as if they were viewed through muslin. The largest size block completely obscured the left half of the images. Superposition of the masks on the discriminanda affected performance in both groups according to a U-shaped function. That is, the difficulty of the discriminations increased as the size of the blocks making up the mask increased until half of the image was occluded by a single block, and the discrimination then became easier (Figure 3). Despite a substantial effect on overall performance, superposition of the masks did not differentially affect the performance of monkeys lacking perirhinal cortex.

These findings show that in the absence of any requirement for new learning, and in the presence of a variety of perceptual challenges, perirhinal cortex is not required for perceptual generalization. Although these manipulations did make the discriminations substantially more difficult for both control and operated monkeys, none of the manipulations differentially affected the performance of monkeys with lesions of perirhinal cortex. Therefore, it would appear that the deficits observed in initial learning of the discrimination by monkeys lacking perirhinal cortex were due to a learning or memory impairment, rather

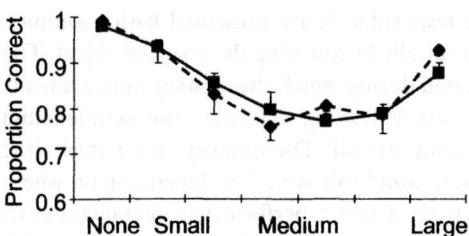

Figure 3.   Accuracy on standard trials and on probe trials in which masks of six different mesh sizes were super-imposed on the discriminanda (Hampton & Murray, 2002, Exp. 6). Control monkeys, solid lines; Monkeys lacking perirhinal cortex, dashed lines. Error bars are standard errors. While superposition of the masks made the discriminations difficult for both groups, there was no difference between the groups.

than an impairment in perception. This hypothesis was tested directly in a second series of experiments, described in the next section, in which new learning was required without any increase in perceptual demand.

Tests of learning free from new perceptual challenge

Monkeys with perirhinal cortex removed were impaired in learning new discrimination problems but were not impaired in generalizing this learning to altered views of the learned discriminations (Hampton & Murray, 2002). One interpretation of these findings is that monkeys with perirhinal cortex lesions are generally impaired in remembering stimulus–reward associations but have intact perception. If so, then they should be impaired under conditions of reversal learning. In reversal learning, the valence of each object discrimination problem is reversed such that the image that was previously unrewarded is now the correct image, while the previously correct image is now unrewarded. The value of reversal learning in the current context is that the monkeys have already demonstrated their ability to discriminate the two stimuli comprising each discrimination problem and thus have demonstrated perceptual mastery of the problems. The new demand placed on them by reversal learning is to learn and remember a new association between an image and food reward. That is, reversal learning provides a measure of learning that is relatively free of new demands on perceptual or representational mechanisms (e.g., Gaffan & Harrison, 1986, p. 7). A perceptual account of perirhinal cortex function predicts that monkeys lacking perirhinal cortex should be unimpaired in reversal learning, while a mnemonic account predicts an impairment. In three replications, monkeys with perirhinal cortex removed were impaired in reversal learning (Hampton & Murray, 2002, Exp. 7, 8, and 9). Reversal learning requires memory, but does not place new demands on perceptual processes. These results show that a memory impairment can be obtained in monkeys lacking perirhinal cortex without taxing perception.

Matching-to-sample

The preceding findings implicate perirhinal cortex in learning or memory, but not in perception. Matching- and nonmatching-to-sample are additional ways of assessing memory. In the primate literature these are generally referred to as recognition memory tests.

In recognition memory tests subjects are presented with a sample stimulus to which they make a study response, typically by touching the image or object. The object is then removed from view for a delay period during which the monkey must remember the stimulus in order to respond correctly at test. Following this delay, the sample stimulus is again presented, with one or more distractor stimuli. The monkeys must then either touch the previously viewed sample stimulus, or avoid this stimulus, depending on whether the rule is matching or nonmatching, respectively. Correct performance demonstrates that the monkey remembers or recognizes the sample stimulus. Manipulation of the delay interval over which monkeys must remember the sample stimulus provides a means of taxing memory—the longer the delay interval, the greater the demand on memory.

The performance of monkeys in recognition memory tests can be diagnostic of both memory impairments and perceptual impairments. When the delay between viewing the sample and presentation of the test is very short, the demand on memory is not great (although some demand does exist because there is a delay, however brief). Monkeys with memory impairments, but with intact perception, would be predicted to perform normally at very short delay intervals, but to perform increasingly poorly in comparison to normal monkeys as the delay interval is increased. Animals with perceptual impairments would be predicted to perform equally poorly relative to controls at all delays because they have difficulty in apprehending or representing the sample stimulus and in discriminating it from the distractor stimuli at test. The patterns of data predicted for animals with perceptual impairment and those with memory impairment are depicted in Figure 4.

A number of experiments have assessed recognition memory performance in monkeys with lesions that include the perirhinal cortex. In the majority of these studies monkeys lacking perirhinal cortex have shown delay-dependent impairments, such that performance compared to controls was worse at long delays than at short delays (Buffalo et al., 1999; Buffalo et al., 2000; Eacott et al., 1994; Meunier et al., 1993; Zola-Morgan et al., 1989; but

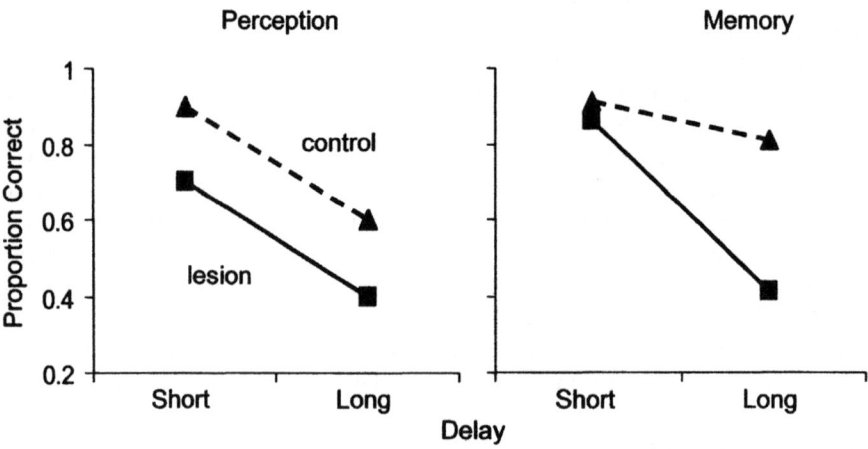

Figure 4.   Patterns of deficit in recognition memory tests predicted by perceptual and mnemonic impairment. Accuracy is plotted on the vertical axis, while the duration of the delay between study and test is plotted on the horizontal axis. The left panel plots the pattern predicted by a perceptual deficit, while the right panel indicates the pattern expected in the case of a mnemonic deficit.

see Eacott et al., 1994, for an exception). Some of these studies also found deficits in performance at the shortest delay used, although often the shortest delay was still considerable. It has been a matter of debate whether or not the performance of normal monkeys and monkeys with perirhinal cortex removed can be equated at very short delays (e.g., Buffalo et al., 2000, pp. 379–380).

The same monkeys used in the experiments described previously (Hampton & Murray, 2002) have produced two relevant results in recognition memory experiments. First, the monkeys with perirhinal cortex removed were shown to have a dramatic delay-dependent memory impairment over the range of 2 to 32 seconds with no significant impairment at the shortest delay (Hampton & Murray, unpublished results). In a second recognition memory experiment, control monkeys were perceptually impaired by masking the sample stimulus with an opaque grid. Performance under these conditions was then compared to that under normal conditions, without the perceptually challenging mask. Masking the sample impaired performance at all delays, but the forgetting curve generated under the conditions of perceptual impairment was parallel to the normal forgetting curve, as in the hypothetical data in the left panel of Figure 4 (Hampton & Murray, unpublished results). The finding of a delay-dependent memory deficit reinforces the view that perirhinal cortex removal causes memory impairment. The second set of results suggests that even if monkeys with perirhinal cortex removed did have perceptual impairments, such impairments would not cause delay-dependent deficits in performance. Perceptually challenging normal monkeys by masking the sample images did not lead to more rapid forgetting than that which occurred under normal conditions. However, it remains for further work to test whether other perceptual manipulations have different effects on recognition memory.

In this section, I have reviewed data showing that monkeys lacking perirhinal cortex were unimpaired in a variety of difficult perceptual tests. The same monkeys were reliably impaired in reversal learning. Data from a variety of sources show that monkeys lacking perirhinal cortex show a delay-dependent impairment in recognition memory tests (e.g., Buffalo et al., 1999; Buffalo et al., 2000; Eacott et al., 1994; Meunier et al., 1993; Zola-Morgan et al., 1989) . It has been argued that delay dependence is an indicator of memory impairment distinct from perceptual impairment. Taken together, these findings demonstrate a learning or memory impairment in the absence of a perceptual impairment. In the next section I present a critique of the influential perceptual-mnemonic model of perirhinal cortex.

What is perceptual-mnemonic?

The perceptual-mnemonic model of perirhinal cortex function posits that perihinal cortex plays a critical role in the binding of individual features making up whole objects (Bussey, Saksida, & Murray, 2002, 2003; Bussey et al., this issue; Lee et al., this issue; Murray & Bussey, 1999; Rolls, Franco, & Stringer, this issue; Saksida & Bussey, 2002). It is based on the observation that neurons located more rostrally in inferior temporal cortex have more complex response properties than do more caudally located neurons (e.g., Desimone & Ungerleider, 1989), suggesting that these rostral neurons represent complex conjunctions of features (Bright, Moss, Stamatakis, & Tyler, this issue). Applied to discrimination learning the perceptual-mnemonic theory is essentially a migration of configural learning theory out of the hippocampus and into subjacent cortex (Rudy & Sutherland, 1995).

The perceptual-mnemonic model presents the greatest challenge to the view put forward in this paper that the perirhinal cortex is involved in memory but not in perception.

I take up four criticisms of the perceptual-mnemonic model: (a) As currently stated, the model is almost invulnerable to direct falsification; (b) a central prediction of the model was shown not to obtain in the first section of this paper; (c) the model has not been evaluated with direct tests of perceptual function in monkeys, free of a requirement of learning; and (d) the model does not account for delay-dependent memory impairments following perirhinal cortex removal.

The perceptual-mnemonic model predicts that monkeys lacking perirhinal cortex will be impaired in discrimination learning only when the images being discriminated are similar such that some features are rewarded when part of one image, but not rewarded when part of another. This situation is referred to as feature overlap or feature ambiguity. When feature ambiguity is high, a configural representation of the discriminanda that binds together component features of an image is expected to greatly facilitate learning. The inability of monkeys lacking perirhinal cortex to form these configural representations would therefore lead to an impairment in discrimination learning only under conditions of feature ambiguity. This prediction has been evaluated in a creative set of experiments employing images that are blended with each other so that they are ambiguous and share many features (Bussey et al., 2002). All monkeys rapidly learned discriminations in which few or no features were shared between discriminanda, and monkeys with perirhinal cortex removed performed nearly as well as control monkeys. In contrast, when feature ambiguity was high, all monkeys learned comparatively slowly, and those lacking perirhinal cortex learned especially slowly (Bussey et al., 2002, 2003). These findings support the perceptual-mnemonic model.

When is feature ambiguity present?

The perceptual-mnemonic model predicts deficits only when feature ambiguity complicates the discrimination of stimuli (Bussey et al., 2002, 2003; Murray & Bussey, 1999; Saksida & Bussey, 2002). In their experiments testing the model, Bussey and colleagues have used novel and creative ways of creating feature ambiguity, most notably the use of software that allows blending of images such that they share many features (Bussey et al., 2003). By using these techniques they have been able to parametrically manipulate feature overlap in an objective manner. But it is difficult to apply the notion of feature ambiguity outside of the confines of these experiments. Further complicating the application of feature ambiguity is the fact that the model predicts that discrimination ability will depend on the perirhinal cortex only if the ambiguity is at a certain level in the hierarchy of visual representation proposed to exist in the ventral visual processing stream. Ambiguity at early stages in visual processing can be resolved at those early stages. It can therefore be argued post hoc whenever deficits in discrimination learning are observed that there must have been feature ambiguity, and when such deficits have not been observed there must have been no such ambiguity. It is difficult or impossible to independently assess the degree and level of feature ambiguity in a given set of stimuli.

Monkeys were trained on discrimination problems that minimized feature ambiguity in an effort to address the argument that feature ambiguity could explain the discrimination learning deficits observed in monkeys with perirhinal cortex removed (Hampton & Murray, 2002). This test limited feature ambiguity in two ways. First, the images used in each discrimination problem were selected because they were extremely distinctive to human observers, subjectively more so than were the low ambiguity images used by Bussey et al. (2003). Second, the problems were presented one at a time (set size = 1), rather than concurrently, so that difficulty with feature overlap between problems would be minimized. Monkeys lacking perirhinal cortex were nonetheless found to be impaired in learning and in reversal of these discrimination problems (Hampton & Murray, 2002, Exp. 8). Given the finding presented in the first section of this paper that no set size effect occurs in discrimination learning, it is not surprising that a deficit was observed even when pairs were learned individually. But the impairment in learning despite a lack of obvious feature ambiguity presents a challenge to the perceptual-mnemonic model. Nonetheless, it can be argued that feature ambiguity was present but was not detected by the human observers. If this were the case, however, one would think that feature ambiguity must be a pervasive quality that would complicate learning of virtually all discrimination problems.

False prediction?

A connectionist model was developed by Saksida and Bussey (2002) in an effort to formalize the perceptual-mnemonic model of perirhinal cortex function. The model instantiates a hierarchical organization of the ventral visual processing stream, with a feature conjunction layer representing perirhinal cortex. The feature conjunction layer can be removed from the model to simulate the effects of removing the perirhinal cortex from a monkey. The validity of the connectionist model, and thus the perceptual-mnemonic model, was evaluated by comparing the performance of a complete system to one with the feature conjunction layer removed. Removing the feature conjunction layer was predicted to have similar effects to those following perirhinal cortex removal in monkeys.

Indeed, the connectionist model performed in a way that closely matched the behaviour hypothesized to occur in real monkeys. Most notably, the connectionist model lacking the feature conjunction layer was impaired in learning a large set of discrimination problems, but was not impaired in learning a small set (Saksida & Bussey, 2002, Exps. 1 and 3), replicating the result reported by Buckley and Gaffan (1997). But this set size effect was reexamined at length in the first section of this paper and was shown not to occur in monkeys! Thus, the model predicts a major result that does not occur in the real world.

Tests of perception confounded with learning

Evaluation of the hypothesis that perirhinal cortex makes a critical contribution to perception requires use of tests that measure perceptual performance free of a requirement for learning. Such tests have not been conducted with stimuli in which feature ambiguity has been systematically manipulated. The perceptual-mnemonic model is supported by the finding of a functional relationship between the degree of feature ambiguity and the magnitude of the impairment in discrimination learning in monkeys lacking perirhinal cortex. That is, Bussey et al. (2002, 2003) report that under conditions of low feature ambiguity monkeys lacking perirhinal cortex show little or no impairment in learning, whereas under conditions of high feature ambiguity the impairment is substantial. It is the apparent functional relationship between perceptual demand and degree of impairment that makes the case that the impairment is due to perceptual demand.

Interpretation of this functional relationship is made difficult by the confounding of learning and perception in discrimination learning. While feature ambiguity does appear to be a challenge to perception, it can also be interpreted as a challenge to learning. Discriminations involving feature ambiguity cannot be solved on the basis of single stimulus features (Saksida & Bussey, 2002). Thus to solve these problems monkeys must remember multiple features of the discriminanda. This means that in order to solve the discriminations, monkeys have to remember more about the stimuli than they would under conditions of low feature ambiguity. This knowledge must be acquired through learning. If indeed more learning is required to solve discriminations high in feature ambiguity, then the observed functional relationship between degree of feature ambiguity and rate of learning can be equally well predicted by mnemonic load as by perceptual load.

Fortunately, this issue can be easily resolved with the techniques described by Hampton and Murray (2002). The same types of blended stimuli used previously by Bussey et al. (2003), which provide an objective and elegant means for manipulating feature ambiguity, could again be used. Monkeys would be trained on a large set of object discriminations using low-ambiguity stimuli. Once monkeys have mastered these discriminations, they would be presented with a single trial involving each discrimination problem, but now the stimuli would be blended together to create feature ambiguity. Because only a single presentation of each pair would be provided, the monkeys could not relearn the altered discriminations. Thus, the tests would be a pure measure of perceptual generalization. The perceptual-mnemonic model predicts that monkeys lacking perirhinal cortex would be impaired on these perceptual generalization tests. A purely mnemonic account of perirhinal cortex predicts that there would be no such impairment, as was observed in the perceptual tests of Hampton and Murray (2002).

In their work with blended stimuli, Bussey and colleagues have come closest to a pure perceptual test in an experiment using performance tests (Bussey et al., 2003, Exp. 2). In this experiment, monkeys were trained on four discrimination problems with low feature ambiguity. In performance tests monkeys were given blocks of trials with the familiar discriminanda now blended together to create feature ambiguity. The problems were presented at two levels of feature ambiguity, and in the originally learned form, in three blocks of 32 test trials. A percentage correct measure was generated by averaging performance in the 32 trial blocks and was taken as a measure of perceptual generalization. Monkeys lacking perirhinal cortex were impaired relative to controls in the two ambiguity conditions. This would provide strong evidence for a perceptual impairment, except for that fact that monkeys had a combined total of 64 trials with the altered stimuli in each performance test. After the first test trial, the monkeys have the opportunity to relearn the discriminations, and any deficit observed could therefore be one in learning rather than in perception. Bussey et al. (2003) reported that no significant learning occurred during the performance test. Given the rate at which control monkeys learned high feature ambiguity discriminations in the first experiment in this series (Bussey et al., 2003, Exp. 1), this is a surprising result. In that experiment, normal monkeys mastered high feature ambiguity discriminations in 96 trials and showed considerable learning over any given block of 64 trials. The comparison performed by Bussey et al. may have lacked sufficient statistical power to detect the improvement in performance across trials within a block, but when all trials were averaged together, the overall difference may have been large enough to detect statistically. In any case, using many discrimination problems and only a single perceptual generalization test for each problem would provide the most direct test of perceptual function in monkeys lacking perirhinal cortex.

Failure to account for amnesia

Delay-dependent recognition memory impairments have been repeatedly demonstrated in monkeys lacking perirhinal cortex (Buffalo et al., 1999; Buffalo et al., 2000; Eacott et al., 1994; Meunier et al., 1993; Zola-Morgan et al., 1989, but see Eacott et al., 1994, for an exception). The perceptual-mnemonic model does not readily offer an account for this pervasive finding. The challenge is to explain how an impairment in coping with feature ambiguity could produce a greater memory impairment at long delays than at short delays. Successful application of the perceptual-mnemonic model to the impairments in recognition memory would be an advance for the model.

GENERAL DISCUSSION

In this paper I have critically evaluated the evidence implicating perirhinal cortex in perceptual processes. Overuse of discrimination learning has complicated interpretation of the function of perirhinal cortex because discrimination learning necessarily confounds learning or memory with perceptual processes. Perceptual processes are best assessed using tests that require subjects to make judgements about stimuli that are presented in only a single test trial. After the first presentation, performance can be influenced by learning and memory.

In the first section of this paper, previously published findings were subjected to reanalysis and reconsideration. It was shown that the oft-cited set size effect that has been used to implicate perirhinal cortex in perception does not obtain. There is no functional relationship between discrimination problem set size and the degree of impairment in monkeys lacking perirhinal cortex, as has been reported (Buckley & Gaffan, 1997). Concomitantly, there is little evidence that impairment in relearning familiar discrimination problems under difficult conditions is exacerbated relative to initial learning. In fact, monkeys lacking perirhinal cortex make a remarkably consistent two errors for every one error made by normal monkeys across a variety of tests (Buckley & Gaffan, 1997, 1998b). There is general agreement that monkeys lacking perirhinal cortex are impaired in discrimination learning, under at least some conditions. But if monkeys lacking perirhinal cortex were perceptually impaired, then the magnitude of the learning deficits in this group would increase as perceptual demand is increased. This was not observed.

In the second section of this paper I reviewed a set of experiments in which perceptual function and learning were assessed independently in distinct tests (Hampton & Murray, 2002). Most significantly, the perceptual assessments involved just one or two presentations of each test. Monkeys lacking perirhinal cortex were unimpaired in the face of a variety of perceptual challenges, despite the fact that the tests were sufficiently difficult to cause significant drops in performance in both normal and operated monkeys. Conversely, when new learning was required that involved familiar and readily discriminated images, monkeys lacking perirhinal cortex were impaired, demonstrating a deficit in learning or memory in these monkeys. Finally, this set of experiments also showed that the impairment in acquisition of new discrimination problems was not affected by the complexity of the images used.

In the final section of this paper, I critically assessed the perceptual-mnemonic model of perirhinal cortex function. While this model and the data supporting it provide the greatest challenge to a purely mnemonic account of perirhinal cortex function, several questions were raised. First, in most situations feature ambiguity is difficult to quantify. This makes the perceptual-mnemonic model difficult to assess outside a narrow set of procedures developed by Bussey and colleagues (Bussey et al., 2002, 2003; Murray & Bussey, 1999; Saksida & Bussey, 2002). Second, the model predicts the set size effect. Buckley and Gaffan (1997) tested for this effect, but reanalysis shows that it did not occur. Third, the perceptual-mnemonic model has yet to be subjected to test by pure measures of perceptual function. The completion of such tests would be a major advance in the assessment of perirhinal cortex function. Finally, the perceptual-mnemonic model has yet to be applied to the most widely reported effect of perirhinal cortex removal—delay-dependent memory impairment.

Because of its location in the brain, it is reasonable to expect that perirhinal cortex is involved in memory, perception, or both. Resolving the functional contribution of this region can only be accomplished through behavioural studies in which perception and memory are clearly defined and operationalized. Here, I have raised the concern that what appears at first to be a body of converging evidence implicating monkey perirhinal cortex in perception may instead be nearly universally flawed by the confounding of perception and memory. Given the ambiguity of the evidence for a perceptual role, students of human neuro-psychology might best regard with scepticism the idea that there is widespread agreement that monkey perirhinal cortex is critical for perception in monkeys. At present, the preponderance of evidence supports a role for perirhinal cortex in learning and memory, whereas the evidence for a role in perception is comparatively weak.

REFERENCES

Bright, P. J., Moss, H. E., Stamatakis, E. A., & Tyler, L. K. (this issue). The anatomy of object processing: The role of anteromedial cortex. *Quarterly Journal of Experimental Psychology, 58B*, 361–377.

Buckley, M. J. (this issue). The role of the perirhinal cortex and hippocampus in learning, memory and perception. *Quarterly Journal of Experimental Psychology, 58B*, 246–268.

Buckley, M. J., Booth, M. C. A., Rolls, E. T., & Gaffan, D. (2001). Selective perceptual impairments after perirhinal cortex ablation. *Journal of Neuroscience, 21*(24), 9824–9836.

Buckley, M. J., & Gaffan, D. (1997). Impairment of visual object-discrimination learning after perirhinal cortex ablation. *Behavioral Neuroscience, 111*, 467–475.

Buckley, M. J., & Gaffan, D. (1998a). Perirhinal cortex ablation impairs configural learning and paired-associate learning equally. *Neuropsychologia, 36*(6), 535–546.

Buckley, M. J., & Gaffan, D. (1998b). Perirhinal cortex ablation impairs visual object identification. *Journal of Neuroscience, 18*(6), 2268–2275.

Buffalo, E. A., Ramus, S. J., Clark, R. E., Teng, E., Squire, L. R., & Zola, S. M. (1999). Dissociation between the effects of damage to perirhinal cortex and area TE. *Learning & Memory, 6*, 572–599.

Buffalo, E. A., Ramus, S. J., Squire, L. R., & Zola, S. M. (2000). Perception and recognition memory in monkeys following lesions of area TE and perirhinal cortex. *Learning & Memory, 7*(6), 375–382.

Bussey, T. J., Saksida, L. M., & Murray, E. A. (2002). Perirhinal cortex resolves feature ambiguity in complex visual discriminations. *European Journal of Neuroscience, 15*(2), 365–374.

Bussey, T. J., Saksida, L. M., & Murray, E. A. (2003). Impairments in visual discrimination after perirhinal cortex lesions: Testing 'declarative' vs. 'perceptual-mnemonic' views of perirhinal cortex function. *European Journal of Neuroscience, 17*(3), 649–660.

Bussey, T. J., Saksida, L. M., & Murray, E. A. (this issue). The perceptual-mnemonic/feature conjugation model of perirhinal function. *Quarterly Journal of Experimental Psychology, 58B*, 269–282.

Desimone, R., & Ungerleider, L. G. (1989). Neural mechanisms of visual processing in monkeys. In F. Boller & J. Grafman (Eds.), *Handbook of neuropsychology* (Vol. 2, pp. 267–299). Amsterdam: Elsevier.

Eacott, M. J., & Gaffan, E. A. (this issue). The roles of the perirhinal cortex, postrhinal cortex, and the fornix in memory for objects, cortexts, and events in the rat. *Quarterly Journal of Experimental Psychology, 58B*, 202–217.

Eacott, M. J., Gaffan, D., & Murray, E. A. (1994). Preserved recognition memory for small sets, and impaired stimulus identification for large sets, following rhinal cortex ablations in monkeys. *European Journal of Neuroscience, 6*(9), 1466–1478.

Eacott, M. J., & Heywood, C. A. (1995). Perception and memory: Action and interaction. *Critical Reviews in Neurobiology, 9*, 311–320.

Gaffan, D., & Harrison, S. (1986). Visual identification following inferotemporal ablation in the monkey. *The Quarterly Journal of Experimental Psychology, 38B*, 5–30.

Hampton, R. R., & Murray, E. A. (2002). Learning of discriminations is impaired, but generalization to altered views is intact, in monkeys (Macaca mulatta) with perirhinal cortex removal. *Behavioral Neuroscience, 116*(3), 363–377.

Jagadeesh, B., Chelazzi, L., Mishkin, M., & Desimone, R. (2001). Learning increases stimulus salience in anterior inferior temporal cortex of the macaque. *Journal of Neurophysiology, 86*, 290–303.

Lee, A. C. H., Barense, M. D., & Graham, K. S. (this issue). The contribution of the human medial temporal lobe to perception: Bridging the gap between animal and human studies. *Quarterly Journal of Experimental Psychology, 58B*, 300–325.

Logothetis, N. (1998). Object vision and visual awareness. *Current Opinion in Neurobiology, 8*, 536–544.

Meunier, M., Bachevalier, J., Mishkin, M., & Murray, E. A. (1993). Effects on visual recognition of combined and separate ablations of the entorhinal and perirhinal cortex in rhesus-monkeys. *Journal of Neuroscience, 13*(12), 5418–5432.

Murray, E. A. (2000). Memory for objects in nonhuman primates. In M. S. Gazzaniga (Ed.), *The new cognitive neurosciences* (2nd ed., pp. 753–764). Cambridge, MA: MIT Press.

Murray, E. A., & Bussey, T. J. (1999). Perceptual-mnemonic functions of the perirhinal cortex. *Trends in Cognitive Sciences, 3*, 142–151.

Murray, E. A., Gaffan, D., & Mishkin, M. (1993). Neural substrates of visual stimulus stimulus association in rhesus-monkeys. *Journal of Neuroscience, 13*(10), 4549–4561.

Murray, E. A., Malkova, L., & Goulet, S. (1998). Crossmodal associations, intramodal associations, and object identification in macaque monkeys. In A. Milner (Ed.), *Comparative neuropsychology* (pp. 51–69). Oxford, UK: Oxford University Press.

Murray, E. A., & Mishkin, M. (1986). Visual recognition in monkeys following rhinal cortical ablations combined with either amygdalectomy or hippocampectomy. *Journal of Neuroscience, 6*(7), 1991–2003.

Murray, E. A., & Mishkin, M. (1998). Object recognition and location memory in monkeys with excitotoxic lesions of the amygdala and hippocampus. *Journal of Neuroscience, 18*(16), 6568–6582.

Rolls, E. T., Franco, L., & Stringer, S. M. (this issue). The perirhinal cortex and long-term familiarity memory. *Quarterly Journal of Experimental Psychology, 58B*, 234–245.

Rudy, J. W., & Sutherland, R. J. (1995). Configural association theory and the hippocampal formation: An appraisal and reconfiguration. *Hippocampus, 5*, 375–389.

Saksida, L. M., & Bussey, T. J. (2002). The organization of visual object representations: A connectionist model of effects of lesions in perirhinal cortex. *European Journal of Neuroscience, 15*, 355–364.

Suzuki, W. A. (1996). Neuroanatomy of the monkey entorhinal, perirhinal and parahippocampal cortices: Organization of cortical inputs and interconnections with amygdala and striatum. *Seminars in the Neurosciences, 8*, 3–12.

Zola-Morgan, S., Squire, L. R., Amaral, D. G., & Suzuki, W. A. (1989). Lesions of perirhinal and parahippocampal cortex that spare the amygdala and hippocampal-formation produce severe memory impairment. *Journal of Neuroscience, 9*(12), 4355–4370.

THE QUARTERLY JOURNAL OF EXPERIMENTAL PSYCHOLOGY
2005, 58B (3/4), 300–325

The contribution of the human medial temporal lobe to perception: Bridging the gap between animal and human studies

Andy C. H. Lee, Morgan D. Barense, and Kim S. Graham

MRC Cognition and Brain Sciences Unit, Cambridge, UK

The medial temporal lobe (MTL) has been considered traditionally to subserve declarative memory processes only. Recent studies in nonhuman primates suggest, however, that the MTL may also be critical to higher order perceptual processes, with the hippocampus and perirhinal cortex being involved in scene and object perception, respectively. The current article reviews the human neuropsychological literature to determine whether there is any evidence to suggest that these same views may apply to the human MTL. Although the majority of existing studies report intact perception following MTL damage in human amnesics, there have been recent studies that suggest that when scene and object perception are assessed systematically, significant impairments in perception become apparent. These findings have important implications for current mnemonic theories of human MTL function and our understanding of human amnesia as a result of MTL lesions.

Large lesions to the medial temporal lobes (MTL), including the hippocampus proper, subiculum, parahippocampal cortex, enthorinal cortex, and perirhinal cortex, are known to produce severe long-term memory impairments in both nonhuman primates (Alvarez, Zola-Morgan, & Squire, 1994; Gaffan, Parker, & Easton, 2001; Mishkin, 1982) and humans (Scoville & Milner, 1957). On the basis of these observations, it has been proposed that the primate MTL structures function as a single memory system important for the acquisition of new facts and events (termed declarative memory), with the degree of memory deficit correlating positively with the extent of MTL damage (Squire & Zola-Morgan, 1991; Zola-Morgan, Squire, & Ramus, 1994). A modification of this theory is that the different MTL areas subserve distinct mnemonic processes: for example, with the hippocampus mediating the recollection of episodic and contextual information, particularly in the

---

Correspondence should be addressed to Dr Andy Lee, MRC Cognition and Brain Sciences Unit, 15 Chaucer Road, Cambridge CB2 2EF, UK. Email: andy.lee@mrc-cbu.cam.ac.uk

The authors would like to thank Drs M. Buckley, T. Bussey, D. Gaffan, E. Murray, and L. Saksida for feedback on this and related work. The authors are funded by the Alzheimer's Research Trust, UK, and the Medical Research Council, UK.

DOI:10.1080/02724990444000168

domain of space, and the perirhinal cortex playing a unique role in object memory and recognition memory based on familiarity (Aggleton & Brown, 1999; Brown & Aggleton, 2001; Mishkin, Suzuki, Gadian, & Vargha-Khadem, 1997). According to this account, the profile of memory impairment following MTL lesions will depend on the exact location of any damage.

A more recent and controversial suggestion, which has risen primarily from the animal literature, is that the primate MTL may not mediate declarative memory processes exclusively but may, in fact, be involved in higher order perceptual processes (e.g., formation of visual representations; Buckley, Booth, Rolls, & Gaffan, 2001; Bussey, Saksida, & Murray, 2002; Murray & Bussey, 1999). The majority of evidence for this view comes from studies that have examined the role of the perirhinal cortex in monkeys in object perception and have found that lesions to this structure can result in severe object discrimination deficits (see Buckley, this issue; Bussey, Saksida, & Murray, this issue, full discussion). In brief, it has been proposed that the ventral "what" visual processing stream may culminate in the perirhinal cortex (Murray & Bussey, 1999). Thus, while simple features (e.g., colour, size) may be represented in more caudal visual regions such as V4 and TE/TEO, more complex conjunctions of features (e.g., objects) may be represented in rostral regions including the perirhinal cortex.

A nonmnemonic role in spatial perception has also been proposed for the hippocampus (Gaffan, 2001; Horel, 1978; O'Keefe, 1999). While there has been, to date, no direct experimental evidence to support this hypothesis, a large body of existing animal hippocampal data appear to be consistent with a role for the hippocampus in processing spatial information. For instance, nonhuman primate lesion studies have demonstrated that the hippocampus is essential for tasks on which performance is dependent upon spatial memory (Buckley, Charles, Browning, & Gaffan, 2004; Hampton, Hampstead, & Murray, 2004; Murray, Davidson, Gaffan, Olton, & Suomi, 1989), and, furthermore, there is an abundance of evidence that animals (Hori et al., 2003; O'Keefe, 1976; O'Keefe & Burgess, 1996; O'Keefe, Burgess, Donnett, Jeffery, & Maguire, 1998; Ono, Nakamura, Fukuda, & Tamura, 1991; Ono, Nakamura, Nishijo, & Eifuku, 1993; Wilson & McNaughton, 1993) possess hippocampal place cells that may signal aspects of spatial location and navigation.

One extension to a perceptual view of the primate MTL is that the severe declarative memory impairments that often result from lesions to MTL structures may be explained, at least in part, by a primary deficit in higher order perception (Gaffan, 2001; Horel, 1978): for instance, difficulties with object perception following perirhinal cortex damage or an impairment in spatial perception after lesions to the hippocampus. If true, then perceptual theories of MTL function will have a profound impact on our current understanding of the functional organization of declarative memory in the brain, as well as profiles of amnesia in animals and humans following MTL damage.

The aim of the current review, therefore, is to explore whether there is any evidence for the human MTL subserving the same perceptual processes that have been proposed for the perirhinal cortex and hippocampus in animals. Given that functional neuroimaging studies of MTL function have already been discussed in detail elsewhere (see Henson, this issue) this paper concentrates primarily on the patient neuropsychological literature. Furthermore, due to the large volume of studies that have been conducted on MTL function, discussion is restricted primarily to patients with static focal lesions, as opposed to cases with progressive diseases that impact the MTL (e.g., Alzheimer's Disease). As a starting point, a brief overview of existing studies of mnemonic function in amnesic patients with MTL damage is provided, in particular focusing on those investigations that have assessed perceptual function alongside memory performance. This shows that, to date, there has been little evidence for the human MTL subserving perceptual processing (although this has not been tested systematically) and that, as a result, the existing literature supports the general conclusion that the human MTL mediates declarative memory processes exclusively.

Following this, a more detailed discussion is carried out on a handful of recent neuropsychological studies that were specifically conducted to investigate the perceptual abilities of amnesic patients (Buffalo, Reber, & Squire, 1998; Holdstock, Gutnikov, Gaffan, & Mayes, 2000a; Lee et al., 2005a; Lee et al., 2005c; Stark & Squire, 2000). In contradiction with the existing literature, two of these studies have demonstrated that MTL damage in humans can lead to perceptual deficits (Lee et al., 2005a; Lee et al., 2005c). The implications of these new findings are considered, in terms of (a) how they may be reconciled with studies that have failed to support a role for MTL structures in perception (Buffalo et al., 1998; Holdstock et al., 2000a; Stark & Squire, 2000) and (b) how deficits in perception may interact with declarative memory function and underlie aspects of the human amnesic syndrome. With respect to the latter, specific mention is made of a recent study that has examined the impact that deficits in object perception may have on a form of declarative learning (Barense et al., 2005). This review then finishes by considering how more recent perceptual theories of MTL function may be reconciled with existing mnemonic theories of the human MTL, and possible avenues of further research are discussed.

A role for the human MTL in memory only?

Existing studies of mnemonic function and the human MTL

Since the initial description of severe amnesia in patient H.M. and other patients with large bilateral lesions to the MTL (Scoville & Milner, 1957), there have been a myriad of studies that have investigated various aspects of mnemonic functioning and profiles of retrograde and anterograde memory loss following damage to MTL structures (for a recent review, see Spiers, Maguire, & Burgess, 2001a). These include investigations into new semantic memory learning (e.g., Bayley & Squire, 2004; Gadian et al., 2000; Manns, Hopkins, & Squire, 2003b; Vargha-Khadem et al., 1997), differences in recollection versus familiarity-based recognition (e.g., Holdstock et al., 2002; Manns, Hopkins, Reed, Kitchener, & Squire, 2003a; Mayes, Holdstock, Isaac, Hunkin, & Roberts, 2002; Mayes et al., 2004; Yonelinas et al., 2002), working memory (e.g., Cave & Squire, 1992b; Fujii, Yamadori, Endo, Suzuki, & Fukatsu, 1999), and different facets of nondeclarative memory: for instance, priming (e.g., Cave & Squire, 1992a; Levy, Stark, & Squire, 2004; Musen & Squire, 1992; Postle & Corkin, 1998), perceptual learning (e.g., Kartsounis, Rudge, & Stevens, 1995; Manns & Squire, 2001), and motor learning (e.g., Gabrieli, Corkin, Mickel, & Growdon, 1993; Muramoto, Kuru, Sugishita, & Toyokura, 1979). Although a detailed discussion of the findings of these studies is beyond the scope and aims of the present review (for full review see Spiers, Maguire, & Burgess, 2001a; Squire, Stark, & Clark, 2004), there are a couple of observations that can be made with respect to the MTL and mnemonic function. First, there is an abundance of evidence to suggest that the MTL is essential for long-term declarative memory, but often not for nondeclarative memory (e.g., Cave & Squire, 1992a; Kartsounis et al., 1995; Levy et al., 2004; Manns & Squire, 2001; Musen & Squire, 1992; but see Chun & Phelps, 1999) or working memory (e.g., Cave & Squire, 1992b; Rempel-Clower, Zola, Squire, & Amaral, 1996). Second, as mentioned in the Introduction, the data on functional fractionation within the MTL are less clear, with some reports arguing for functional specialization within different MTL structures (Aggleton & Brown, 1999; Brown & Aggleton, 2001; Mishkin et al., 1997) and others concluding that all MTL structures function in concert as a unitary declarative memory system (Squire, 2004; Squire et al., 2004; Squire & Zola-Morgan, 1991; Zola-Morgan et al., 1994).

In favour of a single memory system, there have been numerous studies showing that selective lesions to the human hippocampus can result in mnemonic deficits across multiple domains of declarative memory function, including impaired acquisition of new semantic memory and deficits in episodic memory retrieval on the basis of both recollection and familiarity (e.g., Bayley & Squire, 2004; Gabrieli, Cohen, & Corkin, 1988; Manns et al., 2003a; Manns et al., 2003b; for review see Squire et al., 2004). Contradictory evidence, however, comes from studies that have shown that circumscribed hippocampal damage does leave some memory functions intact. For example, in contrast to those studies that support a unitary view, Vargha-Khadem et al. (1997) and Gadian et al. (2000) found that the hippocampal formation is not essential for acquiring new semantic information in developmental amnesic cases. Furthermore, there have been a number of studies that have found preservation of object recognition memory in hippocampal lesion patients (Baddeley, Vargha-Khadem, & Mishkin, 2001; Holdstock et al., 2002; Mayes et al., 2002; Yonelinas et al., 2002).

In addition to the above, there is also a considerable collection of neuropsychological data indicating that the human hippocampus, as well as the parahippocampal gyrus, may have a significant role in spatial memory and navigation. Patients with hippocampal and/or parahippocampal damage are known to be significantly impaired in spatial memory and navigation (Bohbot, Iaria, & Petrides, 2004; Bohbot et al., 1998; Burgess, Maguire, & O'Keefe, 2002; Feigenbaum & Morris, 2004; Holdstock et al., 2000b; King, Burgess, Hartley, Vargha-Khadem, & O'Keefe, 2002; King, Trinkler, Hartley, Vargha-Khadem, & Burgess, 2004; Spiers, Burgess, Hartley, Vargha-Khadem, & O'Keefe, 2001b). Additionally, there is now evidence that there are cells within the human MTL, analogous to those found in the rat and monkey hippocampus, that signal aspects of space processing, with cells sensitive to location in the hippocampus and cells sensitive to views of landmarks in the parahippocampal gyrus (Ekstrom et al., 2003).

There are currently two main contrasting theories that attempt to make sense of these data. In brief, one line of thought posits that the hippocampus in animals and humans operates as a spatial module or cognitive map (O'Keefe & Nadel, 1978), with additional involvement in temporal information processing and linguistic processes critical for the formation of episodic memory in humans (O'Keefe, 1999; O'Keefe & Burgess, 1996). In contrast, it has been proposed that the hippocampus is involved in "relational" memory— that is, memory for relationships among perceptually distinct items (Cohen et al., 1999; Eichenbaum & Cohen, 2002; Eichenbaum, Otto, & Cohen, 1994). Thus, according to this view, the hippocampus is not specialized for spatial processing per se but, rather, plays a broader role in processing a wide range of relational associations, including spatial relations.

Existing studies of perceptual function and the human MTL

If the human MTL does subserve nonmnemonic perceptual processes, as suggested by recent data in the animal literature (see Buckley, this issue; Bussey et al., this issue), then one would expect there to have been some support for this view from the wealth of neuropsychological studies that have investigated the effects of human MTL damage on cognitive function over the past 50 years. This, however, does not appear to be the case. First, the fact that patients with MTL damage generally show normal motor and perceptual learning has been taken as evidence that perception is intact in these patients (e.g., Gage, 1985; Iverson, 1977). For example, amnesic cases can demonstrate an improvement in performance accuracy and response times if they are required to do certain tasks repeatedly (e.g., mirror drawing, visual search), even in the absence of any explicit memory for previous experiences (e.g., Gabrieli et al., 1993; Manns & Squire, 2001; Muramoto et al., 1979; but see Chun & Phelps, 1999). Second, there has been little evidence from studies of perceptual function that damage to the human MTL alone can lead to perceptual difficulties.

Notably, however, a brief review of the existing literature reveals that only a proportion of studies has assessed perceptual abilities along with performance on memory tasks. Table 1 (pp. 306–309) lists these studies, giving details of the patients involved (e.g., location and extent of their cortical damage), which perceptual assessments were carried out, and, lastly, the patients' performance on these tests. Although a few of these studies assessed visual perception by using experimental tasks that placed a low demand on mnemonic processes—for example tasks of visual matching (Buffalo et al., 1998; Holdstock et al., 2000a; Squire, 1993)—more commonly standardized neuropsychological tests of perception were utilized. These include the copying condition of the Rey–Osterrieth figure (Osterrieth, 1944), which requires subjects to make a copy of a complex drawing using pen and paper, and the Visual Object Space Perception (VOSP; Warrington & James, 1991) battery (Table 1), which is composed of a series of tests that assess a variety of perceptual abilities, including identifying incomplete pictures of letters, noting the positions and quantities of items on a page, and analysing three-dimensional line drawings of blocks composed of cubes. Due to space limitations, studies that have included patients with amnesia due to fornix transection have been omitted. Moreover, only those studies that have used perceptual tasks that do not contain a memory component (e.g., semantic/episodic) have been included. For example, while the Gollin incomplete figures test (Gollin, 1960) and the object silhouette task of the VOSP (Warrington & James, 1991) assess the ability to perceive common objects from incomplete drawings and silhouettes, respectively, they both require semantic knowledge of these objects.

The overall consensus from the studies listed in Table 1 is that patients with MTL damage usually perform within the normal range on tasks of perception. Interestingly, however, there have been isolated cases where MTL patients have demonstrated impaired performance on tasks that do not appear to be mnemonic in nature. For instance, while patient HM was able to accurately perceive facial features and cartoon drawings, he was, however, impaired on a version of Gottschaldt's (Gottschaldt, 1929) hidden figure task, in which geometric patterns have to be identified and traced out within a network of embedding and overlapping lines (Milner, Corkin, & Teuber, 1968). In addition to this, a number of studies have found MTL lesion patients to be impaired on standard neuropsychological tasks, such as the Rey–Osterrieth figure copy condition (e.g., Benson, Marsden, & Meadows, 1974; Woods, Schoene, & Kneisley, 1982). Importantly, however, these deficits have often not been attributed to damage to the MTL, but rather to cortical damage beyond this region, in particular to occipital and more posterior lateral temporal lobe regions (e.g., Benson et al., 1974). In keeping with this, there have been, to our knowledge, no previous reports of perceptual deficits in patients who have cortical damage that is restricted to MTL structures bilaterally (see Table 1).

Two recent studies that specifically attempted to investigate a possible role for the perirhinal cortex in object perception are Buffalo et al. (1998) and Holdstock et al. (2000a). In brief, both of these investigations employed variations of a matching to sample (MTS) paradigm to assess a range of amnesic patients, including those with focal damage encompassing the perirhinal cortex bilaterally. In Buffalo et al. (1998), the participants were presented with four successive multicoloured complex visual designs, each for 1 s and with a 1 s interstimulus interval per trial. Following delays of 0 to 40 s, the participants were presented with a single test image and were required to decide whether this stimulus was one of the designs that had been shown previously. Similarly, Holdstock et al. (2000a) presented participants with a single monochromatic abstract pattern for 2.5 s per trial. In a simultaneous MTS condition, this pattern remained on the screen with 14 possible targets, and the subjects were instructed to match the two identical stimuli. In contrast, in the delayed MTS conditions, the target item was removed from the screen, and after delays of 0 to 30 s, the participants were required to select the pattern that they had seen earlier from 14 possible choices.

In the aforementioned studies, it was found that amnesic patients with perirhinal cortex damage were not impaired on MTS of abstract visual stimuli when there was an absent or minimal memory demand (e.g., simultaneous to 5-s MTS in Holdstock et al., 2000a; 0–2 s MTS in Buffalo et al., 1998). Significant deficits in performance were, however, evident when the delay was increased between stimulus presentation and recognition (10 s and above in Holdstock et al., 2000a; 6 s and above in Buffalo et al., 1998), suggesting that these patients suffered from recognition memory, but not perceptual, difficulties. On the basis of these findings, both Buffalo et al. (1998) and Holdstock et al. (2000a; this issue) concluded that the human perirhinal cortex does not appear to be critical for visual perception, and that this MTL region may subserve long-term mnemonic processes, especially recognition memory for objects, exclusively.

While the absence of a perceptual deficit following human MTL damage is contrary to recent reports from the animal literature (Buckley et al., 2001; Bussey et al., 2002; Bussey, Saksida, & Murray, 2003; Eacott, Gaffan, & Murray, 1994), it is possible that this discrepancy may be explained by considering the stimuli that were employed by Buffalo et al. (1998) and Holdstock et al. (2000a). For example, in these experiments two-dimensional abstract patterns, composed of varying shapes, colours, and lines, were used, with few overlapping features (or stimulus components) across the different stimuli. It is possible, therefore, to argue that these stimuli did not place a large demand on the types of perceptual process that have been attributed to the perirhinal cortex in monkeys (e.g., perception of *conjunctions* of object features; Buckley et al., 2001; Bussey & Saksida, 2002; Bussey et al., 2002; Bussey et al., 2003; Eacott et al., 1994; Murray & Bussey, 1999) and that the patients in these studies

TABLE 1

Neuropsychological studies that have assessed perception in human patients with static lesions to the medial temporal lobes

Study	Year	Patient/s	Lesion details	Perceptual assessment	Finding
Beatty, Salmon, Bernstein, & Butters	1987	MRL	Bilateral HC.	Rey copy.	Intact performance.
Benson et al.	1974	10 cases	All medial temporal-occipital damage.	Copying complex figures.	One patient impaired.
Benzing & Squire	1989	LM	Bilateral HC.	Binocular depth perception from random dot stereograms.	Intact performance.
Bohbot et al.	1998	BS, FL, KoA, MH, KP, MJ, PV, PM, KrA, FA, KS, SV, VP	**BS, MH, KoA:** R. HC, Amyg & PRh; **KP, MJ:** R. HC & Amyg; **FL:** R. HC & bilateral Amyg; **PV:** R. PHG, HC & PRh; **PM:** R. PHG, HC & Amyg; **KRA:** R. PHG, ERh & PRh; **VP, FA:** L. HC & Amyg; **KS:** L. HC, Amyg, ERh, PRh; **SV:** L. HC, PRh & bilateral Amyg; **SI:** L. PHG, Amyg & PRh.	Rey copy.	All demonstrated intact performance, but 4 unidentified patients showed signs of "perceptual defragmentation".
Buffalo et al.	1998	EP, GT, AB, LJ, RC, NF, PN, JW	**EP:** Bilateral HCF, Amyg, PRh, PHG, fusiform gyrus; **GT:** Bilateral HC, Amyg, PRh, PHG, inferior, middle & superior temporal gyri; **NF:** Mamillary bodies, bilateral HC; **LJ:** bilateral HC; **RC, JW, PN:** Mamillary bodies; **AC:** unknown.	Simultaneous matching for novel visual patterns.	All demonstrated intact performance.
Chun & Phelps	1999	5 cases	All bilateral HC.	Visual search task (initial assessment).	All demonstrated intact performance.
Cipolotti et al.	2001	VC	Bilateral HC, L PHG.	VOSP incomplete letters, cube analysis.	Intact performance.
Cummings, Tomiyasu, Read, & Benson	1984	Unknown	Bilateral HC; R. frontal-parietal damage, L. thalamus.	Copying complex 3D figures.	Intact performance.

Author(s)	Year	Case(s)	Anatomy	Test	Performance
Dejong, Itabashi, & Olson	1969	1 case	Bilateral HC; PHG and fusiform gyrus extending into occipital lobes.	Copying Gestalt designs.	Intact performance.
Duyckaerts et al.	1985	1 case	Bilateral HC, Amyg.	Rey copy.	Intact performance.
Eslinger	1998	DR, MR, PD	**DR**: R. HCF, temporal pole, anterior lateral temporal lobe; Amyg, medial occipitotemporal gyri; basal forebrain, insula; **MR**: bilateral HC; **PD**: bilateral HC; Amyg.	Not stated.	Intact perception and spatial cognition.
Fujii et al.	1999	1 case	Bilateral HCF & Amyg.	Rey copy.	Intact performance.
Henke et al.	1999	DF	Bilateral HCF.	Mental rotation; Spatial subtest of differential aptitude test (Bennett, Seashore, & Wesman, 1972).	Intact performance.
Holdstock et al.	2000a	CF, RS, YW, NM, CW, RT, JT	**CF**: R. HC, Amyg, PHG, PRh & ERh, L. insula, superior & middle temporal gyrus, occipitotemporal gyrus; **RS**: Bilateral HC, R. PRh, L. PHG, PRh, ERh. **YW**: R. Amyg, HC, PHG, PRh, ERh, inferior & middle temporal gyri, L. inferior temporal gyrus; **NM**: Bilateral Amyg, HC, PHG, PRh, ERh, mammillary bodies, cerebellum, superior frontal and parietal lobes; **CW**: Posterior temporal lobe, occipital lobe, R. MTL, **RT**, JT: unknown.	Simultaneous matching to sample of novel visual patterns.	All demonstrated intact performance.
Holdstock et al.	2000b	YR	Bilateral HC.	Complete VOSP battery; Judgement of line orientation (Benton, Hamsher, Varney, & Spreen, 1983); 'Little Nen' test of mental rotation (Ratcliff, 1979).	Intact performance.

(continued)

TABLE 1 Continued

Study	Year	Patient/s	Lesion details	Perceptual assessment	Finding
Kapur et al.	1994	TJ	Bilateral temporal pole, inferior, middle & superior temporal gyri & R. HC, Amyg.	Mooney visual "closure" task (organizing face from a chaotic black and white pattern with incomplete contour).	Intact performance.
Kartsounis et al.	1995	1 case	Bilateral HC, L. Amyg.	Not stated.	Intact perceptual skills and learning.
Kitchener, Hodges, & McCarthy	1998	RS	L. HC, PHG, PRh, ERh, posterior thalamus, medial frontal lobe & R. HC.	VOSP fragmented letters & dot centre; Judgement of line orientation.	Intact performance.
Milner et al.	1968	HM	Bilateral MTL, including HCF & ERh; Cerebellum.	Detection of anomalous cartoon drawing features, tachistocopic letter recognition, masked letter perception, Mooney visual "closure" task, contrast perception.	Intact performance.
				Gottschaldt hidden figure test (detection of geometric patterns within a network of overlapping lines).	Impaired performance.
Oxbury, Oxbury, Renowden, Squier, & Carpenter	1997	CG	L. temporal lobe & R. HCF.	Rey copy.	Intact performance.

Sidman, Stoddard, & Mohr	1968	HM	See Milner et al. (1968)	Discrimination of circles and ellipses.	Intact performance.
Stark & Squire	2000	EP, GP, GT, LJ, PN, AB, PH, MH, JW	**EP, GT, AB, LJ, PN, JW**: see Buffalo et al., 1998. **GP**: Bilateral HC, ERh, PRh, PHG, Amyg; **PN**: Mamillary nuclei, frontal lobes; **MH**: Unknown.	Oddity judgement for colour, size, shape, faces, and objects.	All demonstrated intact performance.
Starr & Phillips	1970	MK	Bilateral temporal lobe.	Unstated.	Intact vision.
Vargha-Khadem et al.	1997	Kate, Beth, Jon	All bilateral HC.	Rey copy.	All demonstrated intact performance.
Von Cramon & Schuri	1992	DC	L. HC, superior parietal lobe, thalamus.	Visual search.	Intact performance.
Woods et al.	1982	1 case	Bilateral HCF; L. posterior inferior temporal lobe extending to occipital pole.	Rey copy.	Impaired performance.
Zola-Morgan, Squire, & Amaral	1986	RB	Bilateral HCF.	Rey copy.	Intact performance.

Note: Lesion details key: **R.** = Right; **L.** = Left; HC = Hippocampus; HCF = Hippocampal formation (includes hippocampus, subiculum and entorhinal cortex); Amyg = Amygdala; PHG = Parahippocampal gyrus; PRh = Perirhinal cortex; ERh = Entorhinal cortex.

solved the tasks primarily on the basis of single-feature discrimination (which is not thought to be dependent upon perirhinal cortex). A similar argument may be applied to existing standardized neuropsychological tests of perception (Rey–Osterrieth figure copy, VOSP battery), in that none of these tasks adequately assess the ability to discriminate conjunctions of features. Consequently, it may not be surprising that, to date, there has been no convincing evidence that lesions to the human MTL can result in perceptual difficulties.

This issue was addressed in a study by Stark and Squire (2000), in which the ability of amnesic patients to discriminate visual stimuli was assessed using an adaptation of a nonmnemonic oddity judgement task that had been previously utilized in monkeys with perirhinal cortex lesions (Buckley et al., 2001). In this paradigm, the participants were required to select the odd stimulus from an array of six images taken from a variety of stimulus categories. While some of these discriminations could be made on the basis of single features (e.g., oddity judgement for shape, colour, and size), other discriminations placed a greater demand on perceiving conjunctions of features (e.g., oddity judgement for objects, masked with varying degrees of visual noise, and faces). In the original monkey study, Buckley et al. (2001) found that simple single-feature discriminations were not dependent on the perirhinal cortex, while judgements for objects and faces were impaired following selective perirhinal cortex damage. Contrary to these observations, however, Stark and Squire (2000) found that their amnesic patients with perirhinal cortex damage performed within the normal range across all stimulus conditions, even on those that were sensitive to perirhinal cortex damage in monkeys. Given that this difference could not be attributed to a lack of difficulty in the tasks used (both patients and healthy controls performed around the 60% mark in the hardest oddity conditions), it was concluded that the human perirhinal cortex is "functionally different from the perirhinal cortex in monkeys with respect to visual perception" (Stark & Squire, 2000) and that this region must serve mnemonic processes exclusively.

New evidence for a role of the human MTL in perception

Deficits in visual discrimination following human MTL damage

Although recent studies have investigated the role of the human perirhinal cortex in perception, no studies have, to our knowledge, examined a possible role for the human hippocampus in spatial perception as suggested recently in the animal literature (e.g., Gaffan, 2001). In order to address this issue and also to replicate and extend the findings of Stark and Squire (2000), we recently assessed fine visual discrimination in amnesic patients with either selective hippocampal damage (HC group) or more extensive medial temporal damage including the perirhinal cortex (MTL group). Importantly, both patient groups were tested initially on general neuropsychological tasks of perception, including the complete VOSP battery (Warrington & James, 1991), the Rey–Osterrieth figure copy condition (Osterrieth, 1944), and the Benton face task (Benton & Van Allen, 1968). It was found, in agreement with the existing literature, that all patients performed within the normal range on these tests, suggesting intact perceptual abilities as they are traditionally assessed.

With respect to the experimental tasks, different conditions of two simple computerized visual discrimination tasks were administered (Lee et al., 2005c). In brief, the two

discrimination tasks were both based on the same experimental paradigm, but whereas successful performance on one task was dependent on the participants learning a single visual
image (*discrimination performance task*), the second test placed a minimal demand on
mnemonic processing (*simultaneous matching task*). On the first three trials of each condition in the discrimination performance task (see Figure 1), a pair of unfamiliar images from
one of four stimulus categories (faces, objects, spatial scenes, or colour) was presented on a
touchscreen monitor, and the participants were instructed to identify the "correct" picture
by touching it. Selecting the correct stimulus produced a high tone, while the incorrect
image was associated with a low tone. From Trials 4–53, the same pictures were then
blended together to create 50 new trial-unique pairs with five different levels of overlapping features: 0–9%, 10–19%, 20–29%, 30–39%, and 40–49% of shared features. There
were 10 trials for each level of blending, and these were pseudorandomly ordered such that
2 trials from each level were presented during each block of 10 trials. The participants were
asked to select the image that they perceived to contain a greater proportion of the original

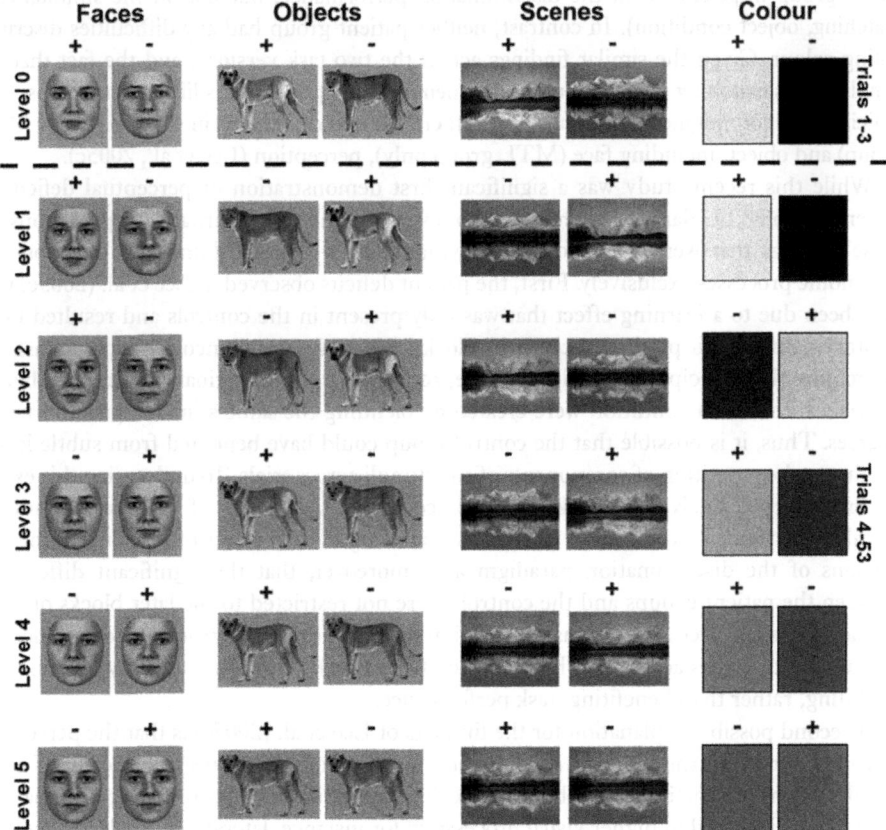

Figure 1.   One trial from each level of feature overlap from the discrimination performance task used in Lee
et al. (2005c). (+) indicates correct stimulus; (−) indicates incorrect stimulus. The simultaneous matching task was
identical, but on each trial the original correct stimulus (i.e., Level 0) was displayed above the two choice stimuli.

correct stimulus, and auditory feedback was provided for all trials. Thus, this first task required the participants to hold onto the representation of the original correct picture throughout the 50 trials of each condition, and it assessed different difficulties of fine visual discrimination as the two test stimuli on each trial shared varying levels of features.

In the second simultaneous matching task (administered a few months later), the identical paradigm was used but with one critical difference: On each trial the original correct stimulus was always shown above the two choice stimuli. Thus, there was a minimal mnemonic component in this task, and successful performance was not dependent on the participants having to remember the original correct stimulus throughout the duration of each task condition.

In support of the idea that the human hippocampus and perirhinal cortex may be critical to scene and object perception, respectively, it was found that on both the discrimination performance and simultaneous matching tasks, the two patient groups were impaired in the discrimination of spatial scenes, while the MTL group patients demonstrated an additional impairment in discriminating faces and, to a lesser extent, objects (there was a significant MTL group impairment on the discrimination performance, but not on the simultaneous matching, object condition). In contrast, neither patient group had any difficulties discriminating colour. Given the similar findings across the two task versions, and the fact that the simultaneous matching task had a reduced mnemonic component, it is likely that the observed deficits were not mnemonic in nature, but rather reflected difficulties in scene (HC and MTL group) and object, including face (MTL group only), perception (Lee et al., 2005c).

While this recent study was a significant first demonstration of perceptual deficits in patients with MTL damage, there were, however, a number of alternative explanations for these findings that were not inconsistent with the view that the human MTL subserves mnemonic processes exclusively. First, the patient deficits observed in Lee et al. (2005c) may have been due to a learning effect that was only present in the controls and resulted in an improvement in their performance across blocks. While the simultaneous matching task did not require the participants to hold onto a representation of the original correct stimulus, all 50 trials within each condition were created by blending the same stimulus pair to varying degrees. Thus, it is possible that the control group could have benefited from subtle learning due to the repetition of components of the stimuli across trials. To undermine this explanation, however, analyses of task performance across the blocks of each task condition revealed that none of the subject groups showed a significant effect of learning in the two versions of the discrimination paradigm and, moreover, that the significant differences between the patient groups and the controls were not restricted to the later blocks of trials on the scenes and faces conditions. In addition to this, the control groups made an increasing number of errors across trial blocks for certain stimuli, indicating that any learning was inhibiting, rather than benefiting, task performance.

A second possible explanation for the findings of Lee et al. (2005c) is that the perceptual deficits observed in the HC and MTL patient groups could have been the result of additional atrophy to cortical areas beyond the hippocampus and perirhinal cortex that are known to be involved in higher visual processing: for instance, lateral temporal lobe regions, including area TE/TEO. One line of evidence against this idea, however, is the structural ratings of the MRI scans of the patients. These ratings, based on a scale validated against volumetric methods (Galton et al., 2001), revealed that the hippocampal patients did not

have any cortical damage beyond the hippocampus, suggesting that any perceptual impairment in this group was likely to be the result of this selective damage. Similarly, whereas the
MTL patients had larger lesions that encompassed the hippocampus and surrounding
regions (including the perirhinal cortex, amygdala, and anterior temporal lobe), the majority
of these patients did not have significant damage to lateral temporal lobe regions. In addition to this, both HC and MTL patients performed within the normal range when making
difficult colour discriminations, an ability that has been previously shown to be dependent
upon area TE/TEO in nonhuman primates (Buckley, Gaffan, & Murray, 1997).

Deficits in oddity judgement following human MTL damage

In an attempt to replicate the findings of Lee et al. (2005c) and to reconcile the findings of
this study with that of Stark and Squire (2000), Lee et al. (2005a) assessed oddity judgement
in the same amnesic patients that were assessed in Lee et al. (2005c). Versions of the original
oddity tasks that were used in perirhinal lesioned monkeys (Buckley et al., 2001) were administered, as well as a number of new oddity tests to assess spatial perception (see Figure 2):

1. *Reassessment of Stark and Squire (2000).*   To investigate the findings of Stark and
Squire (2000), oddity judgement for colour, faces, and objects was investigated. As in this pre-
vious study, participants were required to select the odd stimulus from an array of six images.
In our experiment, however, the stimulus set sizes of the faces and objects were significantly

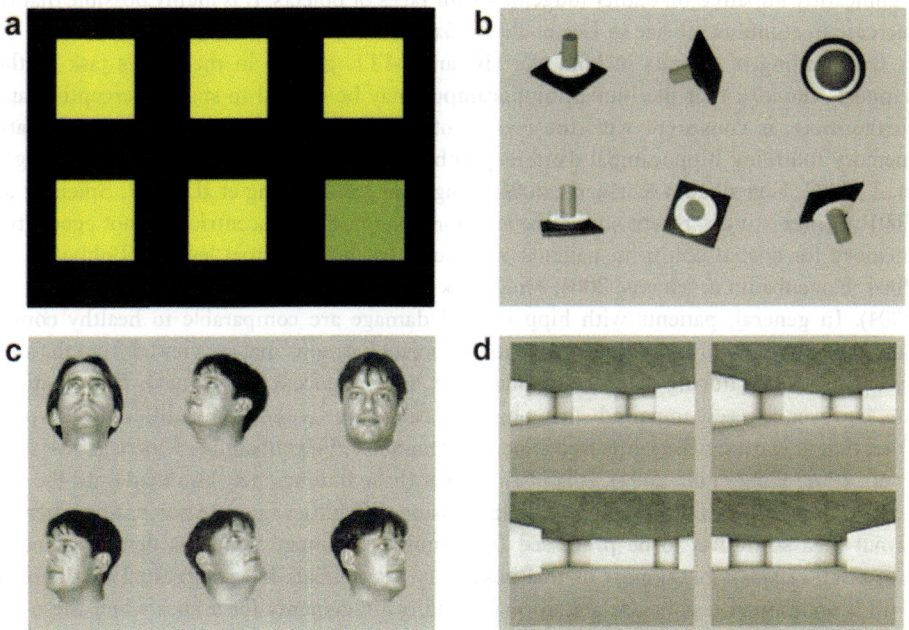

Figure 2.   One trial from the (a) colour; (b) object; (c) face; and (d) scene oddity tasks used in Lee et al. (2005a).

increased (sets of 20 items were used, in comparison to 10 in Stark & Squire, 2000). This change was made since studies in monkeys have shown that large set sizes may be required to observe visual concurrent discrimination deficits following perirhinal cortex lesions (Buckley & Gaffan, 1997; although see Hampton, this issue), presumably by increasing feature "overlap" across trials.

   2. *Assessing spatial perception and the possible role of learning.*   To investigate whether spatial perception deficits could be observed in the context of an oddity paradigm, we administered a test that assessed oddity judgement for virtual reality rooms. On any given trial, patients were presented with four images on the screen and were required to select the odd stimulus. Three of these scenes were different vantage points of the same virtual reality room, while one was a fourth view of a different room that differed slightly in the arrangement and/or size of certain aspects (e.g., wall, pillar, or room cavity). Thus, in order to solve this task, the participants had to use a complete three-dimensional representation of the rooms within each trial. It is important to note that in order to rule out possible effects of learning, trial-unique stimuli were used, with no room being shown more than once in each condition.

   Consistent with recent suggestions that the hippocampus is involved in space, and not object, perception, the HC group was significantly impaired on oddity judgements for virtual reality rooms, but not for faces, objects, or colour. The MTL patients were also impaired on the virtual reality rooms task, but additionally demonstrated significant impairments when they were required to make oddity judgements for faces as well as objects. They were, however, not significantly impaired in the colour condition. Thus, the profile of performance observed by Lee et al. (2005a) in the MTL patients contradicted the findings of Stark and Squire (2000), who found, using similar tasks, that patients with MTL damage were not significantly impaired on oddity judgements for faces or objects. It is highly possible that the increase in stimulus set size in Lee et al., 2005a, contributed to this difference.

   The finding of deficits in both the HC and MTL groups in the scenes task further supports the idea that the human hippocampus may be critical to spatial perception and, furthermore, is consistent with the myriad of evidence demonstrating deficits in spatial memory following hippocampal damage (Bohbot et al., 2004; Bohbot et al., 1998; Burgess et al., 2002; Feigenbaum & Morris, 2004; King et al., 2002; King et al., 2004; Spiers et al., 2001b). Interestingly, recent studies have reported impaired allocentric, but not egocentric, memory for spatial scenes in patients with selective hippocampal lesions (Bohbot et al., 2004; Feigenbaum & Morris, 2004; Holdstock et al., 2000b; King et al., 2002; King et al., 2004). In general, patients with hippocampal damage are comparable to healthy control subjects or only mildly impaired at recognizing object locations that are viewed from the same viewpoint as that during the learning phase (i.e., intact egocentric memory). These patients are, however, severely impaired when object locations are viewed from a different viewpoint from that at learning (i.e., impaired allocentric memory; Feigenbaum & Morris, 2004; King et al., 2002; King et al., 2004). It is possible that these findings may also map onto the perceptual domain, with hippocampal damage causing difficulties in scene perception when the layout of a scene has to be processed from multiple vantage points as demonstrated by Lee et al. (2005a). In support of this possibility, recent data have suggested that patients with hippocampal lesions are able to make oddity judgements for scenes when these discriminations can be solved on the basis of egocentric processing. The amnesic patients from Lee et al. (2005a) were presented with three identical images of the same room taken from the same viewpoint with a fourth image of a separate room from a different view. It was found that irrespective of the location of their cortical damage (i.e., selective hippocampal damage versus larger MTL lesions) all the patients were able to select successfully the odd stimulus (Lee et al., 2005a).

Convergence of findings in visual discrimination and oddity judgement

The consistent findings of perceptual deficits in two separate paradigms (Lee et al., 2005a; Lee et al., 2005c), as well as their convergence with data from the nonhuman primate literature (e.g., Buckley et al., 2001; Bussey et al., 2002; Bussey et al., 2003), have provided the strongest evidence to date that the human MTL may play a critical role in higher order perception (see Figure 3), with the hippocampus and perirhinal cortex involved in spatial and object perception, respectively. The fact that the amnesic patients in these studies performed within the normal range on general tests of perception (e.g., the copying condition of the Rey–Osterrieth figure, Osterrieth, 1944, and the VOSP, Warrington & James, 1991) highlights again the fact that these tests do not place an adequate demand on the perceptual processes that the human MTL have been proposed to subserve. For example, making a copy of the Rey–Osterrieth figure with pen and paper can be done on a feature-by-feature basis, while many of the tests of the VOSP (e.g., the counting of dots and letter identification) do not involve complex object or spatial scene perception, which are dependent on processing conjunctions of features. Perceptual deficits following MTL damage may, therefore, only be evident when tasks that cannot be solved on the basis of single visual features (i.e., that place a demand on the perception of multiple features within objects and scenes) are employed.

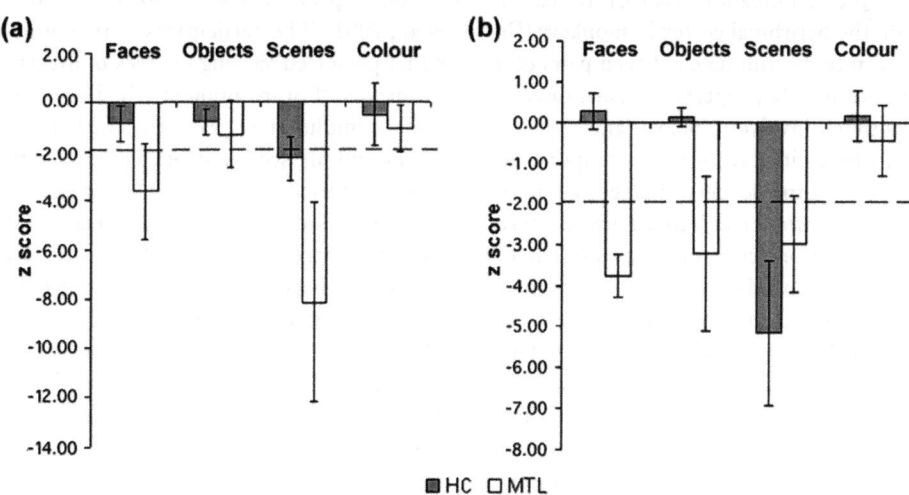

Figure 3. Z score plots for the two patient groups when compared to matched controls on different conditions of (a) simultaneous matching (Lee et al., 2005c) and (b) oddity judgement (Lee et al., 2005a). Scores beyond the dashed line ($z = -1.96$) indicate significant impairment.

Perceptual deficits as a cause of amnesia?

Given the recent evidence for perceptual deficits in humans following MTL damage (Lee et al., 2005a; Lee et al., 2005c), it is important to consider how perceptual and mnemonic deficits may interact with each other. One controversial proposal stemming primarily from the animal literature proposes that the perceptual deficits documented following MTL damage may, in fact, be the underlying cause of the memory deficits seen after such injury (Gaffan, 2001; Horel, 1978; O'Keefe, 1999). When considering the existing neuropsychological literature, however, it is difficult to conceive that the human amnesic syndrome may result primarily from perceptual deficits. For instance, impairments in spatial perception in patients with selective hippocampal damage (Lee et al., 2005a; Lee et al., 2005c) cannot account easily for difficulties in mnemonic tasks that do not, at least overtly, contain a spatial component. These include remembering studied pairs of words (e.g., Giovanello, Verfaellie, & Keane, 2003) as well as recalling a previously presented prose passage after a delay.

It is important to note that although these types of task do not place an explicit demand on processing spatial information, it is possible, nevertheless, that spatial perception may still be essential to task performance. One argument is that spatial memory may be spontaneously recruited during the encoding and retrieval of an experienced episode (Gaffan, 2001). For instance, the accuracy of recalling an event in life may benefit from the reconstruction of the entire event, a critical part of which may be the spatial context in which the event took place. Subsequently, an impairment in spatial processing may impede this process and produce difficulties in mnemonic processing (Gaffan, 2001).

Although further research is required to understand a possible interaction between space perception and mnemonic impairments in human amnesics, the impact of object perception difficulties on learning has been investigated recently by Barense et al. (2005). In brief, the amnesic patients seen previously in Lee et al. (2005a, 2005c) were assessed on a visual concurrent discrimination task that was based on a paradigm previously shown to be dependent upon the perirhinal cortex in monkeys (Bussey et al., 2002). The participants were required to learn to discriminate between pairs of objects that possessed varying degrees of overlapping features (a property of visual discrimination termed "feature ambiguity"). Two of the conditions involved "blobs" and "bugs", and each stimulus item was composed of two explicitly defined features, or components (e.g., shape and fill; see Figure 4). For each stimulus type, three levels of perceptual discrimination were used: *minimum*, *intermediate*, and *maximum feature ambiguity*. Each of these levels involved four objects (two correct targets and two incorrect nontargets) presented in pairs (one target randomly paired with one nontarget), and the number of trials that the participants needed to learn the target items was assessed. In the minimum feature ambiguity level, no object features were ambiguous (i.e., both features were unique to target and nontarget objects). In contrast, in the intermediate feature ambiguity level, half of the features were ambiguous (i.e., one feature appeared in both a target and a nontarget, but the other feature was unique to the target), whereas in the maximum feature ambiguity level, all features were ambiguous (i.e., both features of the target also appeared separately in nontargets). Successful discriminations in the maximum condition required the participants to learn the conjunction of the two features that correctly distinguished targets from nontargets (e.g., dot fill and six-pronged shape), and they could not be solved using a single feature (e.g., dot fill) since no individual

Minimum   Intermediate   Maximum
  +   -   +   -   +   -

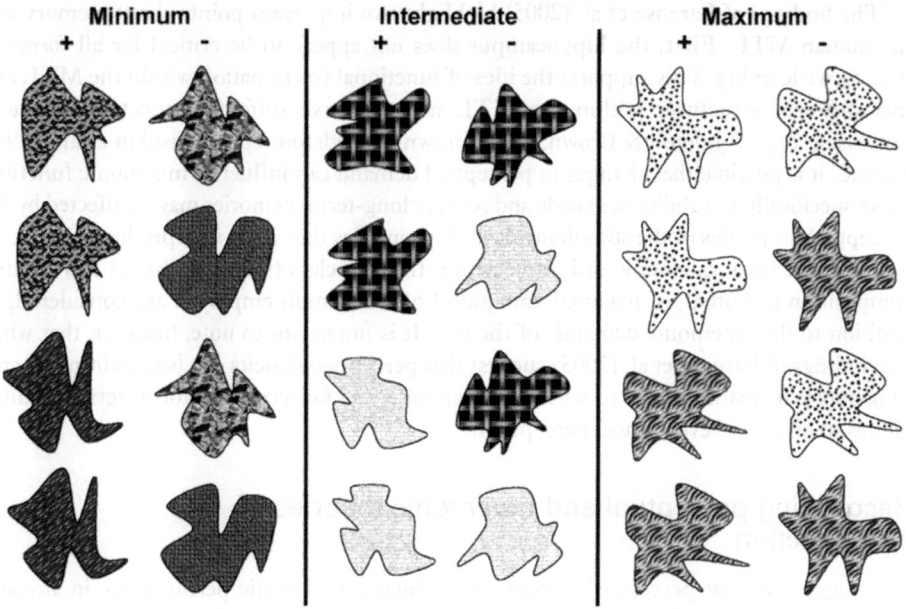

Figure 4. Minimum, intermediate, and maximum feature ambiguity levels for the "blobs" condition in Barense et al. (2005). (+) indicates correct stimulus; (−) indicates incorrect stimulus. The two manipulated feature components were shape and fill and the position of the target on the screen was counter balanced.

features were unique to the targets. Importantly, the number of objects to be remembered was constant for all conditions, but as feature ambiguity was increased, the demand on learning feature conjunctions was increased parametrically.

Contrary to the idea that all MTL structures function as a single mnemonic system, Barense et al. (2005) found that the patients with selective hippocampal lesions (HC group) performed within the normal range on all conditions of the blobs and bugs tasks, thus demonstrating an intact ability to learn object discriminations (i.e., they required a similar number of trials as did healthy control subjects to learn all discriminations). In contrast, whereas the patients with larger MTL lesions (MTL group) generally performed normally on all minimum feature ambiguity conditions, they were significantly impaired on the intermediate and maximum feature ambiguity levels (i.e., when a greater demand was placed on discriminating conjunctions of features, they required a significantly greater number of trials than did healthy control subjects to learn the discrimination). Given that healthy control subjects performed equally across all conditions, it is unlikely that the deficits observed in the MTL group were due to task difficulty. Moreover, although there were an equal number of stimuli in each condition, a greater number of features were presented in the minimum (eight features) than in the maximum ambiguity (four features) levels. Thus, the magnitude of impairment in the MTL group was inversely related to the number of features presented. This suggests that the number of items to be discriminated and any differences in memory load were not critical to eliciting impairments in MTL patients.

The findings of Barense et al. (2005) highlight two important points about memory and the human MTL. First, the hippocampus does not appear to be critical for all forms of declarative learning. This supports the idea of functional fractionation within the MTL and that different structures within the MTL may subserve different aspects of memory processing (e.g., Aggleton & Brown, 1999; Brown & Aggleton, 2001; Mishkin et al., 1997). Second, it is possible that changes in perceptual demand can influence mnemonic function. More specifically, the ability to encode and retrieve long-term memories may be affected by the perceptual properties of the stimuli involved. This implies that when interpreting behavioural performance on a mnemonic task, it is critical that the class (scene or object) and featural composition (minimal vs. maximal ambiguity) of the stimuli employed are considered, in addition to the mnemonic demands of the test. It is important to note, however, that while the findings of Barense et al. (2005) suggest that perceptual deficits can lead to impairments in memory, it is still unclear to what extent amnesia can be accounted for in terms of difficulties in object, or even space, perception.

Reconciling perceptual and mnemonic theories of MTL function

The suggestion that perceptual deficits can influence mnemonic performance in amnesic cases with focal MTL damage (Barense et al., 2005) raises the question as to how perceptual theories of MTL function can be reconciled with existing ideas of the role of MTL structures in mnemonic processes. One likelihood is that perceptual and mnemonic theories of MTL function are not mutually exclusive. For instance, considering the perirhinal cortex, it is conceivable that different populations of neurons within this region mediate familiarity-based recognition memory and object perception. It is known from nonhuman primate electrophysiological studies that there are perirhinal neurons that decrease their firing in response to subsequent presentations of unfamiliar objects (Brown, Wilson, & Riches, 1987; Li, Miller, & Desimone, 1993; Sobotka & Ringo, 1993), and it has been suggested that this mechanism may constitute the neural basis of recognition memory (for review see Brown & Xiang, 1998). There are also, however, neurons that possess stimulus specificity but do not exhibit decremental response patterns (Xiang & Brown, 1998), and, moreover, studies have shown that recognition memory can be dissociated from the decremental response properties of some perirhinal neurons (Sobotka & Ringo, 1996). Thus, it is possible that there may be a variety of neural mechanisms that make separate contributions to the mnemonic and perceptual processes that the perirhinal cortex has been suggested to subserve.

With regard to the hippocampus, the idea that this region may be critical for spatial perception concurs with the existing theory that this structure may function as a spatial module or cognitive map (O'Keefe & Nadel, 1978). It is possible that hippocampal cells that signal aspects of spatial location are important to this role (Ekstrom et al., 2003; Hori et al., 2003; O'Keefe, 1976; O'Keefe & Burgess, 1996; O'Keefe et al., 1998; Ono et al., 1991; Ono, Nakamura, Nishijo, & Eifuku, 1993; Wilson & McNaughton, 1993), although it is still unclear as to whether this same mechanism may contribute to declarative memory encoding and retrieval processes.

A possible role for the hippocampus in spatial, but not object, perception is also not necessarily contradictory to a relational memory view of the hippocampus (i.e., that this structure is involved in the memory for relationships among perceptually distinct items; Cohen et al., 1999; Eichenbaum & Cohen, 2002; Eichenbaum et al., 1994) providing that the hippocampus is believed to mediate the association of relations among elements of scenes or events and not the binding of features of individual objects. Thus, while patients with selective hippocampal lesions are able to discriminate objects that possess overlapping features, they have difficulties when they have to perceive the multiple relations between these objects: for example, in the context of a spatial scene.

Future directions

If the human MTL is involved in higher order perceptual processes, then one would expect the perceptual deficits seen in amnesics with static lesions to be replicable in other patient populations with MTL pathology. Two ideal candidates for this are Alzheimer's Disease (AD), a condition that is characterized by a primary deficit in episodic memory and at least one other significant impairment in a different cognitive domain, such as attention, working memory and language (Grady et al., 1988; Hodges & Patterson, 1995; McKahnn et al., 1984; Perry & Hodges, 1996, 1999; Welsh, Butters, Hughes, Mohs, & Heyman, 1992), and semantic dementia (SD), a neurodegenerative disease that is associated with a progressive, crossmodal loss of semantic knowledge, with relative preservation in other cognitive domains (Garrard, Perry, & Hodges, 1997; Hodges & Patterson, 1995; Hodges, Patterson, Oxbury, & Funnell, 1992; Perry & Hodges, 1996; Snowden, Goulding, & Neary, 1989; Snowden, Neary, & Mann, 1996; Warrington, 1975). Recent volumetric studies have shown differing profiles of atrophy to MTL regions in these two diseases (Chan et al., 2001; Davies, Graham, Xuereb, Williams, & Hodges, 2004; Davies, Xuereb, & Hodges, 2002; Galton et al., 2001), with predominant involvement of the perirhinal cortex in SD compared to other MTL regions, and primary atrophy throughout the hippocampus in AD (Davies, Graham, Xuereb, Williams & Hodges, 2004, although see Bright Moss, Stamatakis, & Tyler, this issue). Consistent with perceptual theories of the MTL, preliminary data have suggested that perceptual deficits may, in fact, also be present in patients with these disorders, with AD patients demonstrating impairments in scene discrimination and SD patients showing deficits in discriminating faces (Lee et al., 2005b).

Another important issue for investigation is defining the role that the hippocampus may play in spatial perception. As discussed earlier, existing studies of spatial memory suggest that the hippocampus may be important for allocentric processes. Recent nonhuman primate work has found, however, that hippocampal dysfunction, by means of fornix transection, can result in deficits in learning conjunctions of spatial information: for instance, the association between tail length, orientation, and spatial location of tadpoles within a two-dimensional scene (Buckley et al., 2004). Similarly, our own investigations suggest that amnesic patients with hippocampal damage have difficulties discriminating images of scenes that share a high proportion of overlapping features but do not necessarily place a demand on allocentric processing (Lee et al., 2005c). Thus, it is currently unclear how perceiving conjunctions of spatial information relates to allocentric processes and what role the hippocampus may play in each of these. In addition to this, given that the parahippocampal gyrus is also likely to play an important role in spatial memory (Bohbot et al., 2004; Bohbot et al., 1998; Ekstrom et al., 2003) as well as the encoding of perceptual information about the layout of scenes

(Epstein, Harris, Stanley, & Kanwisher, 1999), further research will be necessary to deter-mine the distinct contributions made by the parahippocampal gyrus and hippocampus to spatial perception.

Summary

Contrary to traditional views, there is now evidence emerging to suggest that the human MTL may mediate the same nonmnemonic processes that have been proposed for the animal MTL, with the hippocampus and perirhinal cortex subserving aspects of spatial and object perception, respectively. If true, perceptual theories of human MTL function will have a profound impact on the way in which human amnesia is understood, in particular whether deficits in perception are related to impairments in mnemonic processes.

REFERENCES

Aggleton, J. P., & Brown, M. W. (1999). Episodic memory, amnesia and the hippocampal-anterior thalamic axis. *Behavioral and Brain Sciences, 22,* 425–289.

Alvarez, P., Zola-Morgan, S., & Squire, L. R. (1994). The animal model of human amnesia: Long-term memory impaired and short-term memory intact. *Proceedings of the National Academy of Sciences USA, 91,* 5637–5641.

Baddeley, A., Vargha-Khadem, F., & Mishkin, M. (2001). Preserved recognition in a case of developmental amnesia: Implications for the acquisition of semantic memory? *Journal of Cognitive Neuroscience, 13,* 357–369.

Barense, M. D., Bussey, T. J., Lee, A. C. H., Rogers, T. T., Davies, R. R., Saksida, L. M., et al. (2005). *Functional specialization in the human medial temporal lobe.* Manuscript submitted for publication.

Bayley, P. J., & Squire, L. R. (2004). Failure to acquire new semantic knowledge in patients with large medial temporal lobe lesions. *Hippocampus, 15,* 273-280.

Beatty, W. W., Salmon, D. P., Bernstein, N., & Butters, N. (1987). Remote memory in a patient with amnesia due to hypoxia. *Psychological Medicine, 17,* 657–665.

Bennett, G. K., Seashore, H. G., & Wesman, A. G. (1972). *Differential aptitude test manual* (5th ed.). New York: Psychological Corporation.

Benson, D. F., Marsden, C. D., & Meadows, J. C. (1974). The amnesic syndrome of posterior cerebral artery occlusion. *Acta Neurologica Scandinavica, 50,* 133–145.

Benton, A. L., Hamsher, K., Varney, N. R., & Spreen, O. (1983). *Judgement of line orientation: Contributions to neuropsychological assessment.* New York: Oxford University Press.

Benton, A. L., & Van Allen, M. W. (1968). Impairment in facial recognition in patients with cerebral disease. *Cortex, 4,* 344–358.

Benzing, W. C., & Squire, L. R. (1989). Preserved learning and memory in amnesia: Intact adaptation-level effects and learning of stereoscopic depth. *Behavioral Neuroscience, 103,* 538–547.

Bohbot, V. D., Iaria, G., & Petrides, M. (2004). Hippocampal function and spatial memory: Evidence from functional neuroimaging in healthy participants and performance of patients with medial temporal lobe resections. *Neuropsychology, 18,* 418–425.

Bohbot, V. D., Kalina, M., Stepankova, K., Spackova, N., Petrides, M., & Nadel, L. (1998). Spatial memory deficits in patients with lesions to the right hippocampus and to the right parahippocampal cortex. *Neuropsychologia, 36,* 1217–1238.

Bright, P., Moss, H. E., Stamatakis, E. A., & Tyler, L. K. (this issue). The anatomy of object processing: The role of anteromedial temporal cortex. *Quarterly Journal of Experimental Psychology, 58B,* 361–377.

Brown, M. W., & Aggleton, J. P. (2001). Recognition memory: What are the roles of the perirhinal cortex and hippocampus? *Nature Reviews Neuroscience, 2,* 51–61.

Brown, M. W., Wilson, F. A., & Riches, I. P. (1987). Neuronal evidence that inferomedial temporal cortex is more important than hippocampus in certain processes underlying recognition memory. *Brain Research, 409,* 158–162.

Brown, M. W., & Xiang, J. Z. (1998). Recognition memory: Neuronal substrates of the judgement of prior occurrence. *Progress in Neurobiology, 55*, 149–189.

Buckley, M. J. (this issue). The role of the perirhinal cortex and hippocampus in learning, memory, and perception. *Quarterly Journal of Experimental Psychology, 58B*, 246–268.

Buckley, M. J., Booth, M. C., Rolls, E. T., & Gaffan, D. (2001). Selective perceptual impairments after perirhinal cortex ablation. *Journal of Neuroscience, 21*, 9824–9836.

Buckley, M. J., Charles, D. P., Browning, P. G., & Gaffan, D. (2004). Learning and retrieval of concurrently presented spatial discrimination tasks: Role of the fornix. *Behavioral Neuroscience, 118*, 138–149.

Buckley, M. J., & Gaffan, D. (1997). Learning and transfer of object–reward associations and the role of the perirhinal cortex. *Behavioral Neuroscience, 112*, 1–9.

Buckley, M. J., Gaffan, D., & Murray, E. A. (1997). Functional double dissociation between two inferior temporal cortical areas: Perirhinal cortex versus middle temporal gyrus. *Journal of Neurophysiology, 77*, 587–598.

Buffalo, E. A., Reber, P. J., & Squire, L. R. (1998). The human perirhinal cortex and recognition memory. *Hippocampus, 8*, 330–339.

Burgess, N., Maguire, E. A., & O'Keefe, J. (2002). The human hippocampus and spatial and episodic memory. *Neuron, 35*, 625–641.

Bussey, T. J., & Saksida, L. M. (2002). The organization of visual object representations: A connectionist model of effects of lesions in perirhinal cortex. *European Journal of Neuroscience, 15*, 355–364.

Bussey, T. J., Saksida, L. M., & Murray, E. A. (2002). Perirhinal cortex resolves feature ambiguity in complex visual discriminations. *European Journal of Neuroscience, 15*, 365–374.

Bussey, T. J., Saksida, L. M., & Murray, E. A. (2003). Impairments in visual discrimination after perirhinal cortex lesions: Testing 'declarative' vs. 'perceptual-mnemonic' views of perirhinal cortex function. *European Journal of Neuroscience, 17*, 649–660.

Bussey, T. J., Saksida, L. M., & Murray, E. A. (this issue). The perceptual-mnemonic/feature conjunction model of perirhinal cortex function. *Quarterly Journal of Experimental Psychology, 58B*, 269–282.

Cave, C. B., & Squire, L. R. (1992a). Intact and long-lasting repetition priming in amnesia. *Journal of Experimental Psychology: Learning, Memory, and Cognition, 18*, 509–520.

Cave, C. B., & Squire, L. R. (1992b). Intact verbal and nonverbal short-term memory following damage to the human hippocampus. *Hippocampus, 2*, 151–163.

Chan, D., Fox, N. C., Scahill, R. I., Crum, W. R., Whitwell, J. L., Leschziner, G., et al. (2001). Patterns of temporal lobe atrophy in semantic dementia and Alzheimer's disease. *Annals of Neurology, 49*, 433–442.

Chun, M. M., & Phelps, E. A. (1999). Memory deficits for implicit contextual information in amnesic subjects with hippocampal damage. *Nature Neuroscience, 2*, 844–847.

Cipolotti, L., Shallice, T., Chan, D., Fox, N., Scahill, R., Harrison, G., et al. (2001). Long-term retrograde amnesia...the crucial role of the hippocampus. *Neuropsychologia, 39*, 151–172.

Cohen, N. J., Ryan, J., Hunt, C., Romine, L., Wszalek, T., & Nash, C. (1999). Hippocampal system and declarative (relational) memory: Summarizing the data from functional neuroimaging studies. *Hippocampus, 9*, 83–98.

Cummings, J. L., Tomiyasu, U., Read, S., & Benson, D. F. (1984). Amnesia with hippocampal lesions after cardiopulmonary arrest. *Neurology, 34*, 679–687.

Davies, R. R., Graham, K. S., Xuereb, J. H., Williams, G. B., & Hodges, J. R. (2004). The human perirhinal cortex and semantic memory. *European Journal of Neuroscience, 20*, 2441–2446.

Davies, R. R., Xuereb, J. H., & Hodges, J. R. (2002). The human perirhinal cortex in semantic memory: An in vivo and postmortem volumetric magnetic resonance imaging study in semantic dementia, Alzheimer's disease and matched controls. *Neuropathology and Applied Neurobiology, 28*, 167–168.

Dejong, R. N., Itabashi, H. H., & Olson, J. R. (1969). Memory loss due to hippocampal lesions. *Archives of Neurology, 20*, 339–348.

Duyckaerts, C., Derouesne, C., Signoret, J. L., Gray, F., Escourolle, R., & Castaigne, P. (1985). Bilateral and limited amygdalohippocampal lesions causing a pure amnesic syndrome. *Annals of Neurology, 18*, 314–319.

Eacott, M. J., Gaffan, D., & Murray, E. A. (1994). Preserved recognition memory for small sets, and impaired stimulus identification for large sets, following rhinal cortex ablations in monkeys. *European Journal of Neuroscience, 6*, 1466–1478.

Eichenbaum, H., & Cohen, N. J. (2002). *From conditioning to conscious recollection: Memory systems of the brain.* Oxford, UK: Oxford University.

Eichenbaum, H., Otto, T., & Cohen, N. J. (1994). Two functional components of the hippocampal memory system. *Behavioral and Brain Science, 17*, 449–518.

Ekstrom, A. D., Kahana, M. J., Caplan, J. B., Fields, T. A., Isham, E. A., Newman, E. L., et al. (2003). Cellular networks underlying human spatial navigation. *Nature, 425*, 184–188.

Epstein, R., Harris, A., Stanley, D., & Kanwisher, N. (1999). The parahippocampal place area: Recognition, navigation, or encoding? *Neuron, 23*, 115–125.

Eslinger, P. (1998). Autobiographical memory after temporal lobe lesions. *Neurocase, 4*, 481–495.

Feigenbaum, J. D., & Morris, R. G. (2004). Allocentric versus egocentric spatial memory after unilateral temporal lobectomy in humans. *Neuropsychology, 18*, 462–472.

Fujii, T., Yamadori, A., Endo, K., Suzuki, K., & Fukatsu, R. (1999). Disproportionate retrograde amnesia in a patient with herpes simplex encephalitis. *Cortex, 35*, 599–614.

Gabrieli, J. D., Cohen, N. J., & Corkin, S. (1988). The impaired learning of semantic knowledge following bilateral medial temporal-lobe resection. *Brain and Cognition, 7*, 157–177.

Gabrieli, J. D., Corkin, S., Mickel, S. F., & Growdon, J. H. (1993). Intact acquisition and long-term retention of mirror-tracing skill in Alzheimer's disease and in global amnesia. *Behavioral Neuroscience, 107*, 899–910.

Gadian, D. G., Aicardi, J., Watkins, K. E., Porter, D. A., Mishkin, M., & Vargha-Khadem, F. (2000). Developmental amnesia associated with early hypoxic-ischaemic injury. *Brain, 123, (3)*, 499–507.

Gaffan, D. (2001). What is a memory system? Horel's critique revisited. *Behavioral Brain Research, 127*, 5–11.

Gaffan, D., Parker, A., & Easton, A. (2001). Dense amnesia in the monkey after transection of fornix, amygdala and anterior temporal stem. *Neuropsychologia, 39*, 51–70.

Gage, P. (1985). Preserved and impaired information processing in human bitemporal amnesiacs and their intrahuman analogues. *Journal of Mind and Behaviour, 6*, 515–551.

Galton, C. J., Gomez-Anson, B., Antoun, N., Scheltens, P., Patterson, K., Graves, M., et al. (2001). Temporal lobe rating scale: Application to Alzheimer's disease and frontotemporal dementia. *Journal of Neurology, Neurosurgery, and Psychiatry, 70*, 165–173.

Galton, C. J., Patterson, K., Graham, K., Lambon-Ralph, M. A., Williams, G., Antoun, N., et al. (2001). Differing patterns of temporal atrophy in Alzheimer's disease and semantic dementia. *Neurology, 57*, 216–225.

Garrard, P., Perry, R., & Hodges, J. R. (1997). Disorders of semantic memory. *Journal of Neurology, Neurosurgery, and Psychiatry, 62*, 431–435.

Giovanello, K. S., Verfaellie, M., & Keane, M. M. (2003). Disproportionate deficit in associative recognition relative to item recognition in global amnesia. *Cognitive, Affective and Behavioral Neuroscience, 3*, 186–194.

Gollin, E. S. (1960). Developmental studies of visual recognition of incomplete objects. *Perceptual and Motor Skills, 11*, 289–298.

Gottschaldt, K. (1929). Über den Einfluss der Erfahrung auf die Wahrnehmung von Figuren. *Psychologische Forschung, 12*, 1–87.

Grady, C. L., Haxby, J. V., Horwitz, B., Sundaram, M., Berg, G., Schapiro, M., et al. (1988). Longitudinal study of the early neuropsychological and cerebral metabolic changes in dementia of the Alzheimer type. *Journal of Clinical and Experimental Neuropsychology, 10*, 576–596.

Hampton, R. R. (this issue). Monkey perirhinal cortex is critical for visual memory, but not for visual perception: Re-examination of the behavioural evidence from monkeys. *Quarterly Journal of Experimental Psychology, 58B*, 283–299.

Hampton, R. R., Hampstead, B. M., & Murray, E. A. (2004). Selective hippocampal damage in rhesus monkeys impairs spatial memory in an open-field test. *Hippocampus, 14*, 808–818.

Henke, K., Kroll, N. E., Behniea, H., Amaral, D. G., Miller, M. B., Rafal, R., et al. (1999). Memory lost and regained following bilateral hippocampal damage. *Journal of Cognitive Neuroscience, 11*, 682–697.

Henson, R. (this issue). A mini-review of FMRI studies of human medial temporal lobe activity associated with recognition memory. *Quarterly Journal of Experimental Psychology, 58B*, 340–360.

Hodges, J. R., & Patterson, K. (1995). Is semantic memory consistently impaired early in the course of Alzheimer's disease? Neuroanatomical and diagnostic implications. *Neuropsychologia, 33*, 441–459.

Hodges, J. R., Patterson, K., Oxbury, S., & Funnell, E. (1992). Semantic dementia. Progressive fluent aphasia with temporal lobe atrophy. *Brain, 115 (6)*, 1783–1806.

Holdstock, J. S. (this issue). The role of the human medial temporal lobe in object recognition and object discrimination. *Quarterly Journal of Experimental Psychology, 58B*, 326–339.

Holdstock, J. S., Gutnikov, S. A., Gaffan, D., & Mayes, A. R. (2000a). Perceptual and mnemonic matching-to-sample in humans: Contributions of the hippocampus, perirhinal and other medial temporal lobe cortices. *Cortex, 36*, 301–322.

Holdstock, J. S., Mayes, A. R., Cezayirli, E., Isaac, C. L., Aggleton, J. P., & Roberts, N. (2000b). A comparison of egocentric and allocentric spatial memory in a patient with selective hippocampal damage. *Neuropsychologia, 38*, 410–425.

Holdstock, J. S., Mayes, A. R., Roberts, N., Cezayirli, E., Isaac, C. L., O'Reilly, R. C., et al. (2002). Under what conditions is recognition spared relative to recall after selective hippocampal damage in humans? *Hippocampus, 12*, 341–351.

Horel, J. A. (1978). The neuroanatomy of amnesia. A critique of the hippocampal memory hypothesis. *Brain, 101*, 403–445.

Hori, E., Tabuchi, E., Matsumura, N., Tamura, R., Eifuku, S., Endo, S., et al. (2003). Representation of place by monkey hippocampal neurons in real and virtual translocation. *Hippocampus, 13*, 190–196.

Iverson, S. D. (1977). Temporal lobe amnesia. In C. W. M. Whitty & O. L. Zangwill (Eds.), *Amnesia* (pp. 136–182). Boston: Butterworths.

Kapur, N., Ellison, D., Parkin, A. J., Hunkin, N. M., Burrows, E., Sampson, S. A., et al. (1994). Bilateral temporal lobe pathology with sparing of medial temporal lobe structures: Lesion profile and pattern of memory disorder. *Neuropsychologia, 32*, 23–38.

Kartsounis, L. D., Rudge, P., & Stevens, J. M. (1995). Bilateral lesions of CA1 and CA2 fields of the hippocampus are sufficient to cause a severe amnesic syndrome in humans. *Journal of Neurology, Neurosurgery, and Psychiatry, 59*, 95–98.

King, J. A., Burgess, N., Hartley, T., Vargha-Khadem, F., & O'Keefe, J. (2002). Human hippocampus and viewpoint dependence in spatial memory. *Hippocampus, 12*, 811–820.

King, J. A., Trinkler, I., Hartley, T., Vargha-Khadem, F., & Burgess, N. (2004). The hippocampal role in spatial memory and the familiarity–recollection distinction: A case study. *Neuropsychology, 18*, 405–417.

Kitchener, E. G., Hodges, J. R., & McCarthy, R. (1998). Acquisition of post-morbid vocabulary and semantic facts in the absence of episodic memory. *Brain, 121* (7), 1313–1327.

Lee, A. C. H., Buckley, M. J., Pegman, S. J., Spiers, H., Scahill, V. L., Gaffan, D., et al. (2005a). *Specialisation in the medial temporal lobe for processing of objects and scenes*. Manuscript submitted for publication.

Lee, A. C. H., Bussey, T. J., Murray, E. A., Levi, N., Saksida, L. M., Davies, R. R., et al. (2005b). *Object and scene memory in semantic dementia and Alzheimer's Disease: A double dissociation*. Manuscript in preparation.

Lee, A. C., Bussey, T. J., Murray, E. A., Saksida, L. M., Epstein, R. A., Kapur, N., et al. (2005c). Perceptual deficits in amnesia: Challenging the medial temporal lobe 'mnemonic' view. *Neuropsychologia, 43*, 1–11.

Levy, D. A., Stark, C. E., & Squire, L. R. (2004). Intact conceptual priming in the absence of declarative memory. *Psychology Science, 15*, 680–686.

Li, L., Miller, E. K., & Desimone, R. (1993). The representation of stimulus familiarity in anterior inferior temporal cortex. *Journal of Neurophysiology, 69*, 1918–1929.

Manns, J. R., Hopkins, R. O., Reed, J. M., Kitchener, E. G., & Squire, L. R. (2003a). Recognition memory and the human hippocampus. *Neuron, 37*, 171–180.

Manns, J. R., Hopkins, R. O., & Squire, L. R. (2003b). Semantic memory and the human hippocampus. *Neuron, 38*, 127–133.

Manns, J. R., & Squire, L. R. (2001). Perceptual learning, awareness, and the hippocampus. *Hippocampus, 11*, 776–782.

Mayes, A. R., Holdstock, J. S., Isaac, C. L., Hunkin, N. M., & Roberts, N. (2002). Relative sparing of item recognition memory in a patient with adult-onset damage limited to the hippocampus. *Hippocampus, 12*, 325–340.

Mayes, A. R., Holdstock, J. S., Isaac, C. L., Montaldi, D., Grigor, J., Gummer, A., et al. (2004). Associative recognition in a patient with selective hippocampal lesions and relatively normal item recognition. *Hippocampus, 14*, 763–784.

McKahnn, G., Drachman, D., Folstein, M., Katzman, R., Price, D., & Stadlan, E. M. (1984). Clinical diagnosis of Alzheimer's disease: Report of the NINCDS-ADRDA Work Group under the auspices of Department of Health and Human Services Task Force on Alzheimer's Disease. *Neurology, 34*, 939–944.

Milner, B., Corkin, S., & Teuber, H.-L. (1968). Further analysis of the hippocampal amnesic syndrome: 14-year follow-up of HM. *Neuropsychologia, 6*, 215–234.

Mishkin, M. (1982). A memory system in the monkey. *Philosophical Transactions of the Royal Society of London. Series B: Biological Sciences, 298*, 83–95.

Mishkin, M., Suzuki, W. A., Gadian, D. G., & Vargha-Khadem, F. (1997). Hierarchical organization of cognitive memory. *Philosphical Transactions of the Royal Society of London. Series B: Biological Sciences, 352*, 1461–1467.

Muramoto, O., Kuru, Y., Sugishita, M., & Toyokura, Y. (1979). Pure memory loss with hippocampal lesions: A pneumoencephalographic study. *Archives of Neurology, 36*, 54–56.

Murray, E. A., & Bussey, T. J. (1999). Perceptual-mnemonic functions of the perirhinal cortex. *Trends in Cognitive Sciences, 3*, 142–151.

Murray, E. A., Davidson, M., Gaffan, D., Olton, D. S., & Suomi, S. (1989). Effects of fornix transection and cingulate cortical ablation on spatial memory in rhesus monkeys. *Experimental Brain Research, 74*, 173–186.

Musen, G., & Squire, L. R. (1992). Nonverbal priming in amnesia. *Memory & Cognition, 20*, 441–448.

O'Keefe, J. (1976). Place units in the hippocampus of the freely moving rat. *Experimental Neurology, 51*, 78–109.

O'Keefe, J. (1999). Do hippocampal pyramidal cells signal non-spatial as well as spatial information? *Hippocampus, 9*, 352–364.

O'Keefe, J., & Burgess, N. (1996). Geometric determinants of the place fields of hippocampal neurons. *Nature, 381*, 425–428.

O'Keefe, J., Burgess, N., Donnett, J. G., Jeffery, K. J., & Maguire, E. A. (1998). Place cells, navigational accuracy, and the human hippocampus. *Philosophical Transactions of the Royal Society of London. Series B: Biological Sciences, 353*, 1333–1340.

O'Keefe, J., & Nadel, L. (1978). *The hippocampus as a cognitive map*. Oxford, UK: Clarendon Press.

Ono, T., Nakamura, K., Fukuda, M., & Tamura, R. (1991). Place recognition responses of neurons in monkey hippocampus. *Neuroscience, Letters, 121*, 194–198.

Ono, T., Nakamura, K., Nishijo, H., & Eifuku, S. (1993). Monkey hippocampal neurons related to spatial and nonspatial functions. *Journal of Neurophysiology, 70*, 1516–1529.

Osterrieth, P. A. (1944). Le test de copie d'une figure complexe. *Archives de Psychologie, 30*, 205–220.

Oxbury, S., Oxbury, J., Renowden, S., Squier, W., & Carpenter, K. (1997). Severe amnesia: An usual late complication after temporal lobectomy. *Neuropsychologia, 35*, 975–988.

Perry, R. J., & Hodges, J. R. (1996). Spectrum of memory dysfunction in degenerative disease. *Current Opinion in Neurology, 9*, 281–285.

Perry, R. J., & Hodges, J. R. (1999). Attention and executive deficits in Alzheimer's disease. A critical review. *Brain, 122* (3), 383–404.

Postle, B. R., & Corkin, S. (1998). Impaired word-stem completion priming but intact perceptual identification priming with novel words: Evidence from the amnesic patient H.M. *Neuropsychologia, 36*, 421–440.

Ratcliff, G. (1979). Spatial thought, mental rotation and the right cerebral hemisphere. *Neuropsychologia, 17*, 49–54.

Rempel-Clower, N. L., Zola, S. M., Squire, L. R., & Amaral, D. G. (1996). Three cases of enduring memory impairment after bilateral damage limited to the hippocampal formation. *Journal of Neuroscience, 16*, 5233–5255.

Scoville, W. B., & Milner, B. (1957). Loss of recent memory after bilateral hippocampal lesions. *Journal of Neurochemistry, 20*, 11–21.

Sidman, M., Stoddard, L. T., & Mohr, J. P. (1968). Some additional quantitative observations of immediate memory. *Neuropsychologia, 6*, 245–254.

Snowden, J., Goulding, P., & Neary, D. (1989). Semantic dementia: A form of circumscribed cerebral atrophy. *Behavioural Neurology, 2*, 167–182.

Snowden, J., Neary, D., & Mann, D. (1996). *Frontotemporal lobar degeneration: Frontotemporal dementia, progressive aphasia, semantic dementia*. London: Churchill Livingstone.

Sobotka, S., & Ringo, J. L. (1993). Investigation of long-term recognition and association memory in unit responses from inferotemporal cortex. *Experimental Brain Research, 96*, 28–38.

Sobotka, S., & Ringo, J. L. (1996). Mnemonic responses of single units recorded from monkey inferotemporal cortex, accessed via transcommissural versus direct pathways: A dissociation between unit activity and behavior. *Journal of Neuroscience, 16*, 4222–4230.

Spiers, H. J., Burgess, N., Hartley, T., Vargha-Khadem, F., & O'Keefe, J. (2001b). Bilateral hippocampal pathology impairs topographical and episodic memory but not visual pattern matching. *Hippocampus, 11*, 715–725.

Spiers, H. J., Maguire, E. A., & Burgess, N. (2001a). Hippocampal amnesia. *Neurocase, 7*, 357–382.

Squire, L. R. (1993). Cross-modal matching performance in amnesia. *Neuropsychology, 7*, 375–384.

Squire, L. R. (2004). Memory systems of the brain: a brief history and current perspective. *Neurobiology Learning and Memory, 82,* 171–177.

Squire, L. R., Stark, C. E., & Clark, R. E. (2004). The medial temporal lobe. *Annual Review of Neuroscience, 27,* 279–306.

Squire, L., & Zola-Morgan, S. (1991). The medial temporal lobe memory system. *Science, 253,* 1380–1386.

Stark, C. E., & Squire, L. R. (2000). Intact visual perceptual discrimination in humans in the absence of perirhinal cortex. *Learning & Memory 7,* 273–278.

Starr, A., & Phillips, L. (1970). Verbal and motor learning in the amnestic syndrome. *Neuropsychologia, 8,* 75–88.

Vargha-Khadem, F., Gadian, D. G., Watkins, K. E., Connelly, A., Van Paesschen, W., & Mishkin, M. (1997). Differential effects of early hippocampal pathology on episodic and semantic memory. *Science, 277,* 376–380.

Von Cramon, D. Y., & Schuri, U. (1992). The septo-hippocampal pathways and their relevance to human memory: A case report. *Cortex, 28,* 411–422.

Warrington, E. K. (1975). The selective impairment of semantic memory. *Quarterly Journal of Experimental Psychology, 27,* 635–657.

Warrington, E. K., & James, M. (1991). *The Visual Object and Space Perception battery.* Bury St. Edmunds, UK: Thames Valley Test Company.

Welsh, K. A., Butters, N., Hughes, J. P., Mohs, R. C., & Heyman, A. (1992). Detection and staging of dementia in Alzheimer's disease. Use of the neuropsychological measures developed for the Consortium to Establish a Registry for Alzheimer's Disease. *Archives of Neurology, 49,* 448–452.

Wilson, M. A., & McNaughton, B. L. (1993). Dynamics of the hippocampal ensemble code for space. *Science, 261,* 1055–1058.

Woods, B. T., Schoene, W., & Kneisley, L. (1982). Are hippocampal lesions sufficient to cause lasting amnesia? *Journal of Neurology, Neurosurgery, and Psychiatry, 45,* 243–247.

Xiang, J. Z., & Brown, M. W. (1998). Differential neuronal encoding of novelty, familiarity and recency in regions of the anterior temporal lobe. *Neuropharmacology, 37,* 657–676.

Yonelinas, A. P., Kroll, N. E., Quamme, J. R., Lazzara, M. M., Sauve, M. J., Widaman, K. F., et al. (2002). Effects of extensive temporal lobe damage or mild hypoxia on recollection and familiarity. *Nature Neuroscience, 5,* 1236–1241.

Zola-Morgan, S., Squire, L. R., & Amaral, D. G. (1986). Human amnesia and the medial temporal region: Enduring memory impairment following a bilateral lesion limited to field CA1 of the hippocampus. *Journal of Neuroscience, 6,* 2950–2967.

Zola-Morgan, S., Squire, L. R., & Ramus, S. J. (1994). Severity of memory impairment in monkeys as a function of locus and extent of damage within the medial temporal lobe memory system. *Hippocampus, 5,* 232–239.

THE QUARTERLY JOURNAL OF EXPERIMENTAL PSYCHOLOGY
2005, 58B (3/4), 326–339

The role of the human medial temporal lobe in object recognition and object discrimination

J. S. Holdstock

University of Liverpool, UK

This paper reviews evidence from neuropsychological patient studies relevant to two questions concerning the functions of the medial temporal lobe in humans. The first is whether the hippocampus and the adjacent perirhinal cortex make different contributions to memory. Data are discussed from two patients with adult-onset bilateral hippocampal damage who show a sparing of item recognition relative to recall and certain types of associative recognition. It is argued that these data are consistent with Aggleton and Brown's (1999) proposal that familiarity-based recognition memory is not dependent on the hippocampus but is mediated by the perirhinal cortex and dorso-medial thalamic nucleus. The second question is whether the recognition memory deficit observed in medial temporal lobe amnesia can be explained by a deficit in perceptual processing and representation of objects rather than a deficit in memory per se. The finding that amnesics were impaired at recognizing, after short delays, patterns that they could successfully discriminate suggests that their memory impairment did not result from an object-processing deficit. The possibility remains, however, that the human perirhinal cortex plays a role in object processing, as well as in recognition memory, and data are presented that support this possibility.

The medial temporal lobes are known to play a critical role in declarative memory (memory for facts and events) in humans, but it is currently unresolved whether the hippocampus and the adjacent medial temporal lobe cortices (entorhinal, perirhinal, and parahippocampal cortices) make distinct contributions to memory, and, if so, what these contributions are. Furthermore, recent work with nonhuman primates has suggested that the role of the perirhinal cortex may not be exclusively one of memory. The current paper focuses primarily on two issues: first, whether recognition memory for individual items in humans is dependent on the integrity of the hippocampus or whether it can be mediated by adjacent cortical regions such as the perirhinal cortex; second, whether the object recognition memory deficit observed in amnesics with medial temporal lobe lesions can be explained by

Correspondence should be addressed to Dr J. S. Holdstock, School of Psychology, University of Liverpool, Eleanor Rathbone Building, Bedford Street South, Liverpool, L69 3BX, UK. Email: J.Holdstock@liverpool.ac.uk
I would like to thank Keith A. May and John J. Downes for their comments on an earlier version of this paper.

DOI:10.1080/02724990444000177

a deficit in the perceptual processing and representation of objects rather than a deficit in memory per se.

The human medial temporal lobe and recognition memory

The evidence in relation to the first issue from studies of patients with medial temporal lobe lesions has been conflicting. In some patients, selective hippocampal damage has been reported to impair both recall and recognition (Cipolotti et al., 2001; Manns, Hopkins, Reed, Kitchener, & Squire, 2003; Manns & Squire, 1999; Reed & Squire, 1997). These data have been interpreted as supporting the view that the medial temporal lobe functions as a highly integrated memory system in which both recognition and recall are dependent on the hippocampus (Squire & Zola-Morgan, 1991). However, other patients with selective hippocampal damage have been reported to show a sparing of item recognition relative to recall (Henke et al., 1999; Holdstock, Mayes, Gong, Roberts, & Kapur, 2005b; Holdstock, Mayes, Isaac, Gong, & Roberts, 2002a; Holdstock et al., 2002b; Mayes, Holdstock, Isaac, Hunkin, & Roberts, 2002; Vargha-Khadem, Gadian, Watkins, Connelly, van Paesschen, & Mishkin, 1997). This variability between patients remains to be explained but is not considered in detail here. Possible explanations such as reorganization of function, differences in the tests used by different research groups, differences in location and extent of pathology within and outside of the hippocampus, and differences in strategies used by different patients and different control groups have been discussed in detail elsewhere (see Holdstock et al., 2005b; Mayes et al., 2002; Mayes et al., 2004). While accepting that a number of patients with selective structural hippocampal damage have been reported to show a global amnesia, in this paper I summarize the data from two patients we have tested who have shown a relative sparing of item recognition.

Over a number of years we were able to extensively study the memory of patient YR who had adult-onset selective bilateral hippocampal damage. YR's pathology probably resulted from a vascular incident related to the administration of an opiate drug at the age of 48 years, although this could not be confirmed. Detailed volumetric analysis of YR's structural MRI indicated a bilateral reduction in hippocampal volume of approximately 50%. The volumes of other brain regions including frontal, temporal, and parietal lobes, and perirhinal and entorhinal cortices were comparable to the control mean (see Holdstock et al., 2000b; Mayes et al., 2002; Mayes et al., 2004). As shown in Figure 1, YR performed on average only 0.5 standard deviations below the control mean on a total of 43 item recognition tests, whereas her mean performance was over 3.5 standard deviations below the control mean on a battery of 34 recall tests (Mayes et al., 2002). Taking a criterion of impairment of 1.96 standard deviations below the control mean (Type 1 error probability of .05, two-tailed), YR's item recognition was unimpaired. Although her performance was below average, it cannot be considered to be impaired because, assuming the population is normally distributed, more than 30% of the normal population would be expected to perform worse than her. YR's item recognition performance (measured as number of SDs from the control mean) was unaffected by the difficulty of the tasks for the control subjects, indicating that the relative sparing of item recognition was unlikely to simply reflect better performance on easier tasks (Holdstock et al., 2002b; Mayes et al., 2002). Her performance was also unaffected

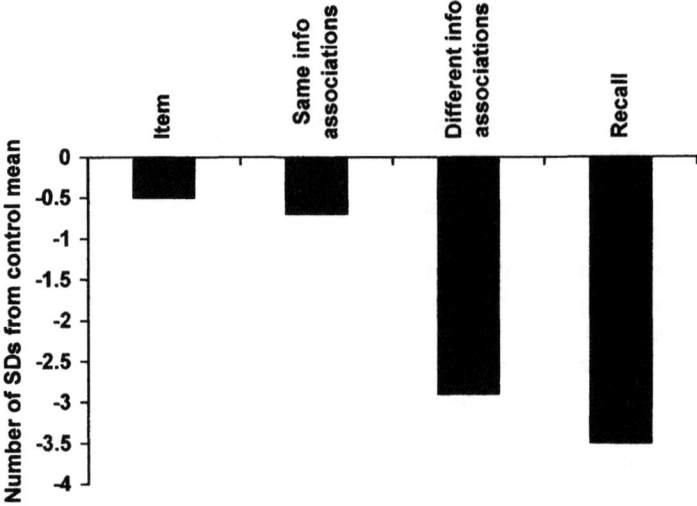

Figure 1.   Performance of patient YR, expressed as the number of *SD*s from the mean of matched control subjects, on tests of item recognition, recognition of associations between information of the same kind, recognition of associations between information of different kinds, and recall. Negative values indicate performance below the control mean. *Key:* Item = mean performance on 43 item recognition tests (data from Mayes et al., 2002). Same info associations = mean performance on four tests of recognition of associations between same types of information (e.g., word pairs and face pairs) (data from Mayes et al., 2004). Different info associations = mean performance on 18 tests of recognition of associations between information of different kinds (data from Mayes et al., 2004). Recall = mean performance on 34 recall tests (data from Mayes et al., 2002).

by length of the retention interval, list length, or number of foils (Mayes et al., 2002). YR's performance on forced-choice and yes/no item recognition tests differed only when targets and their corresponding foils were made very similar (Holdstock et al., 2002b; Mayes et al., 2002); under these conditions her yes/no recognition was impaired whereas her forced-choice recognition was unimpaired (Holdstock et al., 2002b). Like item recognition, YR's recognition of word pairs and face pairs was unimpaired. Her mean performance on four tests of this kind was just 0.7 standard deviations below the control mean. In contrast, her mean performance on 18 tests tapping recognition of associations between information of different kinds (e.g., objects and locations, faces and voices, pictures and sounds, new words and their definitions) was 2.9 standard deviations below the control mean and clearly impaired (see Figure 1; Mayes et al., 2004; see also Holdstock et al., 2002a, 2002b; Mayes et al., 2001). This pattern of spared item recognition and recognition of pairs of items of the same kind but impaired recall and recognition of associations between different kinds of information has also been reported for a group of young patients with bilateral hippocampal lesions (Vargha-Khadem et al., 1997). This suggests that some aspects of recognition memory may be spared by hippocampal damage, although this region is critical for recall and also for recognition of associations between different types of information.

A similar, though less striking, dissociation between relatively spared item recognition and impaired recall and associative recognition was observed in another patient, BE, who also suffered bilateral hippocampal pathology as an adult. BE suffered from herpes simplex encephalitis when he was 45 years old. He was found to have a 37% reduction in right hippocampal volume and a 39% reduction in left hippocampal volume. In contrast, whole temporal lobe volume (which included the hippocampus) was reduced by only 1–2% relative to controls. Although BE's scan was not of sufficient quality for us to estimate the volumes of medial temporal lobe cortices, two independent radiographers who examined the scans considered these regions to be intact. Structural damage therefore appeared to be restricted to the hippocampus bilaterally. However, SPECT and PET scanning revealed that this was accompanied by bilateral hypometabolism in the temporal lobes with SPECT indicating a 25% greater reduction in perfusion in the left than in the right hemisphere (Holdstock et al., 2005b). BE showed deficits of both recognition and recall for verbal material. However, his item recognition for nonverbal material was unimpaired on four of seven forced-choice tests and three of four yes/no tests and above chance on all but one test. In contrast, his recall of nonverbal material and recognition of associations between objects and locations and the temporal order of patterns was impaired and close to chance (Holdstock et al., 2005b). The dissociation between BE's impaired verbal item recognition but relatively spared nonverbal item recognition is unlikely to be explained by his structural damage, as his hippocampus was reduced in volume by a very similar amount on each side (37% vs. 39%). However, it may relate to the difference in severity of hypometabolism in his right and left temporal lobes. Hypometabolism was reported to be less in the right than the left temporal lobe, and, consistent with this, he showed a sparing of nonverbal item recognition relative to recall and associative recognition, but not a similar sparing of verbal item recognition. These data are of considerable interest because, not only do they demonstrate a relative sparing of nonverbal item recognition in this patient, but they also highlight the importance of using functional as well as structural imaging to identify the extent of dysfunction in future case studies. It is possible that the existing conflict in the literature is due, at least in part, to incomplete information about the extent of brain dysfunction in the patients. No functional imaging data have been reported for the majority of patients with selective hippocampal damage including those, reported by Squire and his colleagues (Manns et al., 2003; Manns & Squire, 1999; Reed & Squire, 1997), who have both recall and recognition deficits. It is therefore possible that the patients who have a global amnesia after selective structural hippocampal damage have greater dysfunction outside of the hippocampus, not detected by a structural scan, than those who have a relative sparing of item recognition. The additional information that functional imaging can provide will help to determine whether this is the case.

Although, as discussed above, the reported effects of selective structural hippocampal lesions on memory have been variable, the fact that recognition of items and word and face pairs has been spared by hippocampal damage in at least some patients is consistent with the view that familiarity-based memory decisions are not dependent on the hippocampus but can be supported by neocortical regions (Aggleton & Brown, 1999; Norman & O'Reilly, 2001; see also O'Reilly & Norman, 2002). Electrophysiological studies have shown that neurons in the anterior inferior temporal lobe have the response properties necessary to support familiarity judgements (Brown & Bashir, 2002). Further, lesion studies in monkeys and rats have shown that selective lesions of the perirhinal cortex and dorso-medial nucleus of the thalamus produce deficits in object recognition (Meunier, Bachevalier, Mishkin, & Murray, 1993; Meunier, Hadfield, Bachevalier, & Murray, 1996) whereas object recognition is, at most, mildly impaired by lesions of the hippocampus (Alvarez, Zola-Morgan, & Squire, 1995; Murray & Mishkin, 1986, 1996; Zola-Morgan & Squire, 1986; Zola-Morgan, Squire, & Amaral, 1989; Zola-Morgan, Squire, Clower, & Rempel, 1993), fornix (Aggleton, Hunt, & Shaw, 1990; Bachevalier, Parkinson, & Mishkin, 1985a; Bachevalier, Saunders, & Mishkin, 1985b; Gaffan, Sheilds, & Harrison, 1984; Rothblat & Kromer, 1991; Shaw & Aggleton, 1993; Zola-Morgan et al., 1989) parahippocampal cortex (Meunier et al., 1996; Ramus, Zola-Morgan, & Squire, 1994) and entorhinal cortex (Leonard, Amaral, Squire, & Zola-Morgan, 1995; Meunier et al., 1993). These findings led Aggleton and Brown (1999) to suggest that familiarity-based recognition decisions may be mediated by a system that includes the perirhinal cortex, dorso-medial thalamic nucleus, and prefrontal cortex, whereas recall/recollection is supported by a system that includes the hippocampus, anterior thalamic nucleus, and prefrontal cortex.

The view of Aggleton and Brown (1999) predicts that, after bilateral hippocampal damage, patients will rely primarily on familiarity when making their memory decisions. Consistent with this proposal, estimates of familiarity obtained from eight tests that used the remember/know procedure indicated that familiarity was probably normal in YR (Holdstock et al., 2002b). Mean estimates of familiarity for YR were above the control mean when assumptions of independence and exclusivity were made and 0.17 standard deviations below the control mean when a redundancy relationship was assumed. It was also shown that YR's discrimination of studied line-drawn pictures from very similar foils was unimpaired when a forced-choice paradigm was used but not when a yes/no paradigm was used (Holdstock et al., 2002b). According to the computational model of Norman and O'Reilly (Norman & O'Reilly, 2001; see also O'Reilly & Norman, 2002), with forced-choice tasks, good performance can be achieved using familiarity alone, whereas with yes/no tasks involving very similar targets and foils, good performance relies on the use of recollection. Using the remember/know procedure with a forced-choice paradigm, we also found that patient BE based his forced-choice item recognition decisions on a feeling of familiarity more often than did control subjects (Holdstock et al., 2005b). The data therefore suggest that the two patients rely primarily on familiarity in making their recognition memory decisions.

Many researchers assume that there is an independence relationship between recollection and familiarity (see Yonelinas, 2002)—that is, the occurrence of recollection does not depend on or influence the occurrence of familiarity or vice versa. This means that, for a healthy individual, any item recognition decision may be based on recollection alone, familiarity alone, or both familiarity and recollection. If the patients can only base their recognition decisions on familiarity, it may not be surprising that, although item recognition has been relatively spared in some patients with hippocampal damage, performance has been below the control mean. This is because, although for both patients and controls there would be occasions when they would be able to use familiarity as a basis for their recognition decisions, for the control subjects there would be additional occasions when familiarity is absent but recollection can be used as a basis for their recognition decisions.

Although data from patients such as YR and BE have suggested that the hippocampus is not necessary for familiarity-based item recognition, alone they have not provided any constraints as to which regions may mediate this type of recognition in humans. Of relevance to this issue, Holdstock, Gutnikov, Gaffan, and Mayes (2000a) compared the pattern recognition performance of a mixed group of patients with anterograde amnesia with that of patient YR. The amnesic group comprised four patients with extensive medial temporal lobe damage due to either encephalitis or meningitis, two patients who had suffered Wernicke Korsakoff syndrome and so were likely to have damage to both the anterior and dorso-medial thalamic nuclei (although scan information was not available to confirm this), one patient who had suffered from a posterior communicating artery aneurysm, which had resulted in pathology in the posterior temporal lobe, right medial temporal lobe, and occipital cortex, and two patients who had suffered from anterior communicating artery aneurysms. The patients completed a matching-to-sample task for grey-scale computer-generated patterns. Participants viewed a single pattern, which appeared in the centre of the screen for 2.5 s and then disappeared. After delays ranging from 0 to 30 s, 14 patterns appeared on the screen, and the one that was identical to the sample had to be selected. There was also a simultaneous matching condition in which the sample remained on the screen while participants decided which of the choice stimuli matched it. The data from this study are shown in Figure 2. The amnesic group was unimpaired at simultaneous matching and at matching after unfilled delays of 0, 2, and 5 s. However, after filled delays of 10, 20, and 30 s, which are likely to have been tapping long-term memory, the amnesic group was

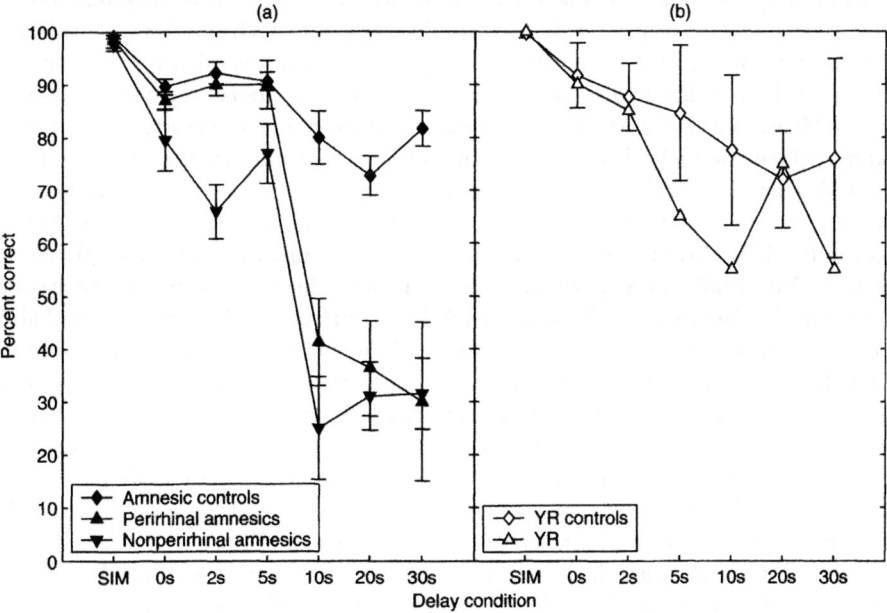

Figure 2. (a) Percentage of correct performance on pattern matching-to-sample of amnesic patients with confirmed damage to the perirhinal cortex (perirhinal amnesics), amnesic patients with no confirmed damage to the perirhinal cortex (nonperirhinal amnesics), and control subjects for the mixed amnesic group (amnesic controls). Error bars indicate the standard error of the mean. (b) Percentage of correct performance on pattern matching-to-sample of patient YR (YR) and YR's matched control group (YR controls). Error bars indicate the standard deviation of the control group. Data replotted from Figure 2 of "Perceptual and mnemonic matching-to-sample in humans: Contributions of the hippocampus, perirhinal and other medial temporal lobe cortices", by J. S. Holdstock, S. A. Gutnikov, D. Gaffan, and P. R. Mayes (2000), *Cortex*, *36*, p. 312. Copyright © 2000, Masson S. p. A. Adapted with permission.

significantly impaired relative to its control group. This impairment was evident at the 10-s delay with little further forgetting after this time. When the patient group was subdivided into those with confirmed perirhinal cortex damage and those without confirmed perirhinal cortex damage, the latter group was found to be impaired from delays of 2 s onwards whereas the former was only impaired after a 10-s delay. However, analysis of variance (ANOVA) showed no significant interaction between amnesic subgroup and delay. In contrast to the memory deficits shown by the amnesic group, patient YR's performance was unimpaired (using a criterion of impairment of 1.96 standard deviations below the control mean) and within the range of her matched control group for all delays. Furthermore, YR's performance was over 2 standard deviations better than the mean of the amnesic group at the 20-s delay and over 1 standard deviation better than the amnesic group mean at delays of 10 and 30 s (Holdstock et al., 2000a). YR achieved this good level of performance even though, in all delayed conditions, her response times were between 2 and 10 standard deviations slower than those of the amnesic patients and controls, suggesting that her memory was effectively being tested after longer delays than those for the other participants, which may have led to her percentage correct performance being underestimated (see Holdstock et al., 2000a).

YR's data suggested that damage to the hippocampus was not sufficient to produce a deficit on this task, suggesting that the deficit shown by the mixed amnesic group was most likely to be due to the damage or dysfunction that these patients had suffered outside of the hippocampus. The patients with medial temporal lobe damage all had MRI-confirmed damage to the perirhinal cortex as well as pathology to additional regions in the medial temporal lobe. The patient who had suffered a posterior communicating artery aneurysm also had medial temporal lobe damage, although there was no confirmation of the exact regions involved. Their data are therefore consistent with Aggleton and Brown's (1999) proposal that a circuit involving perirhinal cortex, dorso-medial thalamus, and prefrontal cortex may mediate familiarity-based item recognition. The data from patients who had suffered from Wernicke Korsakoff syndrome are also consistent with this view, as this disease has been associated with damage to the dorso-medial thalamic nucleus as well as the anterior-thalamic nucleus, which forms part of Aggleton and Brown's hippocampal circuit (Harding, Halliday, Caine, & Kril, 2000). Clipping of anterior communicating artery aneurysms is thought to have a disruptive effect on memory by disconnecting the medial temporal lobe and inferior temporal cortex from cholinergic inputs from the basal forebrain, which is damaged in this patient group (Abe, Inokawa, Kashiwagi, & Yanagihara, 1998; Easton, Ridley, Baker, & Gaffan, 2002). These patients are therefore likely to have extensive medial temporal lobe dysfunction affecting both systems proposed by Aggleton and Brown.

Although the data from Holdstock et al. (2000a) are consistent with the Aggleton and Brown (1999) view, given the widespread brain damage suffered by the mixed amnesic group, stronger evidence is needed concerning the specific contribution of the perirhinal cortex and dorso-medial thalamic nucleus to recognition memory in humans, and this requires studies of patients with more selective damage to these regions. One such study (Isaac et al., 1998) reported both recall and recognition memory deficits in a patient with a bilateral lesion to the dorso-medial thalamic nucleus and some slight atrophy to the mammillary bodies bilaterally. This suggested that the effect of damage to the perirhinal cortex–dorso-medial thalamic system may extend to recall as well as recognition in humans. One possible explanation of this could be that familiarity mediated by this system may be required for normal recall (Isaac et al., 1998). If so, a double dissociation between the effects of hippocampal system damage and perirhinal system damage on recall and item recognition may not be found in humans, although as the input of spatial information into the hippocampal system is thought to be via the parahippocampal cortex rather than the perirhinal cortex, a double dissociation between spatial and item recognition memory may be found (Isaac et al., 1998). The investigation of further cases is required to confirm whether this is the case.

The human medial temporal lobe and object discrimination

The data from Holdstock et al. (2000a) also relate to the second question to be addressed in this paper—that is, whether the recognition memory deficit found in patients with medial temporal lobe amnesia may be attributed to a deficit in object processing and representation rather than a deficit in recognition memory per se. This question has arisen from the findings of studies of nonhuman primates, which have reported object-processing deficits following lesions of the rhinal (perirhinal and entorhinal) or perirhinal cortex (Buckley, this issue; Bussey, Saksida, & Murray, this issue).

The first study to suggest such a deficit showed that monkeys with rhinal lesions were impaired at the simultaneous matching of objects when trial-unique stimuli were used (Eacott, Gaffan, & Murray, 1994). The matching-to-sample task reported by Holdstock et al. (2000a) used stimuli and a task design similar to that used by Eacott et al. Both studies required simultaneous matching and matching after a 0-s delay. Holdstock and colleagues used trial-unique patterns consisting of overlapping complex shapes whereas Eacott et al. used trial-unique patterns composed of overlapping typographical figures. Unlike Eacott and colleagues, we found that the performance of the mixed group of amnesic patients was unimpaired in both the simultaneous and the 0-s delay conditions (see Figure 2). In fact, when the performance of just those patients with MRI-confirmed structural damage to the perirhinal cortex was considered, performance was unimpaired at all delays up to and including the unfilled 5-s delay. Performance was only impaired at the filled delays of 10 s and longer, which are likely to have tapped long-term memory. The patients with medial temporal lobe damage that included damage to the perirhinal cortex were therefore impaired at recognizing abstract patterns after delays of 10 s or longer even though they could accurately discriminate between these stimuli when the task made no demands on memory.

The data from Holdstock et al. (2000a) suggest that the recognition deficit following medial temporal lobe damage in humans is not merely a reflection of a deficit in object processing. However, stronger evidence for this argument would be provided if performance of the participants was not at ceiling in the simultaneous condition. It could be argued that the ceiling effect in the simultaneous condition of Holdstock et al. (2000a) hid an object-processing deficit in the patient group, which only became apparent when the task was made more difficult by making demands on memory. We considered this explanation to be unlikely, but it cannot be excluded until it is demonstrated that patients whose damage includes the perirhinal cortex are unimpaired at simultaneous matching on a task on which performance is below ceiling and yet are impaired at recognition of these kinds of stimuli at longer delays. This is currently being explored with a modified version of the task.

Although the data from Holdstock et al. (2000a) suggest that the recognition deficit following medial temporal lobe damage in humans is unlikely to be attributable to a deficit in object processing, they do not rule out the possibility that the perirhinal cortex plays a role in object processing, in addition to its contribution to recognition memory, but that the discrimination of the stimuli used by Holdstock et al. did not require this process. Although the stimuli were based on those used by Eacott and colleagues (Eacott et al., 1994), humans found the task easier than did the monkeys and were perhaps basing their discriminations on more simple features. This may also have been true of the study reported by Buffalo, Reber, and Squire (1998), which also used trial-unique abstract patterns and found a deficit in yes/no recognition only from delays of 6 s in patients with medial temporal lobe lesions that included perirhinal cortex (see Lee, Barense, & Graham, this issue, for a similar proposal).

The nonhuman primate literature subsequent to the study of Eacott et al. (1994) has suggested that perirhinal cortex lesions impair object processing under some but not all conditions. Impairments have been reported in object discrimination learning when large sets of discriminations were used (Buckley & Gaffan, 1997; although see Hampton & Murray, 2002), large numbers of foils were used (Buckley & Gaffan, 1997), or when the rewarded item was presented in different views on different trials (Buckley & Gaffan, 1998). Deficits were also reported in selecting an odd one out (different object/face) from among different views of a single object (or face) but not from among identical views of a single object (or face; Buckley, Booth, Rolls, & Gaffan, 2001). Impairments were also not seen when detection of the odd one out required fine discriminations of colour, size, or shape (Buckley et al., 2001). It was argued that the tasks on which there were impairments placed high demands on object identification. However, perirhinal cortex damage in monkeys does not impair discrimination under all conditions that increase demands on object identification (Hampton, this issue; Hampton & Murray, 2002).

Bussey and Saksida (2002) have proposed that the key determinant of whether the perirhinal cortex is critical for object identification in a particular task is the extent to which discrimination of objects in that task requires the representation of complex conjunctions of visual features. They argue that the ventral visual processing stream is organized hierarchically so that, although conjunctions of features are represented throughout the ventral visual processing stream, the most complex conjunctions are represented by the perirhinal cortex (Bussey & Saksida, 2002). In support of this model it has been shown that concurrent object discrimination learning was impaired by perirhinal cortex ablations in monkeys when feature ambiguity was high (a photographed scene was rewarded when it was part of one object, but not rewarded when it was part of another object) but that the deficit decreased with decreasing feature ambiguity (Bussey, Saksida, & Murray, 2002, this issue). Further, perirhinal cortex lesions impaired picture discrimination when the stimuli to be discriminated shared a number of features but not when they were perceptually distinct or when difficult colour or shape discriminations were tested (Bright, Moss, Stamatakis, & Tyler, this issue; Bussey, Saksida, & Murray, 2003).

The few studies relevant to this issue in the patient literature have reported conflicting findings. Stark and Squire (2000) found no deficits in three patients with complete bilateral perirhinal cortex damage on tasks modelled on the odd-one-out tasks used by Buckley and colleagues (2001). In contrast, Lee et al. (2005) found deficits in scene and face discrimination and a possible milder deficit in object discrimination in three patients with medial temporal lobe damage that included perirhinal cortex. The issue of whether the human perirhinal cortex is involved in representing complex conjunctions of visual features of objects is therefore unresolved and awaits further studies to explore the generality of the reported deficits to other patients with perirhinal cortex damage and the systematic investigation of the conditions under which such deficits are found in humans (see Lee et al., this issue).

Given the anatomical position of the perirhinal cortex it is plausible to argue that it may not only be involved in representing complex conjunctions of visual features of objects but that it may also be involved in representing conjunctions of object features from different sensory modalities (Goulet & Murray, 2001; Murray & Bussey, 1999). However, there appears to be little evidence directly related to this issue in the literature. Aspiration lesions of the amygdala, which will have directly or indirectly affected the adjacent entorhinal and perirhinal cortices, have been reported to impair delayed-non-matching-to-sample (DNMS) when the sample object was presented tactually and the choices were presented visually, but not when sample and choice items were both presented in the same modality (visual or tactile; Málková & Murray, 1996; Murray & Mishkin, 1985). The importance of perirhinal cortex damage/dysfunction to the occurrence of this deficit is suggested by the finding that an excitotoxic lesion to the amygdala produced only a transient impairment on crossmodal DNMS, whereas a lesion to the rhinal cortex consistently impaired crossmodal DNMS (Goulet & Murray, 2001). The memory component that forms part of this task, however, makes it difficult to determine whether the deficit is one of crossmodal object processing or one of memory.

Evidence from two studies of patients with damage to the medial temporal lobe suggests that this region may indeed be involved in crossmodal object processing. Shaw, Kentridge, and Aggleton (1990) showed that patients who had suffered from herpes simplex encephalitis, which causes pathology in the medial temporal lobe, were impaired at matching a test arc with a comparison circle when the stimuli were presented in different modalities (visual and tactile) but not when they were presented in the same modality (visual or tactile). Although the authors attributed the deficit to damage to the amygdala, given their aetiology these patients are also likely to have had damage to the perirhinal cortex, although MRI was not available to confirmed this.

In recent unpublished work (Holdstock, Blay, Denby, Downes, Roberts, & Mayes, 2005a), crossmodal and intramodal matching were tested in two patients who both had abnormality of the right amygdala and right perirhinal cortex. One patient, a 45-year-old female, JL, had suffered from a head injury at the age of 17 years, and detailed investigation using structural MRI showed that as a result she had damage to medial and lateral orbitofrontal cortex on the right and to bilateral superior, middle, and inferior temporal gyri. Damage to the temporal lobe affected the anterior 60% on the left and the anterior 40% on the right. Volumetric measures showed that the volume of JL's perirhinal cortex was reduced bilaterally but to a greater extent on the right, whereas the volumes of entorhinal cortex and hippocampus were normal bilaterally. She also had partial damage to the right amygdala but volume measures could not be obtained for this structure because of the proximity of the cortical damage (for more detailed information about JL's pathology and neuropsychological test performance see Mayes et al., 2003; and Holdstock et al., 2002a). The second patient, VG, was a 40-year-old male who presented with memory problems but had not suffered from an illness or injury from which this may have resulted. Volumetric analysis of structural MRI revealed a reduction in volume of more than 2 standard deviations relative to the control mean in right hippocampus, amygdala, and perirhinal cortex and a bilateral reduction in entorhinal cortex volume. The volumes of hippocampus, amygdala and perirhinal cortex on the left and the volumes of the left and right temporal lobes were comparable to control mean volumes. The patients were investigated as two separate case studies, and their test performance was compared with that of age-, sex-, and IQ-matched control groups. Both patients showed an identical pattern of performance. The tasks involved selecting which of a number of choice stimuli (abstract shapes) was identical to a simultaneously presented sample stimulus. The sample and choice stimuli were both presented in the visual modality or both presented in the tactile modality, or the sample was presented in the tactile modality, and the choices were presented visually. Consistent with the findings of Shaw and colleagues (1990), the patients were only impaired in the crossmodal matching condition. These data therefore suggest that the role of the perirhinal cortex in object processing may not be restricted to the representation of complex conjunctions of visual features of objects but that it may also be involved in the integration of information about objects from different modalities. However, the data are based on the performance of only a small number of patients, and to be more confident in this conclusion it will be important to demonstrate a similar pattern in larger numbers of patients with damage to this brain region.

Conclusion

In summary, this paper has reviewed the evidence from neuropsychological patient studies relevant to two questions concerning the functions of the medial temporal lobe in humans.

The first question was whether there is evidence to suggest that the hippocampus and the adjacent perirhinal cortex make different contributions to memory. Although the evidence from patients with hippocampal damage is mixed, and further work is required to identify the reasons for the varying patterns of performance that have been reported, the fact that some patients with adult-onset selective hippocampal damage have shown a sparing of item recognition, relative to recognition of associations between different types of information and to recall, suggests that regions other than the hippocampus may be able to support item recognition performance. Given that patients with more extensive medial temporal lobe lesions are impaired on item recognition tests, this finding is consistent with the proposal of Aggleton and Brown (1999) that the hippocampus forms part of a system that supports recall and recollection, whereas the perirhinal cortex and dorso-medial thalamic nucleus support familiarity-based recognition memory.

The second question is whether the recognition memory deficit seen after medial temporal lobe lesions in humans can be attributed to a deficit in object processing rather than memory per se. The evidence presented from the study of Holdstock et al. (2000a) suggests that any perceptual role of the perirhinal cortex is likely to be separable from its role in recognition memory. Although in that study a deficit in object processing was not obtained, studies using tasks in humans that require discriminations on the basis of more complex conjunctions of visual features or the integration of information from different modalities have produced deficits in patients whose damage includes the perirhinal cortex. These data suggest that the perirhinal cortex may contribute to the processing and representation of objects in humans in addition to mediating familiarity-based recognition memory, but further studies are required to strengthen the evidence for this conclusion.

REFERENCES

Abe, K., Inokawa, M., Kashiwagi, A., & Yanagihara, T. (1998). Amnesia after a discrete basal forebrain lesion. *Journal of Neurology, Neurosurgery and Psychiatry, 65*, 126–130.

Aggleton, J. P., & Brown, M. (1999). Episodic memory, amnesia and the hippocampal-anterior thalamic axis. *Behavioral Brain Sciences, 22*, 425–490.

Aggleton, J. P., Hunt, P. R., & Shaw, C. (1990). The effects of mammillary body and combined amygdala-fornix lesions on tests of delayed non-matching-to-sample in the rat. *Behavioural Brain Research, 40*, 145–157.

Alvarez, P., Zola-Morgan, S., & Squire, L. R. (1995). Damage limited to the hippocampal region produces long-lasting memory impairment in monkeys. *Journal of Neuroscience, 15*, 3796–3807.

Bachevalier, J., Parkinson, J. K., & Mishkin, M. (1985a). Visual recognition in monkeys: Effects of separate vs. combined transection of fornix and amygdalofugal pathways. *Experimental Brain Research, 57*, 554–561.

Bachevalier, J., Saunders, R. C., & Mishkin, M. (1985b). Visual recognition in monkeys: Effects of transection of fornix. *Experimental Brain Research, 57*, 547–553.

Bright, P., Moss, H. E., Stamatakis, E. A., & Tyler, L. K. (this issue). The anatomy of object processing: The role of anteromedial temporal cortex. *Quarterly Journal of Experimental Psychology, 58B*, 361–377.

Brown, M. W., & Bashir, Z. I. (2002). Evidence concerning how neurons of the perirhinal cortex may effect familiarity discrimination. *Philosophical Transactions of the Royal Society of London B, 357*, 1083–1095.

Buckley, M. J. (this issue). The role of the perirhinal cortex and hippocampus in learning, memory and perception. *Quarterly Journal of Experimental Psychology, 58B*, 246–268.

Buckley, M. J., Booth, M. C. A., Rolls, E. T., & Gaffan, D. (2001). Selective perceptual impairments after perirhinal cortex ablation. *The Journal of Neuroscience, 21*, 9824–9836.

Buckley, M. J., & Gaffan, D. (1997). Impairment of visual object-discrimination learning after perirhinal cortex ablation. *Behavioral Neuroscience, 111*, 467–475.

Buckley, M. J., & Gaffan, D. (1998). Perirhinal cortex ablation impairs visual object identification. *The Journal of Neuroscience, 18*, 2268–2275.

Buffalo, E. A., Reber, P. J., & Squire, L. R. (1998). The human perirhinal cortex and recognition memory. *Hippocampus, 8*, 330–339.

Bussey, T. J., & Saksida, L. M. (2002). The organization of visual object representations: A connectionist model of effects of lesions in perirhinal cortex. *European Journal of Neuroscience, 15*, 335–364.

Bussey, T. J., Saksida, L. M., & Murray, E. A. (2002). Perirhinal cortex resolves feature ambiguity in complex visual discriminations. *European Journal of Neuroscience, 15*, 365–374.

Bussey, T. J., Saksida, L. M., & Murray, E. A. (2003). Impairments in visual discrimination after perirhinal cortex lesions: testing 'declarative' vs. 'perceptual-mnemonic' views of perirhinal cortex function. *European Journal of Neuroscience, 17*, 649–660.

Bussey, T. J., Saksida, L. M., & Murray, E. A. (this issue). The perceptual-mnemonic/feature conjunction model of perirhinal cortex function. *Quarterly Journal of Experimental Psychology, 58B*, 269–282.

Cipolotti, L., Shallice, T., Chan, D., Fox, N., Scahill, R., Harrison, G. et al. (2001). Long-term retrograde amnesia... the crucial role of the hippocampus. *Neuropsychologia, 39*, 151–172.

Eacott, M. J., Gaffan, D., & Murray, E. A. (1994). Preserved recognition memory for small sets, and impaired stimulus identification for large sets, following rhinal cortex ablations in monkeys. *European Journal of Neuroscience, 6*, 1466–1478.

Easton, A., Ridley, R. M., Baker, H. F., & Gaffan, D. (2002). Unilateral lesions of the cholinergic basal forebrain and fornix in one hemisphere and inferior temporal cortex in the opposite hemisphere produce severe learning impairments in rhesus monkeys. *Cerebral Cortex, 12*, 729–736.

Gaffan, D., Sheilds, C., & Harrison, S. (1984). Delayed matching by fornix-transected monkeys: The sample, the push and the bait. *Quarterly Journal of Experimental Psychology, 36B*, 305–317.

Goulet, S., & Murray, E. A. (2001). Neural substrates of crossmodal association memory in monkeys: The amygdala versus the anterior rhinal cortex. *Behavioral Neuroscience, 115*, 271–284.

Hampton, R. R. (this issue). Monkey perirhinal cortex is critical for visual memory, but not for visual perception: Re-examination of the behavioural evidence from monkeys. *Quarterly Journal of Experimental Psychology*, *58B*, 283–299.

Hampton, R. R., & Murray, E. A. (2002). Learning of discriminations is impaired, but generalization to altered views is intact, in monkeys (macca mulatta) with perirhinal cortex removal. *Behavioral Neuroscience*, *116*, 363–377.

Harding, A., Halliday, G., Caine, D., & Kril, J. (2000) Degeneration of anterior thalamic nuclei differentiates alcoholics with amnesia. *Brain*, *123*, 141–154.

Henke, K., Kroll, N. E. A., Behniea, H., Amaral, D. G., Miller, M. B., Rafal, R. et al. (1999). Memory lost and regained following bilateral hippocampal damage. *Journal of Cognitive Neuroscience*, *11*, 682–697.

Holdstock, J. S., Blay, K., Downes, J. J., Denby, C., Roberts, N., & Mayes, A. R. (2005a). *Intra-modal and cross-modal matching-to-sample following pathology to the right perirhinal cortex and amygdala in humans*. Manuscript in preparation.

Holdstock, J. S., Gutnikov, S. A., Gaffan, D., & Mayes, A. R. (2000a). Perceptual and mnemonic matching-to-sample in humans: Contributions of the hippocampus, perirhinal and other medial temporal lobe cortices. *Cortex*, *36*, 301–322.

Holdstock, J. S., Mayes, A. R., Cezayirli, E., Isaac, C. L., Aggleton, J. P. & Roberts, N. (2000b). A comparison of egocentric and allocentric spatial memory in a patient with selective hippocampal damage. *Neuropsychologia*, *38*, 410–425.

Holdstock, J. S., Mayes, A. R., Gong, Q. Y., Roberts, N., & Kapur, N. (2005b). Item recognition is less impaired than recall and associative recognition in a patient with selective hippocampal damage. *Hippocampus*, *15*, 203–215.

Holdstock, J. S., Mayes, A. R., Isaac, C. L., Gong, Q., & Roberts, N. (2002a). Differential involvement of the hippocampus and temporal lobe cortices in rapid and slow learning of new semantic information. *Neuropsychologia*, *40*, 748–768.

Holdstock, J. S., Mayes, A. R., Roberts, N., Cezayirli, E., Isaac, C. L., O'Reilly, R. et al. (2002b). Under what conditions is recognition spared relative to recall after selective hippocampal damage in humans? *Hippocampus*, *12*, 341–351.

Isaac, C. L., Holdstock, J. S., Cezayirli, E., Roberts, J. N., Holmes, C. J., & Mayes, A. R. (1998). Amnesia in a patient with lesions limited to the dorsomedial thalamic nucleus. *Neurocase*, *4*, 497–508.

Lee, A. C. H., Barense, M. D., & Graham, K. S. (this issue). The contribution of the human medial temporal lobe to perception: Bridging the gap between animal and human studies. *Quarterly Journal of Experimental Psychology*, *58B*, 300–325.

Lee, A. C. H., Bussey, T. J., Murray, E. A., Saksida, L. M., Epstein, R. A., Kapur, N. et al. (2005). Perceptual deficits in amnesia: Challenging the medial temporal lobe 'mnemonic' view. *Neuropsychologia*, *43*, 1–11.

Leonard, B. W., Amaral, D. G., Squire, L. R., & Zola-Morgan, S. (1995). Transient memory impairment in monkeys with bilateral lesions of the entorhinal cortex. *Journal of Neuroscience*, *15*, 5637–5659.

Málková, L., & Murray, E. A. (1996). Effects of partial versus complete lesions of the amygdala on cross-modal associations in cynomolgus monkeys. *Psychobiology*, *24*, 255–264.

Manns, J. R., Hopkins, R. O., Reed, J. M., Kitchener, E. G., & Squire, L. R. (2003). Recognition memory and the human hippocampus. *Neuron*, *37*, 171–180.

Manns, J. R., & Squire, L. R. (1999). Impaired recognition memory on the Doors and People test after damage limited to the hippocampal region. *Hippocampus*, *9*, 495–499.

Mayes, A. R., Holdstock, J. S., Isaac, C. L., Hunkin, N. M., & Roberts, N. (2002). Relative sparing of item recognition memory in a patient with adult-onset damage limited to the hippocampus. *Hippocampus*, *12*, 325–340.

Mayes, A. R., Holdstock, J. S., Isaac, C. L., Montaldi, D., Grigor, J., Gummer, A. et al. (2004). Associative recognition in a patient with selective hippocampal lesions and relatively normal item recognition. *Hippocampus*, *14*, 763–784.

Mayes, A. R., Isaac, C. L., Downes, J. J., Holdstock, J. S., Hunkin, N. M., Montaldi, D. et al. (2001). Memory for single items, word pairs, and temporal order in a patient with selective hippocampal lesions. *Cognitive Neuropsychology*, *18*, 97–123.

Mayes, A. R., Isaac, C. L., Holdstock, J. S., Cariga, P., Gummer, A., & Roberts, N. (2003). Long-term amnesia: A review and detailed illustrative case study. *Cortex*, *39*, 567–603.

Meunier, M., Bachevalier, J., Mishkin, M., & Murray, E. A. (1993). Effects on visual recognition of combined and separate ablations of the entorhinal and perirhinal cortex in rhesus monkeys. *Journal of Neuroscience*, *12*, 5418–5432.

Meunier, M., Hadfield, J., Bachevalier, J., & Murray, E. A. (1996). Effects of rhinal cortex lesions combined with hippocampectomy on visual recognition memory in rhesus monkeys. *Journal of Neurophysiology*, *75*, 1190–1205.

Murray, E. A., & Bussey, T. J. (1999). Perceptual-mnemonic functions of the perirhinal cortex. *Trends in Cognitive Sciences*, *3*, 142–151.

Murray, E. A., Graham, K. S., & Gaffan, D. (this issue). Perirhinal cortex and its neighbours in the medial temporal lobe: Contributions to memory and perception. *Quarterly Journal of Experimental Psychology*, *58B*, 378–396.

Murray, E. A., & Mishkin, M. (1985). Amygdalectomy impairs crossmodal association in monkeys. *Science*, *228*, 604–606.

Murray, E. A., & Mishkin, M. (1986). Visual recognition in monkeys following rhinal cortical ablations combined with either amygdalectomy or hippocampectomy. *Journal of Neuroscience*, *7*, 1991–2003.

Murray, E. A., & Mishkin, M. (1996). 40-Minute visual recognition memory in rhesus monkeys with hippocampal lesions. *Society for Neuroscience Abstracts*, *116*.9.

Norman, K. A., & O'Reilly, R. C. (2001). *Modelling hippocampal and neocortical contributions to recognition memory: A complementary learning systems approach* (ICS Tech. Rep. No. 01-02). Boulder, CO: University of Colorado, Institute of Cognitive Science.

O'Reilly, R. C., & Norman, K. A. (2002). Hippocampal and neocortical contributions to memory: Advances in the complementary learning systems framework. *Trends in Cognitive Sciences*, *6*, 505–510.

Ramus, S. J., Zola-Morgan, S., & Squire, L. R. (1994). Effects of lesions of perirhinal cortex or parahippocampal cortex on memory in monkeys. *Society for Neuroscience Abstracts*, *20*, 1074.

Reed, J. M., & Squire, L. R. (1997). Impaired recognition memory in patients with lesions limited to the hippocampal formation. *Behavioral Neuroscience*, *111*, 667–675.

Rothblat, L. A., & Kromer, L. F. (1991). Object recognition memory in the rat: The role of the hippocampus. *Behavioural Brain Research*, *42*, 25–32.

Shaw, C., & Aggleton, J. P. (1993). The effects of fornix and medial prefrontal lesions on delayed non-matching-to-sample by rats. *Behavioural Brain Research*, *54*, 91–102.

Shaw, C., Kentridge, R. W., & Aggleton, J. P. (1990). Cross-modal matching by amnesic patients. *Neuropsychologia*, *28*, 665–671.

Squire, L. R., & Zola-Morgan, S. (1991). The medial temporal-lobe memory system. *Science*, *253*, 1380–1386.

Stark, C. E. L., & Squire, L. R. (2000). Intact visual perceptual discrimination in humans in the absence of perirhinal cortex. *Learning and Memory*, *7*, 273–278.

Vargha-Khadem, F., Gadian, D. G., Watkins, K. E., Connelly, A., van Paesschen, W., & Mishkin, M. (1997). Differential effects of early hippocampal pathology on episodic and semantic memory. *Science*, *277*, 376–380.

Yonelinas, A. P. (2002). The nature of recollection and familiarity: A review of 30 years of research. *Journal of Memory and Language*, *46*, 441–517.

Zola-Morgan, S., & Squire, L. R. (1986). Memory impairment in monkeys following lesions limited to the hippocampus. *Behavioural Neuroscience*, *100*, 155–160.

Zola-Morgan, S., Squire, L. R., & Amaral, D. G. (1989). Lesions of the hippocampal formation but not lesions of the fornix or the mammillary nuclei produce long-lasting memory impairment in monkeys. *Journal of Neuroscience*, *100*, 155–160.

Zola-Morgan, S., Squire, L. R., Clower, R. P., & Rempel, N. L. (1993). Damage to the perirhinal cortex exacerbates memory impairment following lesions to the hippocampal formation. *Journal of Neuroscience*, *13*, 251–265.

THE QUARTERLY JOURNAL OF EXPERIMENTAL PSYCHOLOGY,
2005, 58B (3/4), 340–360

A mini-review of fMRI studies of human medial temporal lobe activity associated with recognition memory

Richard Henson

MRC Cognition and Brain Sciences Unit, Cambridge, UK

This review considers event-related functional magnetic resonance imaging (fMRI) studies of human recognition memory that have or have not reported activations within the medial temporal lobes (MTL). For comparisons both between items at study (encoding) and between items at test (recognition), MTL activations are characterized as left/right, anterior/posterior, and hippocampus/surrounding cortex, and as a function of the stimulus material and relevance of item/source information. Though no clear pattern emerges, there are trends suggesting differences between item and source information, and verbal and spatial information, and a role for encoding processes during recognition tests. Important future directions are considered.

The recognition memory paradigm has been used with functional neuroimaging to study human long-term memory for over a decade. In this paradigm, participants are exposed to a series of items during a "study" phase, which are later repeated in a "test" phase ("old items") intermixed with further items that were not studied ("new items"). The participant's basic task is to distinguish old items from new items, though this task is often extended to include, for example, confidence judgements, or judgements about some aspect of the context in which an old item was studied (so-called "source information").

Both psychological and neuroanatomical theories of recognition memory have been the subject of considerable debate. For example, some psychologists have proposed that recognition memory is supported by two distinct processes, such as recollection and familiarity (Mandler, 1980; Yonelinas, 2002). Recollection generally refers to retrieval (recall) of source information; familiarity refers to a feeling attributed to recent exposure to an item, in the absence of retrieval of contextual information. Others, however, have argued that behavioural data from the recognition memory paradigm can be explained by assuming only a single continuum of memory strength (usually within the context of signal detection theory,

Correspondence should be addressed to Dr Richard Henson, MRC Cognition and Brain Sciences Unit, 15 Chaucer Road, Cambridge CB2 2EF, UK. Email: rik.henson@mrc-cbu.cam.ac.uk

This paper was written at the invitation of the editors of this special issue. The author would like to thank Kim Graham for her encouragement and help, the reviewers for their comments, and Scott Slotnick for advice.

DOI:10.1080/02724990444000113

Dunn, 2004; Heathcote, 2003). Regarding neuroanatomical theories, some have argued for two subsystems within the medial temporal lobes (MTL)—for example, one involving hippocampus and anterior thalamus and the other involving perirhinal cortex and medial dorsal thalamus—that subserve functions analogous to recollection and familiarity, respectively (Aggleton & Brown, 1999, this issue; Holdstock, this issue). Others, however, have argued that the MTL is an integrated system and that current simple dichotomies do not capture functional differences that might exist between its components (Squire, Stark, & Clark, 2004).

This mini-review focuses on two basic contrasts that can be examined with functional imaging: (a) between items in the study phase that are later recognized ("subsequent hits") and those later forgotten ("subsequent misses"), and (b) between old items in the test phase that are recognized ("hits") and new items that are not recognized ("correct rejections"). The former contrast is called the "subsequent memory effect" and is used to investigate encoding of items into memory. The latter is called the "old–new effect" and is used to investigate successful recognition (see Figure 1). The main interest is whether various experimental manipulations at either study or test have consistently revealed dissociable patterns of activity within MTL associated with such encoding and/or recognition.

The remit of this mini-review is further constrained as follows. First, I only review event-related functional magnetic resonance imaging (fMRI) experiments. The inferior spatial resolution of positron emission tomography (PET) means that it can be difficult to determine the precise location of activations within the MTL. Moreover, the fact that PET normally requires averaging over tens of seconds of a cognitive task (restricting it to so-called "blocked" experimental designs) means that it is difficult to test the two comparisons above. For the same reason, I do not consider blocked fMRI experiments (see Herron, Henson, & Rugg, 2004; Rugg & Henson, 2002, for further discussion). I also only consider group-averaged data from healthy individuals.

I distinguish two types of MTL activity: that in hippocampus and that in surrounding medial temporal cortex (MTC). The latter encompasses both rhinal and parahippocampal cortex. (Although hippocampus is also cortex, I use "MTC" simply to distinguish hippocampal archicortex from surrounding transitional- and neo-cortex.) The reason for such a coarse distinction, even with the superior resolution of fMRI over PET, is that precise localization within MTL is rarely achieved, given the susceptibility-induced distortions associated with echo-planar fMRI (Constable et al., 2000) and the fact that most authors report mean locations within brains that were normalized to a common space and spatially smoothed, which reduce spatial resolution. The reason for choosing this particular distinction is to evaluate theories of functional specialization that distinguish hippocampus from surrounding MTC, like that of Aggleton and Brown (1999).

In the present review, I rely on the authors' anatomical classifications (i.e., I have not attempted to validate them), though I do attempt to characterize activations further as anterior or posterior (as well as left/right), based on a second proposal for a functional dissociation along the anterior–posterior axis (Lepage et al., 1998; Schacter & Wagner, 1999). For this I used an arbitrary division of $y = -20$ in Talairach space (when such coordinates are reported), so this is again a coarse classification (and it should be noted that these coordinates are the maxima of diffuse activations that can extend across the above divisions). I decided not to tabulate Talairach coordinates because they are determined in different ways

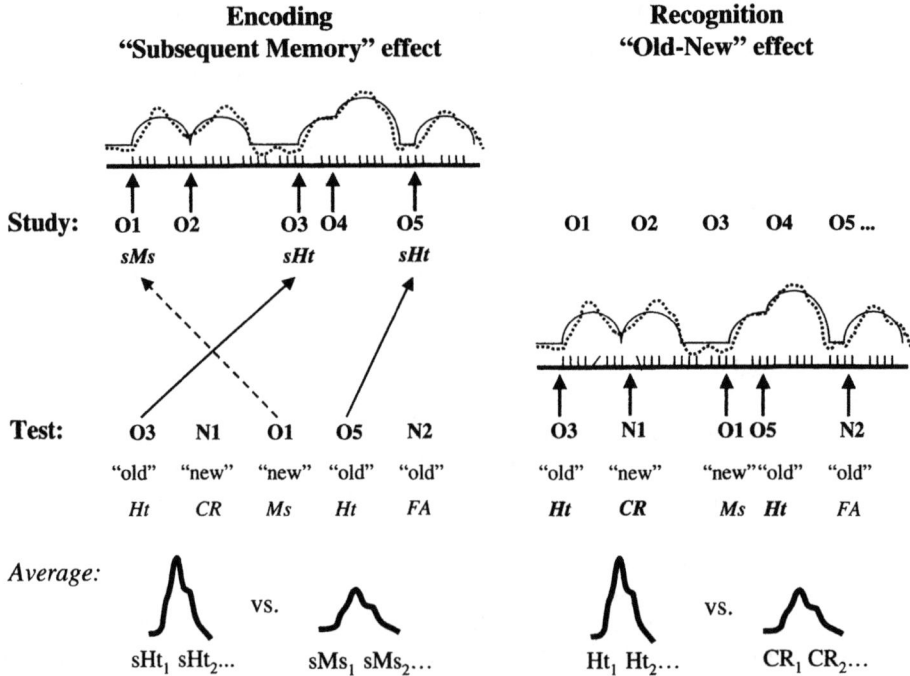

Figure 1. Schematic illustration of subsequent memory effects at study (encoding) and old–new effects at test (recognition). When scanning at study (left) or test (right), dotted and solid lines illustrate actual and fitted timecourses of event-related fMRI responses, with upward arrows indicating onset of an event. "O1", "O2", etc. indicate different "old" (studied) stimuli; "N1", "N2", etc. indicate different "new" (unstudied) stimuli. Events are defined on the basis of the participant's "old" or "new" response at test, which are a correct endorsement of an old item (a "hit", Ht), correct rejection of a new item ("CR"), incorrect rejection of an old item (a "miss", Ms), or incorrect endorsement of a new item (a "false alarm", FA). For the subsequent memory effect, hits and misses at test are used to define events at study as either a "subsequent hit" (sHt) or a "subsequent miss" (sMs). After effective event-locked "averaging", voxels are identified in which the mean response differs for the critical event types.

across experiments and because they can be misleading for anatomically distinct regions within close proximity (Maguire, 1998). Finally, different authors use different statistical criteria for the definition of "reliable" activations, which are difficult to interrelate, particularly when authors focus on a priori regions of interest. The reader should therefore remember that some differences across experiments may reflect different levels of statistical stringency.

In order to see whether any patterns emerge in the conditions under which MTL is activated (in particular, whether those conditions differ for hippocampus and MTC), I also categorize the experiments along different dimensions. There would appear to be only a few such dimensions along which categories can be distinguished that include appreciable numbers of experiments. Apart from the encoding/recognition distinction above, the most obvious is the type of stimulus material (e.g., words, objects, or scenes). The next most obvious is the type of task or instructions used at study and at test. These differ across nearly all experiments, so I grouped the task/instructions into those that engender encoding of, or require retrieval of, source information and those that do not (for which "item information" is sufficient). By source information, I refer to some form of episodic context (e.g., spatiotemporal characteristics of the study episode), but do not distinguish between extrinsic or intrinsic context (Baddeley, 1982), or external and internal source (Johnson, Hashtroudi, & Lindsay, 1993). Though obviously related to the distinction between recollection and familiarity (Mandler, 1980; Yonelinas, 2002), I use the source/item distinction here as an operational distinction based on task requirements, rather than the more theoretically loaded terms of recollection and familiarity. In Tables 1 and 2, experiments are therefore grouped first by stimulus material and then by whether or not the comparison separates source and item information.

As regards further dimensions relevant to animal experiments of recognition memory (like those reported in other articles in this special issue), it is interesting to note that in the present human experiments: (a) the items are invariably trial unique (though they may differ in their level of preexperimental familiarity, particularly for verbal material), (b) only a few experiments explicitly manipulate the similarity between old items and new items, (c) the test format is more often a variant of "yes/no" recognition to single items, rather than "forced choice" between two or more items, and (d) nearly all experiments use visual items, most often words.

Encoding (subsequent memory effects)

Table 1 shows comparisons that examined subsequent memory effects at study. For the basic comparison of subsequent hits (sHt) versus subsequent misses (sMs) according to a yes/no recognition memory test for single words, three experiments reported activation in left anterior hippocampus (Fletcher, Stephenson, Carpenter, Donovan, & Bullmore, 2003; Morcom, Good, Frackowiak, & Rugg, 2003; Otten, Henson, & Rugg, 2001), three reported activation in left or right MTC (Morcom et al., 2003; Otten, Henson, & Rugg, 2002; Wagner et al., 1998), and five failed to find reliable MTL activations (Baker, Sanders, Maccotta, & Buckner, 2001; Buckner, Wheeler, & Sheridan, 2001; Henson, Hornberger, & Rugg, in press; Kirchhoff, Wagner, Maril, & Stern, 2000; Otten & Rugg, 2001). One potentially important factor is whether hits are confined to confident recognition decisions, as is the case in many but not all of the above experiments. This is important because some hits (those typically made with low confidence) can be guesses (Otten et al., 2001) and will therefore weaken the ability to detect true subsequent memory effects. Another potential factor is the type of study task. In all the above experiments, participants were not informed that their memory would be tested later (so-called "incidental" encoding). Nonetheless, the type of incidental task varied, from semantic to orthographic decisions, or even recognition decisions from a previous study phase (Buckner et al., 2001). Of the four experiments that directly compared MTL activity in different study tasks, none found a reliable interaction between study task and subsequent memory effects within the MTL (though such interactions were found elsewhere, e.g., in prefrontal and parietal cortices, Otten et al., 2002), for either semantic and orthographic tasks (Fletcher et al., 2003; Otten et al., 2001) or

TABLE 1
Subsequent memory effects at study (encoding)

Reference	Study material	Item / source	Study task	Test task	Comparison	H		MTC	
						l	r	l	r
Baker et al., 2001	Words	both?	Inc Sem/Ort	Y/N	sHt > sMs, Sem/Ort				
Fletcher et al., 2003	Words	both?	Inc Sem/Ort	Y/N	sHt > sMs, Sem + Ort	ant			
Henson et al., in press	Words	both?	Inc Ort	Y/N	sHt > sMs, Ort				
Wagner et al., 1998	Words	both?	Inc Sem	Y/N, Conf	sHt > sMs (conf)			pos	
Buckner et al., 2001	Words	both?	Inc Recog	Y/N/G	sHt > sMs (no guesses)				
Otten & Rugg, 2001	Words	both?	Inc Sem/Pho	Y/N, Conf	sHt > sMs (conf), Sem/Pho				
Otten et al., 2001	Words	both?	Inc Sem/Ort	Y/N, Conf	sHt > sMs (conf), Sem&Ort	ant			
Otten et al., 2002	Words	both?	Inc Sem/Pho	Y/N, Conf	sHt > sMs (conf), Sem only				
Morcom et al., 2003	Words	both?	Inc Sem	Y/N, Conf	sHt > sMs (conf), Sem	ant		ant	ant
Kirchhoff et al., 2000	Words	both?	Inc Sem	Y/N, Conf	sHt > sMs (conf)				ant
Reber et al., 2002	Words	both?	Rem/For	Y/N	sHt > sMs, Rem&For	pos		pos	
Davachi & Wagner, 2002	3 Words	both?	Inc Elab/Rote	Y/N, Conf	sHt1 to sHt3, Elab only	ant	pos		
Davachi & Wagner, 2002	*3 Words*	*both?*	*Inc Elab/Rote*	*Y/N, Conf*	*sHt3 to sHt1, Elab only*				*pos*
Kensinger et al., 2003	Words	both?	Inc Sem + 2nd	Y/N	sHt > sMs, Hard&Easy			pos	
Henson et al., 1999	*Words*	*item?*	*Inc Lex*	*R/K/N*	*sK > sR*				*pos*
Davachi et al., 2003	Words	item	Image/Read	Y/N, Source2	sHt(Inc source) > sMs			ant	
Ranganath et al., 2004	Words	item?	Inc 2 Sems	Conf 6, Source2	Conf 1 to 5			ant	

344

Reference	Material	Encoded	Encoding task	Test	Contrast			
Kensinger et al., 2003	Words	source?	Inc Sem + 2nd	Y/N	sHt > sMs, Easy only	ant		
Davachi et al., 2003	Words	source	Image/Read	Y/N, Source2	Cor > Inc Source (sHt)	pos	ant	pos
Ranganath et al., 2004	Words	source	Inc, 2 Sems	Conf 6, Source2	Cor > Inc Source (4-6)		pos	
Cansino et al., 2002	Obj-Loc	source	Inc Sem	Y/N, Source4	Ht vs Ms, Cor vs Inc Source			
Jackson & Schacter, 2004	2 Words	source	Inc Assoc	Int/Rer/Sin/New	IntInt > IntRer	ant	ant	ant
Sperling et al., 2003	Face-name	source	Int Match	2AFC name	sHt > sMs, high conf	ant		
Kirwan & Stark, 2004	Face-name	source	Int Assoc	Int/Rer/New	IntInt > IntRer		ant	pos
Kirwan & Stark, 2004	Face-name	item	Int Assoc	Int/Rer/New	IntInt/IntRer > IntNew			both
Kirwan & Stark, 2004	*Face-name*	*both?*	*Inc Recog*	*Y/N*	*sMs > sHt*			*ant*
Brewer et al., 1998	Scenes	both?	Inc in/outdoor	R/K/N	sR > sK > sMs	pos		pos
Kirchhoff et al., 2000	Scenes	both?	Inc in/outdoor	Y/N, Conf	sHt > sMs, high conf	pos		pos
Weis et al., 2004	Scenes	both?	Int building	Y/N, Conf	sHt > sMs, high conf	ant	pos	ant
Stark & Okado, 2003	Scenes	both?	Int	Y/N	sHt > sMs	pos	both	both

Note: Activations for items later "remembered" versus items later "forgotten" (deactivations in *bold italics*). One row per comparison, ordered by study material and then by information encoded (item, source or possibly both). H = hippocampus, MTC = medial temporal cortex, l = left, r = right, ant = anterior (y < −20), pos = posterior (y > −20), sHt = subsequent hit, sMs = subsequent miss, Inc = incidental, Int = intentional, Obj = object, Loc = location, Sem = semantic, Pho = phonological, Lex = lexical, Ort = orthographic, Rem = remember cue, For = forget cue, Elab = elaborate, Assoc = associate, Y/N/G = yes/no/guess, Recog = recognition, Conf = confidence ratings, 2nd = secondary task, SourceN = N-way source decision, Int = intact, Rer = rearranged, Sin = single item recognized, R = remember, K = know, ? = unclear, + = main effect, / = either simple effect, & = both simple effects (see text for further information).

345

semantic and phonological tasks (Otten & Rugg, 2001; Otten et al., 2002). This suggests that the type of processing performed on study items, though it might affect the overall level of memory, does not affect differences between remembered and forgotten items in the MTL.

This apparent independence of MTL subsequent memory effects from participants' intention to memorize or type of task is supported by Reber et al. (2002), who found left posterior hippocampus and MTC activation related to subsequent memory for words, regardless of whether each word was followed by a cue to either remember it or forget it (the "directed forgetting" paradigm, Bjork, 1989). This suggests that the MTL automatically encodes experiences. However, task independence is not supported by Davachi and Wagner (2002), who asked participants to either repeat triplets of words verbatim (rote condition) or reorder them according to semantic attributes (elaborate condition). Only in the elaborate condition did responses in left anterior and right posterior hippocampus increase with the number of words per triplet later recognized (when presented individually at test). This suggests that hippocampal activation only predicts subsequent memory when a certain level of semantic elaboration is performed. Interestingly, Davachi and Wagner (2002) also found right posterior MTC responses that decreased as the number of recognized words increased (a similar decrease in right MTC was found for subsequent remember versus subsequent know judgements to single words, Henson, Rugg, Shallice, Josephs, & Dolan, 1999, and sHt versus sMs for face–name pairs during a recognition test, Kirwan & Stark, 2004; see below).

An experiment by Kensinger, Clarke, and Corkin (2003) examined subsequent recognition as a function of whether an incidental semantic task was performed concurrently with either an "easy" or "hard" secondary task. Left MTC showed activation related to subsequent memory in both easy and hard conditions, whereas left hippocampus only showed subsequent memory effects in the easy condition. Taken together with those above, these findings suggest that, while encoding in MTL may be "automatic" in one sense (i.e., occur under incidental tasks, regardless of instructions to forget or the type of *information* emphasized by the task), it does depend on a certain level of attentional resources and possibly a certain type of *processing* (e.g., elaboration) of that information.

The above experiments did not directly test whether correct subsequent recognition was based on item or source information (though there is evidence that source information is less well encoded under demanding dual-task conditions, Kensinger et al., 2003). Two experiments using words attempted to isolate encoding that leads to later retrieval of source information. In the experiment of Davachi, Mitchell, and Wagner (2003), participants saw words and were randomly cued to either imagine scenes associated with those words, or imagine how they would sound if read backwards. In a later recognition test, participants were asked whether each item was old and, if so, whether it was imaged or read backwards. Left anterior MTC activation predicted subsequent item memory but did not differ according to whether the study task was retrieved correctly. Bilateral hippocampus and left posterior MTC, however, were more responsive to items for which the study task was retrieved correctly than to items that were recognized but their study task not retrieved. These results suggest that anterior MTC (which was identified as perirhinal cortex) encodes item but not source information, whereas hippocampus and posterior MTC (which was identified as parahippocampal cortex) encode only source information.

An experiment by Ranganath et al. (2004) came to similar conclusions. At study, words were presented in two colours, which cued one of two incidental tasks; at test, participants made a six-way confidence judgement for item recognition, followed by a colour/task source judgement. Left anterior MTC activity increased linearly with subsequent confidence levels 1–5 (excluding the most confident level 6). For items recognized with high confidence (rated 4 or more), right posterior hippocampus and right posterior MTC were more active for those associated with correct versus incorrect source judgements. This pattern is generally consistent with the interpretation of Davachi et al. (2003).

An experiment by Cansino, Maquet, Dolan, and Rugg (2002) used objects rather than words, which were presented at study in one of four spatial locations (corners of the screen). They failed, however, to find any subsequent memory effects within MTL, even when comparing subsequent correct versus incorrect retrieval of the spatial source. The reason for this is unknown, though they did find MTL differences at test (see later).

Three experiments looked at associative recognition, in which pairs of items at study were either presented again during test ("intact") or rearranged such that the two items in a test pair came from different study pairs ("rearranged"). Since intact and rearranged pairs cannot be distinguished using item information alone, their comparison is believed to isolate associative or relational information (which I characterize in Tables 1–2 as another example of source information, for simplicity). Jackson and Schacter (2004) used pairs of words and found left anterior hippocampus and bilateral anterior MTC activation to be greater for intact pairs recognized as "intact" than for intact pairs judged to be "rearranged" (pairs in which one or more items were not recognized at all were classified separately). Sperling et al. (2003) presented pairs consisting of a face and a name, followed by two-alternative forced choice (2AFC) for two names presented with a face (one name paired with that face at study, the other paired with a different face at study). Study pairs later attracting high confidence, correct decisions produced greater responses in bilateral anterior hippocampus than did those with later incorrect decisions. Kirwan and Stark (2004) also used face–name pairs but an intact/rearranged/new decision at test (similar to Jackson & Schacter, 2004) rather than 2AFC. They found right hippocampal and right posterior MTC (identified as parahippocampal cortex) activation for intact pairs called "intact" versus intact pairs called "rearranged". Assuming that two items in a study pair were not unitized into a single "item", all three experiments suggest that hippocampus is important for encoding associations between items. If the requirements for encoding associations between items are equivalent to those for encoding associations between an item and contextual information, then these results are consistent with those from source memory tasks considered above.

Kirwan and Stark (2004) also found that more anterior regions in right MTC (including what was identified as perirhinal cortex) showed comparable levels of activation for both intact pairs called "intact" and intact pairs called "rearranged" relative to intact pairs called "new" (by virtue of the face and/or name being forgotten). This is consistent with the suggestion of Davachi et al. (2003) that perirhinal cortex supports subsequent item memory, whereas hippocampus and parahippocampus support subsequent source memory (though see Kirwan & Stark, 2004, for further discussion).

Four experiments used visual scenes rather than words. All four reported bilateral MTC subsequent memory effects (Brewer, Zhao, Desmond, Glover, & Gabrieli, 1998; Kirchhoff et al., 2000; Stark & Okado, 2003; Weis, Klaver, Reul, Elger, & Fernandez, 2004), in additional to bilateral posterior hippocampus in the Kirchhoff et al. (2000) and Stark and Okado (2003) experiments. Only Brewer et al. (1998) made an attempt to distinguish item from source information, by using remember/know (R/K) judgements at test (Tulving, 1985). They found that bilateral posterior MTC responses increased successively from sMs to subsequent K (sK) to subsequent R (sR) judgements. Given that R judgements are likely to involve source retrieval, this is consistent with the above suggestion that posterior MTC (which was identified as parahippocampal cortex) is involved in encoding source information. Though direct, within-experiment comparisons between stimuli are clearly needed, it is noteworthy that MTL subsequent memory effects are more often seen with scenes than with words (and more often seen bilaterally, though see Kirchhoff et al., 2000).

Recognition (old–new effects)

Table 2 (pp. 350–351) shows experiments that examined variants of old–new effects at test. With regard to the basic activation for hits (Ht) relative to correct rejections (CR) in yes/no tests using words, one found left posterior MTC (Daselaar et al., 2001), and one found left posterior hippocampus (Donaldson, Petersen, & Buckner, 2001), but eight did not find activations within MTL (Donaldson, Petersen, Ollinger, & Buckner, 2000; Henson, Rugg, Shallice, & Dolan, 2000; Herron et al., 2004; Jessen et al., 2001; Konishi, Wheeler, Donaldson, & Buckner, 2000; McDermott, Jones, Petersen, Lageman, & Roediger, 2000; Ranganath, Johnson, & D'Esposito, 2000; Rugg, Henson, & Robb, 2003). Four of the latter experiments did, however, find "deactivations" for Ht versus CR. These were generally in anterior MTC rather than hippocampus (Henson et al., in press; Herron et al., 2004), though extending more posteriorly in Rugg et al. (2003) and Jessen et al. (2001) and the laterality varied. Similar deactivations were found in reanalysis of further experiments, with a common locus believed to be perirhinal cortex (Henson, Cansino, Herron, Robb, & Rugg, 2003). Moreover, these MTC deactivations associated with old items (or, conversely, activations associated with new items; see later) did not appear sensitive to source retrieval (see Henson et al., 2003, for further discussion).

Konishi et al. (2002) used a 2AFC recency task, rather than a yes/no task, in which participants saw two old words and judged which was studied more recently. The two words were either close together (high demand condition) or far apart (low demand condition) in the study list. Bilateral posterior MTC activation was found for high versus low demand trials (with accuracy close to ceiling in both).

Three experiments used R/K judgements (Tulving, 1985). Two reported MTL activation for R-hits versus K-hits: in left anterior hippocampus and right posterior MTC (Eldridge, Knowlton, Furmanski, Bookheimer, & Engel, 2000) and in, or near, bilateral anterior hippocampus (Wheeler & Buckner, 2004). The third only found activation in, or near, left posterior hippocampus for R-hits versus CRs (Henson et al., 1999). Two factors are potentially important in these experiments: The first is whether the R/K decision follows an old–new (O/N) decision (as in Eldridge et al., 2000), or whether a single R/K/N decision is made (as in Henson et al., 1999, and Wheeler & Buckner, 2004). The former is better for isolating recollection (Eldridge & Knowlton, 2002) while the latter is better for comparing R/K with N judgements (e.g., CRs). The second factor is whether K responses are more accurate than guesses (as in Henson et al., 1999, and Wheeler & Buckner, 2004, but not necessarily in Eldridge et al., 2000, depending on how R and K judgements are related). The experiment by Kensinger et al. (2003) mentioned earlier did not use R/K judgements during fMRI, but did find more R than K judgements under easy than under hard dual-task study conditions in a separate behavioural experiment, together with greater left anterior hippocampal activation for hits in a yes/no task following easy than following hard dual-task conditions at study.

While R/K judgements might be viewed as a subjective means of distinguishing source information (R) from item information (K), several experiments used objective tests of memory for study context. Wheeler and Buckner (2003) asked participants to study words presented together with either a sound or a picture related to that word. At test, they were asked to indicate whether words were new, or whether they had previously been paired with a sound or picture. Those test words that were repeated 20 times at study (for which accuracy of remembering the associated sound/picture was close to perfect) produced left anterior hippocampal and bilateral posterior MTC activation relative to CRs. A similar experiment was performed by Okado and Stark (2003). At study, participants heard names of objects and then either saw or imagined a picture of that object. At test, they heard old and new object names and were asked whether or not they saw pictures of those objects at study. No MTL differences, however, were found between hits (true memories), false alarms for objects that were only imagined at study (false memories), or correct rejections of objects imagined at study.

Dobbins, Rice, Wagner, and Schacter (2003) cued participants for either a pleasant/unpleasant or a concrete/abstract decision to words. At test, two old words were presented for one of two types of 2AFC: to select the word seen more recently in the study phase (recency task) or to select the word judged for pleasantness in the study phase (source task). Activation for correct versus incorrect decisions was found in left anterior hippocampus and left MTC in the source task, but not in the recency task. Using a similar source task but with yes/no recognition, Kahn, Davachi, and Wagner (2004) found bilateral posterior MTC activation for words whose study task was correctly identified relative to words that were recognized as studied but their study task not identified. Interestingly, this activation was found for words from the visual imagery task, but not those from the read-backwards task (see the earlier description of Davachi et al., 2003, which reported encoding data using the same paradigm). Moreover, the same regions were activated (though to a lesser extent) for false alarms to new items that were (erroneously) judged to have been imaged at study, but not false alarms judged to have been read backwards. This pattern suggests that bilateral posterior MTC activity (particularly in parahippocampal cortex) is associated with episodic retrieval of visual imagery, even if that is cued by a nonstudied item (perhaps reflecting confusion of that item with a different, but related, study item).

Considering the above experiments using source judgements together with those using R/K judgements, it appears that source retrieval is normally correlated with MTL activation (indeed, the likelihood that not all hits are associated with source retrieval in standard yes/no recognition tasks may explain why MTL activation is less common in the basic "old–new" contrast of hits versus correct rejections considered earlier). Whether that activation is in hippocampus or MTC, anterior or posterior, or left or right, varies, however (though the MTC activations are more often in parahippocampus than in rhinal cortex). The differences may depend on the type of source information. Note also that, though

TABLE 2
Old–new and related effects at test (recognition)

Reference	Test material	Item/source	Study task	Test task	Comparison	H l	H r	MTC l	MTC r
Daselaar et al., 2001	Words	both?	Int	Y/N	Ht > CR			pos	
Donaldson et al., 2001	Words	both?	Inc Sem	Y/N	Ht > CR	pos			
Donaldson et al., 2000	Words	both?	Inc Sem	Y/N	Ht vs CR				
Henson et al., 2000	Words	both?	Inc Sem	Y/N Conf	Ht vs CR, high/low conf				
Konishi et al., 2000	Words	both?	Int	Y/N	Ht vs CR				
Ranganath et al., 2000	Words	both?	Inc Sem	Y/N, gen/spec	Ht vs CR				
Henson et al., in press	*Words*	*both?*	*Inc Sem/Ort*	*Y/N*	*CR > Ht*			*ant*	
Herron et al., 2004	*Words*	*both?*	*Inc Sem/Ort*	*Y/N*	*CR > Ht*				*ant*
Rugg et al., 2003	*Words*	*both?*	*Inc Sem*	*In/Exclusion*	*CR > Ht (Inc + Exc)*			*both*	*both*
Jessen et al., 2001	*Words*	*both?*	*Cont recog*	*Y/N Cont*	*CR > Ht*			*pos*	
Konishi et al., 2002	Word pairs	both?	Int	Recency 2AFC	Close > Far (Cor)			pos	pos
Eldridge et al., 2000	Words	source	Int	Y/N, R/K	R(Ht) > K(Ht)/CR	ant			
Henson et al., 1999	Words	source?	Inc Lex	R/K/N	R(Ht) > CR	pos			pos
Wheeler & Buckner, 2004	Words	source	Int	R/K/N/G	R(Ht) > K(Ht)	ant	ant		
Kensinger et al., 2003	Words	source?	Inc Sem + 2nd	Y/N	Ht: Easy > Hard	ant			
Wheeler & Buck, 2003	Words	source	Int, Snd/Pic	Snd/Pic/N	Ht(20) > CR	ant		pos	pos
Okado & Stark, 2003	Aud. word	source	See/Image	Y/N seen	Ht vs FA vs CR				
Dobbins et al., 2003	Word pairs	source	Inc 2 Sem	2AFC: Rec/Sou	Cor > Inc, Source only	ant		both	
Kahn et al., 2004	Words	source	Image/Read	Y/N, source	Cor > Inc Source, Image			pos	pos
McDermott et al., 2000	Words	source	Int	Y/N	CR(sim) vs CR(dis) vs Ht	pos		pos	
von Zerssen et al., 2001	Words	source	Int	Y/N, Conf	Ht/CR(sim) > CR(dis)				
Cabeza et al., 2001	Words	item?	Int	Y/N	Ht/FA(sim) > CR(dis)	pos	pos		

Study	Stimulus	Source	Manipulation	Test	Contrast			
Cabeza et al., 2001	Words	source?	Int	Y/N	Ht > FA(sim)/CR(dis)			pos
Slotnick & Schacter, 2004	Shapes	source?	Int	L/R/N, Conf	Ht/FA(sim) > CR	pos		pos
Maratos et al., 2001	Words	both?	Valence	Y/N	Ht > CR (Pos + Neu + Neg)			pos
Maratos et al., 2001	Words	source?	Valence	Y/N	Neg > Neu (Ht)	ant		
Smith et al., 2004	Objects	source?	Inc assoc scene	Y/N	Neg > Neu (Ht)		pos	pos
Lepage et al., 2003	Object pair	both?	Int Assoc	Int/Rer/New	Int vs Rer vs New			pos
Duzel et al., 2003	Obj/Loc	source	Learn	Y/N Obj/Loc	Old > New pairs, Obj/Loc			
Duzel et al., 2003	*Obj/Loc*	*source*	*Learn*	*Y/N Obj/Loc*	*New > Old pairs, Obj/Loc*		*ant*	
Kirwan & Stark, 2004	Face-name	source	Int Assoc	Int/Rer/New	IntInt > IntRer	ant	ant	ant
Giovanello et al., 2004	Word pairs	source	Inc Sem	Assoc/Item	Assoc > Item (Ht)	ant	ant	
Henson et al., 2002a	Faces	both?	Cont recog	Y/N Cont	Ht vs CR			
Tsivilis et al., 2003	Objects	both	Inc assoc scene	Y/N	Int vs Rer vs New		?	
Slotnick et al., 2003	Shapes	both?	Int	Y/N	Ht > CR	ant		
Stark & Squire, 2001	Names	both?	Int Names	Y/N	Ht > CR (1st test)	both	both	
Stark & Squire, 2001	Names	both?	Int Objects	Y/N	Ht > CR (1st test)		both	
Stark & Squire, 2001	Objects	both?	Int Objects	Y/N	Ht > CR (2nd tes)	both	ant	
Cansino et al., 2002	Objects	source	Inc Sem	Loc 1-4/N	Cor > Inc source (Ht)		ant	pos
Rombouts et al., 2001	*Scenes*	*both?*	*Int*	*Y/N*	*CR > Ht*		*ant*	*ant*
Weis et al., 2004	*Scenes*	*both?*	*Int*	*Y/N, Conf*	*Ms > Ht, high conf*	*ant*	*ant*	*ant*
Stark & Okado, 2003	Scenes	both?	Int	Y/N	Ht > CR(forgotten)	both	both	both
Burgess et al., 2001	Environ	source	Int Assoc	2AFC	Place > Pers/Obj/Width	pos		pos
Burgess et al., 2001	Environ	source	Int Assoc	2AFC	Place > Width	pos	pos?	

Note: Activations for old items versus new items (deactivations in **bold italics**). Ht = hit, CR = correct rejection, Aud = auditory, Environ = virtual environment, Snd = sound, Pic = picture, Cont = continuous, Sim = similar to old item (lures); Dis = dissimilar (unrelated) new items, Neg = negative, Neu = neutral valence, Pers = person. See text and Table 1 for more information.

explicit source judgements can identify whether a specific type of contextual information was retrieved, it is possible that items for which that particular information was not retrieved were nonetheless associated with retrieval of other aspects of study context (so-called "noncriterial" source retrieval). Thus both procedures—objective source decisions and subjective R/K judgements—are valuable.

Four experiments examined recognition memory using new items that were related in some way to studied items. For such "lures", one can look at both false alarms and correct rejections. According to some models (e.g., Jones & Jacoby, 2001), false alarms to lures are attributed to an increased feeling of familiarity in the absence of recollection, whereas correct rejections of lures are attributed to recollection of the related studied item(s) and hence rejection of their increased familiarity (a "recall-to-reject" strategy). Using compound nouns as lures (rearranged from two studied nouns), McDermott et al. (2000) did not find MTL differences between hits, correct rejections of lures, and correct rejections of unrelated new items. Using semantically related lures, von Zerrsen, Mecklinger, Opitz, and Cramonvon (2001) found greater left posterior MTC activation for hits and correct rejection of lures, than for correct rejection of unrelated items. They failed to find MTL differences between false alarms to lures and correct rejections of unrelated items. Cabeza, Rao, Wagner, Mayer, and Schacter (2001) on the other hand found greater bilateral posterior hippocampus activation for both hits and false alarms to lures than for correct rejections of unrelated items, and greater left posterior MTC activation for hits than for both false alarms to lures and correct rejections of unrelated items. This pattern was interpreted as hippocampus being sensitive to retrieval of semantic information, which would not distinguish targets from lures, and parahippocampus being sensitive to retrieval of perceptual information, which would distinguish targets from lures.

However, this hippocampal pattern (hits and false alarms greater than correct rejections) was also found by Slotnick and Schacter (2004) using abstract shapes, which presumably have little semantic content. This suggests that hippocampus is activated by false memory for "gist", even if that gist is a nonverbal visuospatial prototype. Furthermore, Slotnick and Schacter found that it was relatively "early" visual regions, rather than "late" (e.g., parahippocampal) visual regions that showed differences between hits and false alarms to lures. This difference between the two studies may reflect a difference in the types of source information retrieved for true memories (hits): one of two speakers from a video clip (Cabezaet al., 2001) versus presentation left or right of fixation (Slotnick & Schacter, 2004).

These findings using lures raise interesting interpretations (e.g., left posterior MTC activation associated with valid recollection/source retrieval), but more experiments are needed to establish their typicality and robustness. There are also likely to be multiple factors that affect results and their interpretations, such as whether the lures are internally generated at study, whether false recollection occurs in addition to misattributed familiarity, and the type of item and source information (e.g., perceptual or semantic) contributing to the recognition decision.

Two experiments examined recognition of items previously studied in emotionally positive, neutral, or negative contexts (either words within sentences, Maratos, Dolan, Morris, Henson, & Rugg, 2001, or objects within background scenes, Smith, Henson, Dolan, & Rugg, 2004). Maratos et al. (2001) found left posterior MTC activation for hits relative to correct rejections of words, collapsing across study context, and greater left anterior hippocampal activation for negative hits than for neutral hits. Smith et al. (2004) did not find greater MTL activation for hits than for correct rejections of objects, but found greater right posterior hippocampal and left posterior MTC activation for negative hits than for neutral hits. Though these experiments had no objective measure of source retrieval, the difference between negative and neutral hits may reflect differences in the amount (and/or emotional content) of the source information retrieved.

Four experiments looked at variants of the associative recognition task. Using pairs of objects, LePage, Brodeur, and Bourgouin (2003) did not find MTL differences between intact, rearranged, or new pairs. Using pairs consisting of a central face and a peripheral object presented in one of four corners of the screen (for multiple learning trials), Duzel et al. (2003) found greater right posterior MTC activation for studied than for new pairs, but greater right anterior hippocampus activation for new pairs than for studied pairs, regardless of whether recognition decisions were based on the spatial position of the studied object (relative to the face) or the identity of the object (relative to the one studied with that face). In an experiment also described earlier, Kirwan and Stark (2004) found greater bilateral anterior MTC, right posterior MTC, and right anterior hippocampus for intact face–name pairs called "intact" than for intact pairs called "rearranged". Giovanello, Schnyer, and Verfaellie (2004) found greater bilateral anterior hippocampal activation (stronger on left) for intact word-pairs called "intact" in an associative recognition task than for correct recognition for both items of a rearranged pair in an item recognition task. The last three experiments are all consistent with a hippocampal role in retrieval (as well as encoding; see earlier) of associations between pairs of items, though note that different comparisons were used in each case.

Of the experiments using visual objects, Henson, Shallice, Gorno-Tempini, and Dolan (2002a) failed to find MTL old–new effects with either familiar or unfamiliar faces during a continuous recognition memory task. Tsivilis, Otten, and Rugg (2003) also failed to find MTL differences between old or new objects, or between old objects presented in the same or a different background scene as during study. Slotnick, Moo, Segal, and Hart (2003) used unfamiliar, nonverbalizable abstract shapes, and found left anterior MTC activation for hits versus correct rejections in an item recognition task (but no reliable differences in a source recognition task in which participants recollected whether the shapes were studied left or right of fixation).

Stark and Squire (2001) presented objects at study and test, object names at study and test, or objects at study and their names at test. Comparing hits with correct rejections, they found left hippocampal activation in the name–name condition and right hippocampal activation in the object–name condition (probably including both anterior and posterior loci in both cases). Right hippocampal activation was found in the object–object condition only when the recognition test was repeated (which the authors attributed to a high level of encoding-related activation for the new objects; see later). Interestingly, no hippocampal differences were found for hits across the three conditions, suggesting that retrieval of the object associated with a name cue does not engage hippocampus any more than does retrieval associated with an object or word copy cue (Tulving, 1982).

Participants in the Cansino et al. (2002) performed an incidental semantic task on objects presented in one of four corners of the screen (as described in the encoding section), then at test saw a central object and indicated in which of the four positions it appeared at study, or that it was new. Right anterior hippocampus and left posterior MTC were activated for source hits versus source misses. This experiment thus resembles both the source judgement tasks described above using words, and the associative recognition task using objects and locations by Duzel et al. (2003), though the only common activation would appear to be posterior MTC.

Three experiments examined recognition memory for scenes (e.g., landscapes, buildings). Rombouts, Barkhof, Witter, Machielsen, and Scheltens (2001) found deactivations in bilateral anterior MTC and right anterior hippocampus for hits relative to correct rejections, while Weis et al. (2004) found deactivations in left anterior MTC for confident hits relative to confident misses. These "deactivations" associated with recognition, at least with regard to anterior MTC, are analogous to those for words described earlier (and Henson et al., 2003). Furthermore, that the deactivation was found relative to misses by Weis et al. (2004) suggests that it is likely to reflect participants' explicit memory, rather than (implicit memory for) the objective old/new status of the item (see also Henson et al., in press).

Stark and Okado (2003) on the other hand found no MTL differences for the usual comparison of hits versus correct rejections. However, when they restricted correct rejections to those new items that were not recognized in a second subsequent recognition test, they found bilateral activation in hippocampus and MTC. The lack of MTL differences for the usual hits versus correct rejection comparison was therefore attributed to encoding-related activation for those new items that were recognized in the second test (see later).

Finally, one experiment by Burgess, Maguire, Spiers, and O'Keefe (2001) used virtual reality to distinguish retrieval of different types of source information. Participants were told to remember objects given to them by virtual people at specific locations within a virtual environment. During test trials, they were shown two objects, again in the context of a specific person and location, and cued to choose (a) which of an old and a new object was studied ("object" trials), (b) which of two old objects was studied together with that person ("person" trials), (c) which of two old objects was studied in that location ("place" trials), or (d) which of two old objects was "wider" (control trials). Bilateral posterior MTC was activated for hits in the place trials versus hits in the object, person and control trials. Since the person trials required retrieval of source information, this result suggests that these regions are specifically involved in retrieval of (allocentric) spatial information. Left (and right at a lower threshold) posterior hippocampus was also activated in the place versus control trials, but showed intermediate levels of activity (on the left) in the person and object trials, suggesting a more general role in item and/or source retrieval.

Summary

Returning to look at Tables 1 and 2 as a whole, I think there is little doubt that structures within MTL show memory-related differences in haemodynamic activity during both the study and test phases of recognition memory tasks. Unfortunately however, no clear pattern emerges (at least to my eye!) for functional divisions between hippocampus and MTC, between anterior or posterior MTL, or even left versus right MTL. One possibility is that, contrary to proposals such as that by Aggleton and Brown (1999, this issue), there is no functional division of labour within the MTL or, at least, any functional division that exists does not conform to the dichotomies considered here (Squire et al., 2004).

Having said this, some trends do appear to be emerging. One trend is that hippocampus and posterior MTC (specifically, parahippocampal cortex) appear particularly important for encoding and retrieving source information and associations between distinct items. Anterior MTC (most likely perirhinal cortex) on the other hand seems more concerned with item information (see below). Indeed, a clearer functional dissociation might be between (peri)rhinal and parahippocampal cortex rather than hippocampus and MTC (unfortunately not all the experiments reviewed here allowed a clear distinction between perirhinal and parahippocampal activation).

Another trend, regarding the stimulus material, is that scenes seem particularly effective at eliciting memory-related MTL activations (at least relative to words). This may reflect the novelty, complexity or spatial components of scenes. It may even relate to the impression that scenes give of being present in an allocentric environment. It is noteworthy therefore that all three experiments that explicitly examined retrieval of spatial relations found right hippocampal activation, whether those relations were likely to be egocentric (Cansino et al., 2002; Duzel et al., 2003) or allocentric (Burgess et al., 2001). Furthermore, there is a slight trend for left-lateralization of activations associated with verbal information (particularly at test), in contrast with bilateral activation for scenes.

A third trend is for decreased responses in anterior MTC for old relative to new items during recognition tests. This resembles decreased firing rates associated with familiarity in perirhinal neurons in animals (Brown & Xiang, 1998). Alternatively (or perhaps equivalently), this pattern could reflect increased responses to new items (e.g., a novelty response). Indeed, such increases may even reflect encoding-related processes, given that another interesting possibility concerns encoding-related activation during recognition tests: Three experiments examined differences between new items in one test as a function of whether they are recognized in a second test, and two found MTL activation (Kirwan & Stark, 2004; Stark & Okado, 2003). If MTL is involved in encoding new items, as well as recognizing old items, this could explain why MTL activations are less often seen in the basic comparison of hits versus correct rejections during recognition tests. It may only be when old items are associated with high levels of source or associative retrieval (see above) that they produce activity over and above that produced by the encoding of new items.

Future directions

Many future experiments are suggested by the above findings. First there is a need to examine a larger range of stimulus material, given the current dominance of verbal material (probably reflecting the long tradition of verbal learning in human behavioural experiments). One reason is that nonverbal stimuli are easier to compare with those used in animal experiments. Another reason is that words are also confounded with pre-experimental familiarity, making it difficult to separate intra- and extra-experimental sources of information (given that even a simple yes/no recognition memory test can be viewed, at least for familiar items, as a source memory test for whether items were seen in a specific context, viz., the study phase). A related need is to examine recognition memory for nonvisual material, since some MTL activations may be specific to memory for visual information. Future experiments would benefit from using both visual and nonvisual (and/or verbal and nonverbal) material, in order to make direct, within-experiment statistical comparisons (e.g., Duzel et al., 2003; Kirchhoff et al., 2000), rather than relying on comparisons across experiments.

Another need is to further characterize and contrast different types of source information, such as time and space (see above), or conceptual versus perceptual information (see also recent interest in "content effects" at retrieval, e.g., Wheeler, Petersen, & Buckner, 2000). Of particular importance is the theoretical characterization of the commonalities or differences between context information (as tested in source judgements) and associative information (as tested in associative recognition judgements). This may include the role of intrinsic versus extrinsic source (Baddeley, 1982), internal versus external source (Johnson et al., 1993), the difference between unitized versus associated items, and associations between different types of information (Mayes et al., 2004). It is also important to characterize more precisely the theoretical relationship between encoding and novelty (Kirchhoff et al., 2000).

Other factors that may be relevant to MTL activation include the question of explicit (conscious) versus implicit (unconscious) retrieval of information (Henke et al., 2003) and the related question of the relationship between the present findings from recognition tests ("direct" memory tests) and analogous comparisons between novel and repeated items in indirect memory tests (e.g., Donaldson et al., 2001; Henson et al., 2002a; Saykin et al., 1999), where target-related effects are less likely (Herron et al., 2004).

Future recognition experiments would benefit from more comparisons involving misses and false alarms, provided they can be collected in sufficient numbers (e.g., Kahn et al., 2004; Slotnick & Schacter, 2004; Wheeler & Buckner, 2003). This might include the use of lures (see earlier). Future experiments would also benefit from direct statistical comparisons between different MTL regions, such as hippocampus versus perirhinal cortex, or left versus right hippocampus: Current experiments tend to report only reliable findings within each region, with the danger that other regions show similar effects that simply did not reach significance (Henson, 2005). Furthermore, the experiments should test for both increases and decreases in the subtraction of different conditions and might want to consider differences in sustained memory-related "state" effects (Otten et al., 2002; Velanova et al., 2003), as well as between transient "item" effects (as considered here). These tests would benefit from conforming to standardized criteria for allowing for multiple statistical comparisons across voxels within a specified, a priori search space (e.g., MTL), allowing fairer comparison across studies. Most importantly, they should use methods developed for better localization and coregistration of structures within MTL (e.g., Zieneh, Engel, Thompson, & Bookheimer, 2003), including acquisition at higher spatial resolution and efforts to minimize fMRI susceptibility effects. The latter is important not only to prevent geometric distortion and hence mislocation of activations, but also to prevent signal loss or "drop-out", which is particularly common in anterior MTL (another potential reason why some studies in Tables 1–2 failed to find MTL activity). Indeed, it would be valuable if future studies provided information on the degree of signal loss in their fMRI data, in order to evaluate the likelihood of finding MTL activity. Finally, it will be important to compare encoding- and retrieval-related findings from the recognition memory paradigm with those from other paradigms, such as (the technically more challenging) recall paradigms (Henson, Shallice, Josephs, & Dolan, 2002b; Strange, Otten, Josephs, Rugg, & Dolan, 2002).

REFERENCES

Aggleton, J. P., & Brown, M. W. (1999). Episodic memory, amnesia, and the hippocampal-anterior thalamic axis. *Behavioral Brain Sciences, 22*, 425–444.

Aggleton, J. P., & Brown, M. W. (this issue). Contrasting hippocampal and perirhinal cortex function using immediate early gene imaging. *Quarterly Journal of Experimental Psychology, 58B*, 218–233.

Baddeley, A. D. (1982). Domains of recollection. *Psychological Review, 89*, 708–729.

Baker, J. T., Sanders, A. L., Maccotta, L., & Buckner, R. L. (2001). Neural correlates of verbal memory encoding during semantic and structural processing tasks. *Neuroreport, 12*, 1251–1256.

Bjork, R. A. (1989). Retrieval inhibition as an adaptive mechanism in human memory. In H. L. Roediger & F. I. M. Craik (Eds.), *Varieties of memory and consciousness* (pp. 309–330). Hillsdale, NJ: Lawrence Erlbaum, Associates, Inc.

Brewer, J. B., Zhao, Z., Desmond, J. E., Glover, G. H., & Gabrieli, J. D. E. (1998). Making memories: Brain activity that predicts how well visual experience will be remembered. *Science, 281*, 1185–1187.

Brown, M. W., & Xiang, J. Z. (1998). Recognition memory: neuronal substrates if the judgement of prior occurrence. *Progress in Neurobiology, 55*, 149–189.

Buckner, R. L., Wheeler, M. E., & Sheridan, M. A. (2001). Encoding processes during retrieval tasks. *Journal of Cognitive Neuroscience, 13*, 406–415.

Burgess, N., Maguire, E. A., Spiers, H., & O'Keefe, J. (2001). A temporoparietal and prefrontal network for retrieving the spatial context of lifelike events. *Neuroimage, 14*, 439–453.

Cabeza, R., Rao, S. M., Wagner, A. D., Mayer, A. R., & Schacter, D. L. (2001). Can medial temporal lobe regions distinguish true from false? An event-related functional MRI study of veridical and illusory recognition memory. *Proceedings of the National Academy of Sciences, USA, 98*, 4805–4810.

Cansino, S., Maquet, P., Dolan, R. J., & Rugg, M. D. (2002). Brain activity underlying encoding and retrieval of source memory. *Cerebral Cortex, 12*, 1048–1056.

Constable, R. T., Carpenter, A., Pugh, K., Westerveld, M., Oszunar, Y., & Spencer, D. D. (2000). Investigation of the human hippocampal formation using a randomized event-related paradigm and Z-shimmed functional MRI. *Neuroimage, 12*, 55–62.

Daselaar, S. M., Rombouts, S. A. R. B., Veltman, D. J., Raaijmakers, J. G. W., Lazeron, R. H. C., & Jonker, C. (2001). Parahippocampal activation during successful recognition of words: A self-paced event-related fMRI study. *Neuroimage, 12*, 1113–1120.

Davachi, L., Mitchell, J. P., & Wagner, A. D. (2003). Multiple routes to memory: Distinct medial temporal lobe processes build item and source memories. *Proceedings of the National Academy of Sciences, USA, 100*, 2157–2162.

Davachi, L., & Wagner, A. D. (2002). Hippocampal contributions to episodic encoding: insights from relational and item-based learning. *Journal of Neurophysiology, 88*, 982–990.

Dobbins, I. G., Rice, H. J., Wagner, A. D., & Schacter, D. L. (2003). Memory orientation and success: Separable neurocognitive components underlying episodic recognition. *Neuropsychologia, 41*, 318–333.

Donaldson, D. I., Petersen, S. E., & Buckner, R. L. (2001). Dissociating memory retrieval processes using fMRI: Evidence that priming does not support recognition memory. *Neuron, 31*, 1047–1059.

Donaldson, D. I., Petersen, S. E., Ollinger, J. M., & Buckner, R. L. (2000). Dissociating state and item components of recognition memory using fMRI. *Neuroimage, 13*, 129–142.

Dunn, J. C. (2004). Remember–know: A matter of confidence. *Psychological Review, 111*, 524–542.

Duzel, E., Habib, R., Rotte, M., Guderian, S., Tuvling, E., & Heinze, H.-J. (2003). Human hippocampal and parahippocampal activity during visual associative recognition memory for spatial and nonspatial stimulus configurations. *Journal of Neuroscience, 23*, 9439–9444.

Eldridge, L. L., & Knowlton, B. J. (2002). The effect of testing procedure on remember/know judgments. *Psychonomic Bulletin and Review, 9*, 139–145.

Eldridge, L. L., Knowlton, B. J., Furmanski, C. S., Bookheimer, S. Y., & Engel, S. A. (2000). Remembering episodes: A selective role for the hippocampus during retrieval. *Nature Neuroscience, 3*, 1149–1152.

Fletcher, P. C., Stephenson, C. M. E., Carpenter, T. A., Donovan, T., & Bullmore, E. T. (2003). Regional brain activations predicting subsequent memory success: An event-related fMRI study of the influence of encoding tasks. *Cortex, 39*, 1009–1026.

Giovanello, K. S., Schnyer, D. M., & Verfaellie, M. (2004). A critical role for the anterior hippocampus in relational memory: Evidence from an fMRI study comparing associative and item recognition. *Hippocampus, 14*, 5–8.

Heathcote, A. (2003). Item recognition memory and the receiver operating characteristic. *Journal of Experimental Psychology: Learning, Memory & Cognition, 29*, 1210–1230.

Henke, K., Treyer, V., Nagy, E. T., Kneifel, S., Dursteler, M., Nitsch, R. M., et al. (2003). Active hippocampus during nonconscious memories. *Consciousness and Cognition, 12*, 31–48.

Henson, R. N. A. (2005). What can functional imaging tell the experimental psychologist? *Quarterly Journal of Experimental Psychology, 58A*, 193–233.

Henson, R. N. A., Cansino, S., Herron, J. E., Robb, W. G. K., & Rugg, M. D. (2003). A familiarity signal in human anterior medial temporal cortex? *Hippocampus, 13*, 259–262.

Henson, R. N. A., Hornberger, M., & Rugg, M. D. (in press). Further dissociating the processes involved in recognition memory: An fMRI study. *Journal of Cognitive Neuroscience.*

Henson, R. N. A., Rugg, M. D., Shallice, T., & Dolan, R. J. (2000). Confidence in recognition memory for words: Dissociating right prefrontal roles in episodic retrieval. *Journal of Cognitive Neuroscience, 12*, 913–923.

Henson, R. N. A., Rugg, M. D., Shallice, T., Josephs, O., & Dolan, R. (1999). Recollection and familiarity in recognition memory: An event-related fMRI study. *Journal of Neuroscience, 19*, 3962–3972.

Henson, R. N., Shallice, T., Gorno-Tempini, M. L., & Dolan, R. J. (2002a). Face repetition effects in implicit and explicit memory tests as measured by fMRI. *Cerebral Cortex, 12*, 178–186.

Henson, R., Shallice, T., Josephs, O., & Dolan, R. (2002b). Functional magnetic resonance imaging of proactive interference during spoken cued recall. *Neuroimage, 17*, 543–558.

Herron, J. E., Henson, R. N., & Rugg, M. D. (2004). Probability effects on the neural correlates of retrieval success: An fMRI study. *Neuroimage, 21*, 302–310.

Holdstock, J. S. (this issue). The role of the human medial temporal lobe in object recognition and object discrimination. *Quarterly Journal of Experimental Psychology, 58B*, 326–339.

Jackson, O., III, & Schacter, D. L. (2004). Encoding activity in anterior medial temporal lobe supports subsequent associative recognition. *Neuroimage, 21*, 456–462.

Jessen, F., Flacke, S., Granath, D-O., Manka, C., Scheef, L., Papassotiropoulus, A., et al. (2001). Encoding and retrieval related cerebral activation in continuous verbal recognition. *Cognitive Brain Research, 12*, 199–206.

Johnson, M. K., Hashtroudi, S., & Lindsay, D. S. (1993). Source monitoring. *Psychological Bulletin, 114*, 3–28.

Jones, T. C., & Jacoby, L. L. (2001). Feature and conjunction errors in recognition memory: Evidence for dual-process theory. *Journal of Memory and Language, 45*, 82–102.

Kahn, I., Davachi, L., & Wagner, A. D. (2004). Functional-neuroanatomic correlates of recollection: Implications for models of recognition memory. *Journal of Neuroscience, 24*, 4172–4180.

Kensinger, E. A., Clarke, R. J., & Corkin, S. (2003). What neural correlates underlie successful encoding and retrieval? A functional magnetic resonance imaging study using a divided attention paradigm. *Journal of Neuroscience, 23*, 2407–2415.

Kirchhoff, B. A., Wagner, A. D., Maril, A., & Stern, C. E. (2000). Prefrontal-temporal circuitry for episodic encoding and subsequent memory. *Journal of Neuroscience, 20*, 6173–6180.

Kirwan, C. B., & Stark, C. E. L. (2004). Medial temporal lobe activation during encoding and retrieval of novel face–name pairs. *Hippocampus, 14*, 919–930.

Konishi, S., Uchida, I., Okuaki, T., Machida, T., Shirouzu, I., & Miyashita, Y. (2002). Neural correlates of recency judgment. *Journal of Neuroscience, 22*, 9549–9555.

Konishi, S., Wheeler, M. E., Donaldson, D. I., & Buckner, R. L. (2000). Neural correlates of episodic retrieval success. *Neuroimage, 12*, 276–286.

Lepage, M., Brodeur, M., & Bourgouin, P. (2003). Prefrontal cortex contribution to associative recognition memory in humans: An event-related functional magnetic resonance imaging study. *Neuroscience Letter, 346*, 73–76.

Lepage, M., Habib, R., & Tulving, E. (1998). Hippocampal PET activations of memory encoding and retrieval: The HIPER model. *Hippocampus, 8*, 313–322.

Maguire, E. (1998). The hippocampus and human navigation. *Science, 282*, 2151–2152.

Mandler, G. (1980). Recognizing: The judgement of previous occurrence. *Psychological Review, 87*, 252–271.

Maratos, E. J., Dolan, R. J., Morris, J. S., Henson, R. N. A., & Rugg, M. D. (2001). Neural activity associated with episodic memory for emotional context. *Neuropsychologia, 39*, 910–920.

Mayes, A. R., Holdstock, J. S., Isaac, C. L., Montaldi, D., Grigor, J., Gummer, A., et al. (2004). Associative recognition in a patient with selective hippocampal lesions and relatively normal item recognition. *Hippocampus, 14*, 763–784.

McDermott, K. B., Jones, T. C., Petersen, S. E., Lageman, S. K., & Roediger, H. L. (2000). Retrieval success is accompanied by enhanced activation in anterior prefrontal cortex during recognition memory: An event-related fMRI study. *Journal of Cognitive Neuroscience, 12,* 424–432.

Morcom, A. M., Good, C. D., Frackowiak, R. S. J., & Rugg, M. D. (2003). Age effects on the neural correlates of successful memory encoding. *Brain, 126,* 213–229.

Okado, Y., & Stark, C. E. (2003). Neural processing associated with true and false memory retrieval. *Cognitive, Affective & Behavioral Neuroscience, 3,* 323–334.

Otten, L. J., Henson, R. N., & Rugg, M. D. (2001). Depth of processing effects on neural correlates of memory encoding: Relationship between findings from across- and within-task comparisons. *Brain, 124,* 399–412.

Otten, L. J., Henson, R. N., & Rugg, M. D. (2002). State-related and item-related neural correlates of successful memory encoding. *Nature Neuroscience, 5,* 1339–1344.

Otten, L. J., & Rugg, M. D. (2001). Task-dependency of the neural correlates of episodic encoding as measured by fMRI. *Cerebral Cortex, 11,* 1150–1160.

Ranganath, C., Johnson, M. K., & D'Esposito, M. (2000). Left anterior prefrontal activation increases with demands to recall specific perceptual information. *Journal of Neuroscience, 20,* 1–5.

Ranganath, C., Yonelinas, A. P., Cohen, M. X., Dy, C. J., Tom, S. M., & D'Esposito, M. (2004). Dissociable correlates of recollection and familiarity within the medial temporal lobes. *Neuropsychologia, 42,* 2–13.

Reber, P. J., Siwiec, R. M., Gitleman, D. R., Parrish, T. B., Mesulam, M.-M., & Paller, K. A. (2002). Neural correlates of successful encoding identified using functional magnetic resonance imaging. *Journal of Neuroscience, 22,* 9541–9548.

Rombouts, S. A. R. B., Barkhof, F., Witter, M. P., Machielsen, W. C. M., & Scheltens, P. (2001). Anterior medial temporal lobe activation during attempted retrieval of encoded visuospatial scenes: An event-related fMRI study. *Neuroimage, 14,* 67–76.

Rugg, M. D., & Henson, R. N. A. (2002). Episodic memory retrieval: An (event-related) functional neuroimaging perspective. In A. E. Parker, E. L. Wilding, & T. Bussey (Eds.), *The cognitive neuroscience of memory encoding and retrieval* (pp. 3–37). Hove, UK: Psychology Press.

Rugg, M. D., Henson, R. N., & Robb, W. G. (2003). Neural correlates of retrieval processing in the prefrontal cortex during recognition and exclusion tasks. *Neuropsychologia, 41,* 40–52.

Saykin, A. J., Johnson, S. C., Flashman, L. A., McAllister, T. W., Sparling, M., Darcey, T. M., et al. (1999). Functional differentiation of medial temporal and frontal regions involved in processing novel and familiar words: An fMRI study. *Brain, 122,* 1963–1971.

Schacter, D. L., & Wagner, A. D. (1999). Medial temporal lobe activations in fMRI and PET studies of episodic encoding and retrieval. *Hippocampus, 9,* 7–24.

Slotnick, S. D., Moo, L. R., Segal, J. B., & Hart, J. (2003). Distinct prefrontal cortex activity associated with item memory and source memory for visual shapes. *Cognitive Brain Research, 17,* 75–82.

Slotnick, S. D., & Schacter, D. L. (2004). A sensory signature that distinguishes true from false memories. *Nature Neuroscience, 7,* 664–672.

Smith, A. P. R., Henson, R. N. A., Dolan, R. J., & Rugg, M. D. (2004). fMRI correlates of the episodic retrieval of emotional contexts. *Neuroimage, 22,* 868–878.

Sperling, R., Chua, E., Cocchiarella, A., Rand-Giovannetti, E., Poldrack, R., Schacter, D. L., et al. (2003). Putting names to faces: Successful encoding of associative memories activates the anterior hippocampal formation. *Neuroimage, 20,* 1400–1410.

Squire, L. R., Stark, C. E. L., & Clark, R. E. (2004). The medial temporal lobe. *Annual Review of Neuroscience, 27,* 279–306.

Stark, C. E. L., & Okado, Y. (2003). Making memories without trying: Medial temporal lobe activity associated with incidental memory formation during recognition. *Journal of Neuroscience, 23,* 6748–6753.

Stark, C. E. L., & Squire, L. R. (2001). Simple and associative recognition memory in the hippocampal region. *Learning & Memory, 8,* 190–197.

Strange, B. A., Otten, L. J., Josephs, O., Rugg, M. D., & Dolan, R. J. (2002). Dissociable human perirhinal, hippocampal, and parahippocampal roles during verbal encoding. *Journal of Neuroscience, 22,* 523–528.

Tsivilis, D., Otten, L. J., & Rugg, M. D. (2003). Repetition effects elicited by objects and their contexts: An fMRI study. *Human Brain Mapping, 19,* 145–154.

Tulving, E. (1982). *Elements of episodic memory.* Oxford: OUP.

Tulving, E. (1985). Memory and consciousness. *Canadian Journal of Psychology, 26,* 1–12.

Velanova, K., Jacoby, L. J., Wheeler, M. E., McAvoy, M. P., Petersen, S. E., & Buckner, R. L. (2003). Functional-anatomic correlates of sustained and transient processing components engaged during controlled retrieval. *Journal of Neuroscience, 23,* 8460–8470.

von Zerssen, G. C., Mecklinger, A., Opitz, B., & von Cramon, D. Y. (2001). Conscious recollection and illusory recognition: An event-related fMRI study. *European Journal of Neuroscience, 13,* 2148–2156.

Wagner, A. D., Schacter, D. L., Rotte, M., Koustaal, W., Maril, A., Dale, A. M., et al. (1998). Building memories: Remembering and forgetting of verbal experiences as predicted by brain activity. *Science, 21,* 188–191.

Weis, S., Klaver, P., Reul, J., Elger, C. E., & Fernandez, G. (2004). Temporal and cerebellar brain regions that support both declarative memory formation and retrieval. *Cerebral Cortex, 14,* 256–267.

Wheeler, M. E., & Buckner, R. L. (2003). Functional dissociation among components of remembering: Control, perceived oldness, and content. *Journal of Neuroscience, 23,* 3869–3880.

Wheeler, M. E., & Buckner, R. L. (2004). Functional-anatomic correlates of remembering and knowing. *Neuroimage, 21,* 1337–1349.

Wheeler, M. E., Petersen, S. E., & Buckner, R. L. (2000). Memory's echo: Vivid remembering reactivates sensory-specific cortex. *Proceedings of the National Academy of Sciences, USA, 97,* 11125–11129.

Yonelinas, A. P. (2002). The nature of recollection and familiarity: A review of 30 years of research. *Journal of Memory and Language, 46,* 441–517.

Zieneh, M. M., Engel, S. A., Thompson, P. M., & Bookheimer, S. Y. (2003). Dynamics of the hippocampus during encoding and retrieval of face–name pairs. *Science, 299,* 577–580.

THE QUARTERLY JOURNAL OF EXPERIMENTAL PSYCHOLOGY
2005, 58B (3/4), 361–377

The anatomy of object processing: The role of anteromedial temporal cortex

Peter Bright, Helen E. Moss, Emmanuel A. Stamatakis, and Lorraine K. Tyler

University of Cambridge, UK

How objects are represented and processed in the brain remains a key issue in cognitive neuroscience. We have developed a conceptual structure account in which category-specific semantic deficits emerge due to differences in the structure and content of concepts rather than from explicit divisions of conceptual knowledge in separate stores. The primary claim is that concepts associated with particular categories (e.g., animals, tools) differ in the number and type of properties and the extent to which these properties are correlated with each other. In this review, we describe recent neuropsychological and neuroimaging studies in which we have extended our theoretical account by incorporating recent claims about the neuroanatomical basis of feature integration and differentiation that arise from research into hierarchical object processing streams in nonhuman primates and humans. A clear picture has emerged in which the human perirhinal cortex and neighbouring anteromedial temporal structures appear to provide the neural infrastructure for making fine-grained discriminations among objects, suggesting that damage within the perirhinal cortex may underlie the emergence of category-specific semantic deficits in brain-damaged patients.

Understanding the nature and organization of conceptual knowledge in the brain requires the integration of a cognitive theoretical account of conceptual representations with a theory of the neural substrate for the processing of objects. Such a framework should provide the basis for an explanation of conceptual representation, the nature of impairments following damage to the conceptual system, and how processing concepts map onto specific neural circuits. Within this general context we have developed the conceptual structure account (CSA; Tyler & Moss, 2001; Tyler, Moss, Durrant-Peatfield, & Levy, 2000), a theoretical cognitive account of the structure of concepts and categories. A major focus has been on the dissociation commonly observed between the domains of living and nonliving things within which the typical pattern is for a disproportionate deficit for living things, or for a

Correspondence should be addressed to Peter Bright, Department of Experimental Psychology, University of Cambridge, Cambridge CB2 3EB, UK. Email: pbright@csl.psychol.cam.ac.uk

The research was supported by an MRC (UK) programme grant and Newton Trust (Cambridge University, UK) grant to L.K.T. We thank Olivia Longe for her help with the correlational studies, and Andrew Mayes and Roy Jones for referring patients to us.

particular category of living things, such as animals or fruit (e.g., Caramazza & Shelton, 1998; De Renzi & Lucchelli, 1994; Hillis & Caramazza, 1991; Moss, Tyler, & Jennings, 1997; Warrington & Shallice, 1984), although the opposite pattern has occasionally been described (Hillis & Caramazza, 1991; Sacchett & Humphreys, 1992). In this review we describe a series of investigations in which we have tested key predictions of the CSA with respect to recent claims about the role of the perirhinal cortex in object processing. With this approach we have built on our earlier research by incorporating neuropsychological and neuroimaging evidence concerning the neural correlates of object processing at different levels of specificity.

The primary claim of the CSA is that concepts associated with particular categories (e.g., animals, tools) differ in the number and type of properties and the extent to which these properties are correlated with each other. Within this framework we explain the emergence of selective semantic deficits in terms of differences in the internal structure of concepts in different domains of knowledge. Concepts within living-things categories, such as animals and fruits, tend to be similar in their internal structure. They are characterized by many overlapping and intercorrelated shared features (e.g., animals tend to have legs, eyes, teeth), but relatively few distinctive, distinguishing features (e.g., mane, hump, pouch; Durrant-Peatfield, Tyler, Moss, & Levy, 1997; Gaffan & Heywood, 1993; Greer et al., 2001; Humphreys, Riddoch, & Quinlan, 1988; Randall, Moss, Rodd, Greer, & Tyler, 2004; Tranel, Damasio, & Damasio, 1997; Tyler & Moss, 2001; Tyler et al., 2000). Moreover, the distinctive features of living things tend to be only weakly correlated with one another (for example, there is little correlation between having a mane, being fierce, having a pouch, and so on). In contrast, nonliving concepts (such as tools and vehicles) tend to be less visually similar, with a relatively greater proportion of distinctive to shared properties, and with a greater degree of correlation among distinctive features (e.g., the specific form and function of an artifact are often strongly correlated; Greer et al., 2001; Moss, Tyler, & Devlin, 2002; Tyler et al., 2000). In other words, nonliving concepts are less confusable. Such differences among conceptual domains are supported by an analysis of our property norms (Tyler et al., 2000). We have suggested that it is these differences in the internal structure of concepts within different categories that underpin the emergence of category-specific semantic deficits (Moss, Tyler, Durrant-Peatfield, & Bunn, 1998; Tyler & Moss, 2001; Tyler et al., 2000). As weakly correlated, distinctive properties are more susceptible to damage than are intercorrelated, shared properties (which are mutually reinforcing by their coactivation), this will result in a disproportionate deficit for living things (Tyler et al., 2000), although this may critically depend upon the task. For example, if the task requires knowledge of properties of concepts shared with other category members (rather than their distinctive properties), we would predict similar performance for living and nonliving things—a pattern that we have reported for several patients with category-specific impairments for living things (Moss et al., 1998; Moss et al., 2002; Tyler & Moss, 2001). It is only when the task requires access to the distinctive, distinguishing properties of objects that a living-things deficit will emerge, due to a greater degree of similarity and the weak correlations among distinctive features within living-things categories. Thus, in contrast to theories that explain category-specific deficits in terms of damage to anatomically distinct, content-specific stores (e.g., Caramazza & Shelton, 1998; Shelton & Caramazza, 1999), the CSA provides a framework in which deficits may occur due to damage to a unitary, distributed semantic system.

In this review we present findings from a series of studies integrating the central claims of the CSA with recent theoretical accounts concerning the neural substrates of object processing in the brain.

Although hierarchical organization within the visual system has been a fundamental theme in the study of primate visual cortex, evidence has only recently been brought to bear on whether such a hierarchy of object-processing stages extends throughout the ventral occipital and temporal cortex. For example, recent studies in nonhuman primates have suggested that different stages along the anterior-posterior axis of the ventral processing stream code different types of visual feature, leading from simple combinations of features in posterior regions to complex feature configurations in more anterior and medial temporal cortices (e.g., Buckley, this issue; Bussey & Saksida, 2002; Bussey, Saksida, & Murray, this issue; Hampton, this issue; Lerner, Hendler, Ben-Bashat, Harel, & Malach, 2001; Murray & Bussey, 1999; Murray & Richmond, 2001).

A hierarchical theory of object processing in humans has also been proposed by Damasio (1989) and later developed by Simmons and Barsalou (2003). The claim is that, when processing an object, the visual features associated with it (colour, shape, etc.) are bound together in convergence zones (CZs). In the latter model CZs are arranged hierarchically, such that those in posterior regions code simple combinations of features, and those in more anterior regions bind features into increasingly complex conjunctions of features. Such a structure provides a basis for processing objects at different levels of specificity, with posterior CZs representing more coarse-grained configurations of features (e.g., colour or shape or motion), converging on anteromedial CZs, which enable the coding of more complex feature conjunctions underlying object identification (e.g., colour and shape and motion). These CZs, rather than storing conjunctions of features, function by pointing back or indexing those sensorimotor areas where the features are stored. In this way the CZs represent the co-occurrence of features necessary for the retrieval of a holistic representation.

Evidence that the perirhinal cortex may function as the endpoint of a hierarchically organized object-processing network providing the basis for fine-grained discrimination among objects is supported by a number of lesion studies in monkeys and other nonhuman species (e.g., Buckley, this issue; Buckley, Booth, Rolls, & Gaffan, 2001; Buckley & Gaffan, 1998; Bussey, Muir, & Aggleton, 1999; Eacott, Gaffan, & Murray, 1994; Gilbert & Kesner, 2003; Sato & Nakamura, 2003; Zola-Morgan, Squire, Amaral, & Suzuki, 1989). Although the evidence in humans is more controversial (e.g., Aggleton & Shaw, 1996; Reed & Squire, 1997), it strongly supports the premise that perirhinal cortex damage, on its own or in combination with other medial temporal lobe structures, will produce a severe object-recognition impairment (Holdstock, this issue). However, the nature of this object-processing deficit is unclear with respect to what aspect of object processing is compromised by anteromedial temporal damage. Studies in lesioned animals have suggested that anteromedial temporal cortices (and the perirhinal cortex in particular) may function to enable the identification of a particular configuration of features specific to an object to facilitate object identification and discrimination. For example, Bussey, Saksida, and Murray (2002, this issue), demonstrated that monkeys with perirhinal cortex lesions showed unimpaired visual discrimination among complex object pairs in conditions where there were no ambiguous features, moderately impaired where 50% of features were explicitly ambiguous, and severely impaired where all features were ambiguous. Complementary evidence has also been reported by Buckley and colleagues in which macaques with perirhinal ablation were severely impaired on perceptual tasks requiring discrimination among complex combinations of object features, but not on tasks requiring discrimination of single or less complex features (Buckley, this issue; Buckley et al., 2001; Buckley & Gaffan, 1998). However, human studies have produced inconsistent results, with Barense et al. (Barense, 2005; Lee, Barense, & Graham, this issue) providing evidence that novel object discriminations are influenced by feature ambiguity in amnesic patients with perirhinal damage, but Holdstock, Gutnikov, Gaffan, and Mayes (2000; Holdstock, this issue) suggesting that the perirhinal cortex does not have a specific role in the representation of complex visual stimuli (although their data are based on patients with residual preservation of this area).

Given its reciprocal connections with different sensory cortical areas, it has been claimed that, in monkeys, the perirhinal cortex involvement in object processing is not restricted to visual analysis, but is also critically involved in linking diverse aspects of information about objects across modalities (Murray & Richmond, 2001). A similar role for the perirhinal cortex in the human object-processing system has recently been proposed by Simmons and Barsalou (2003), in which, by nature of its reciprocal connections with inferior temporal cortex and other polymodal brain areas, it is able to integrate information about an object across modalities.

We have argued that one implication of a hierarchically organized object-processing system is that the nature of cortical involvement may not be an invariant process, but is determined, at least in part, by the level or type of processing demands required rather than being driven purely by the visual stimulus itself (Moss, Rodd, Stamatakis, Bright, & Tyler, 2005; Tyler et al., 2004). Evidence consistent with process-driven object processing in anterior temporal cortex has been provided by Grabowski et al. (2001), in which retrieval of proper names of very different complex visual stimuli (persons and landmarks) engaged the same region of the left temporal pole to a similar degree, which these authors suggested was consistent with this area reflecting the level of specificity of retrieval rather than the specific properties of the stimuli. A process-driven view of anteromedial temporal cortex is also consistent with evidence from the memory literature that these regions are involved either in the time-limited consolidation of newly acquired memory representations prior to their storage elsewhere (e.g., Alvarez & Squire, 1994) or else act as indexers or pointers to information held in other regions of the brain (e.g., Nadel & Moscovitch, 1997). Recent evidence has emerged that other regions of the inferior temporal cortex may reflect the nature of visual object-processing demands rather than specific stimulus attributes. Gauthier and colleagues, for example, describe a number of studies suggesting that the "fusiform face area", rather than being a face-specific processing area (Kanwisher, McDermott, & Chun, 1997; Kanwisher, Stanley, & Harris, 1999), may be associated with a large variety of objects, but will be more active under conditions requiring classification at a subordinate level rather than at a more general, categorical level (Gauthier, Curran, Curby, & Collins, 2003; Gauthier, Skudlarski, Gore, & Anderson, 2000; Gauthier, Tarr, Anderson, Skudlarski, & Gore, 1999; Tarr & Gauthier, 2000). Together, these findings are most consistent with a nonmodular object-processing system in which neuroanatomical differentiation associated with the processing of different types of object reflects the level of processing demands necessary to identify or discriminate among objects rather than signalling stimulus or domain-specific recruitment.

The hierarchical object-processing model, in combination with the CSA, provides a framework for explaining the frequent association of category-specific deficits for living things and damage to anterior ventromedial temporal cortex (including perirhinal and entorhinal cortices, parahippocampal gyrus, and hippocampus), usually as a consequence of herpes simplex encephalitis (HSE). Within the CSA framework we have argued that, as living things are more similar (i.e., they have more shared features—they have legs, eyes, etc.) than artefacts, they are therefore more difficult to differentiate. This places greater demands on those processes that are involved in the integration of complex conjunctions of features and fine-grained discrimination. On the assumption that anteromedial temporal cortex is the most plausible candidate region for integration and differentiation, we predicted that fine-grained discrimination among living things, relative to nonliving things, will place greater demands on anterior temporal cortical areas (Moss et al., 2005).

We have tested these claims in a series of fMRI studies with healthy participants and behavioural studies with patients, arguing that, if the same kind of hierarchical object-processing system suggested by primate studies also operates in humans, we would expect differential activation of ventral temporal cortex as a function of the level or type of object processing required. Specifically, to the extent that fine-grained discrimination among similar objects is required, we would predict anteromedial temporal cortex involvement. However, where the task does not require the extraction of such detailed information about an object, involvement would be primarily restricted to more posterior object-processing areas of the temporal cortex. Moreover, these differential patterns will be modulated by the internal properties of concepts, with living-things concepts generating a relatively greater involvement of anteromedial temporal cortex than that for nonliving things, but only when fine-grained discrimination among objects is required (Moss et al., 2005; Tyler et al., 2004).

In a first study, we employed fMRI to explore the neural correlates of object processing, by presenting healthy participants with pictures of common objects and asking them to name the objects either at the domain level (i.e., whether the object was living or man-made) or at a basic level (to name the object itself, for example, "dog", "lion", "table"). We argued that naming at a basic level requires fine-grained differentiation among objects in order that, for example, a picture can be differentiated from similar objects—thus, a lion can be differentiated from a tiger or a leopard (Tyler et al., 2004). On the assumption that antero-medial temporal cortex is critically involved in processes of fine-grained discrimination, we predicted greater anteromedial temporal cortex involvement when participants named the picture at a basic level than when they named it at a domain level. The activation of complex conjunctions of features is not necessary for domain-level naming, which can be achieved on the basis of relatively coarse-grained information (there is no requirement for discriminating among objects within a category); thus we predicted that this condition would only activate intermediate object-processing areas and not anteromedial temporal cortex.

In the study, participants silently named the same set of coloured photos of common objects at the basic and domain level (in two separate sessions). We included baseline events (a fixation cross) in each session. The results were unambiguous (see Figure 1, panel A). At the conventional fMRI statistical threshold of .001 (uncorrected for multiple comparisons), basic-level naming produced predominantly left-sided temporal activation, which included the entire extent of the fusiform gyrus (BA 19/37/36), extending into anteromedial temporal

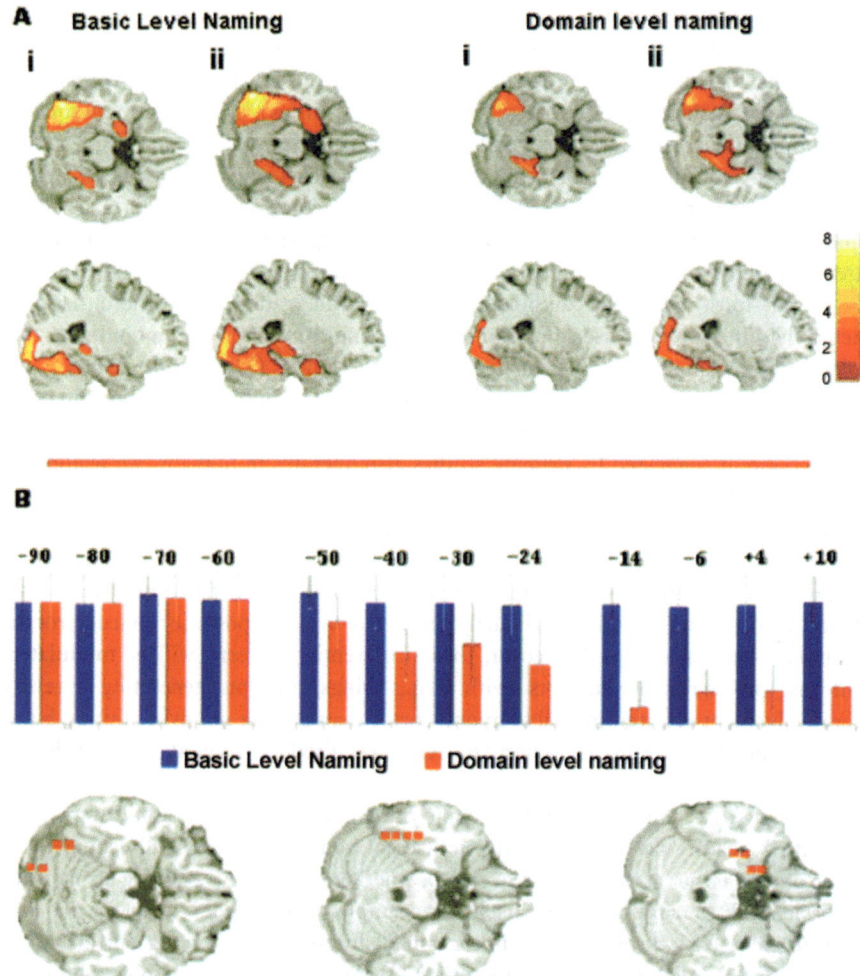

Figure 1. (A) Significant activations for the contrast of basic-level naming and domain-level naming minus baseline are shown superimposed on a T1 anatomical image transformed into the standard stereotactic space of Talairach and Tournoux (1988). The colour bars indicate strength of activation (voxel level T values). The areas shown survived *p* < .05 correction for multiple comparisons at the cluster level and were thresholded at (i) .001 (with small volume correction [SVC] for the left temporal cortex), and (ii) .01. We use neurological convention where left = left. (B) Plots of signal change (with standard error bars) for basic- and domain-level naming relative to baseline at 12 points along the posterior to anterior extent of the left inferior occipital and temporal cortices (values shown in arbitrary units). The approximate location of each point is shown superimposed on axial slices of a normalized brain. Figures above each plot indicate the position on the *y* (coronal) axis in MNI space.

cortex (including entorhinal cortex, BA 28). We also investigated the pattern of activation at a more liberal threshold (.01), which, although increasing the likelihood of false positives, provides information regarding the spatial profile of activation below the standard statistical cut-off. At this threshold, activation extended medially to involve perirhinal cortex, amygdala and hippocampus. Domain-level naming produced activations restricted to the same posterior temporal regions (BA 19, 37), but there was no activation of anterior temporal cortex even when the lower statistical threshold was applied. A direct comparison of basic-against domain-level naming produced significantly more activation in entorhinal and perirhinal cortices (Tyler et al., 2004).

These results show that anteromedial temporal cortex was significantly involved in object processing only when the information-processing demands required relatively fine-grained discriminations among objects. Thus, although participants saw exactly the same stimuli in domain- and basic-level naming conditions, anteromedial involvement appeared to be driven by the processing demands of the task, rather than by the stimulus itself (Figure 1, panel B). We argued on the basis of these data that the specific processes associated with this aspect of object processing relate to the activation of complex configurations of features of the stimulus necessary for basic-level naming. As domain-level naming does not require processing at this level of specificity, activation was restricted to more posterior regions of the inferior temporal cortex (Tyler et al., 2004).

This account also helps to explain the nature of some of the cognitive impairments observed in patients with HSE and semantic dementia (SD). Typically, these patients are unable to consistently name objects at a basic level (e.g., to name a picture of a cat as a cat), but can identify the category to which the object belongs (e.g., "animal"). A hierarchical processing model combined with the CSA suggests that damage at different points along the object-processing pathway should affect a patient's ability to process an object at particular levels of conceptual specificity. Where a patient's damage predominantly involves anteromedial temporal cortex, we predicted a deficit in fine-grained discrimination among objects (involved in basic-level naming), but preserved coarse-grained discriminations (involved in category or domain-level naming). The extent to which lower level discriminations are affected would relate to the extent to which the lesion extends into more posterior areas.

To test this hypothesis, we examined picture-naming performance in four HSE patients, all of whom have extensive left hemisphere anterior temporal cortex damage, but with relatively preserved posterior left inferior temporal cortex. In other words, the same areas found to be active during domain-level naming in healthy adults were largely preserved, while those additional areas active during basic-level naming were damaged in all patients (Tyler et al., 2004). In all four cases, when asked to identify the pictures, a large proportion of responses were either category coordinate errors (e.g., to name a picture of a cat as a dog), or superordinate errors (e.g., to name a cat as an animal), consistent with a particular difficulty in making relatively fine-grained discriminations among objects. On another picture-naming task, directly comparable with that used in the fMRI study, patients were severely impaired on basic-level naming (with a similar profile of error types as found in the first naming study), but unimpaired on domain-level naming (requiring a living thing or man-made response). The results of the fMRI and neuropsychological data together suggest that it may be the left anteromedial regions (including perirhinal and entorhinal cortices), routinely damaged in HSE, that underlie these patients failure to correctly identify objects, and that preservation of more coarse-grained, domain-level naming may be attributed to the relatively unaffected posterior areas of the inferior temporal cortices.

The CSA framework with its claims about variation in the internal structure of concepts enables us to make the further claim for differentiation of activation within anteromedial temporal cortex as a function of the structure of concepts. Living things, with their high degree of within-category similarity, should place relatively greater demands on those object-processing regions—anteromedial temporal cortex—associated with fine-grained discrimination. Thus, basic-level naming for living things should place greater demands on anteromedial cortex (including entorhinal and perirhinal cortices) than should artefacts since they are less confusable. In contrast, domain-level naming, which does not require access to fine-grained detail but relies on shared information, should not produce differential activation for concepts within different categories of knowledge, and activation for both should be confined to more posterior temporal regions.

We tested these claims in a second fMRI event-related study in which healthy participants silently named pictures of common objects from the categories of animals, fruits and vegetables, tools, and vehicles (see Moss et al., 2005, for further details). We also included baseline events (a fixation cross) in each session. As in our earlier study, participants named the objects at a basic level in one block and the same objects at a domain level in a second block (counterbalanced). Living and nonliving trials were randomly presented. The results are shown in Figure 2 (panel A).

The key results are the comparisons of the two visually complex categories (animals and vehicles) and the two visually simple categories (fruit/vegetables and tools). These pairs of categories were also matched on concept agreement, exemplarity, familiarity, and age of acquisition of the basic-level name. The analysis focused on the anteromedial temporal cortex region, which was activated in our earlier study (Tyler et al., 2004) and which included the left perirhinal and entorhinal cortices. As predicted, there was significantly more activation during the basic-level naming of animals relative to vehicles and for fruits and vegetables relative to tools. In both cases the peak was found in the region of the entorhinal cortex (BA 28/34). For completeness, we also compared the nonmatched living and nonliving categories (animals vs. tools, and fruits and vegetables vs. vehicles). Once again, and in both cases, we found significant living over nonliving activations in the entorhinal/dorsal entorhinal cortex (BA 28/34) during basic-level naming. Importantly, there were no differences in any of the comparisons for domain-level naming. Thus, only during basic-level naming was activation in anteromedial temporal cortex modulated as a function of the category from which the objects were drawn (Moss et al., 2005).

These findings clearly suggest an important role for anteromedial temporal cortex in visual object processing at high levels of conceptual specificity. Consistent with the CSA, different objects appear to place greater or lesser demands on these processes according to variations in their internal structure. The greater similarity of objects within living-things categories renders them less distinguishable from each other, and therefore they require the involvement of anteromedial temporal cortex in addition to more posterior regions, which are associated with the processing of objects from both living and nonliving categories.

These results also provide a framework within which to account for patterns of semantic deficits seen in various types of brain-damaged patients. Patients with HSE, associated primarily with ventromedial temporal cortical damage, typically show disproportionate deficits for living things over artefacts. In contrast, semantic dementia patients commonly have anterolateral temporal cortical damage, but not usually the extensive anteromedial temporal lesions associated with HSE, although the typical lateral-medial spread of disease remains unclear (Lambon Ralph, Patterson, Garrard, & Hodges, 2003; Levy, Bayley, & Squire,

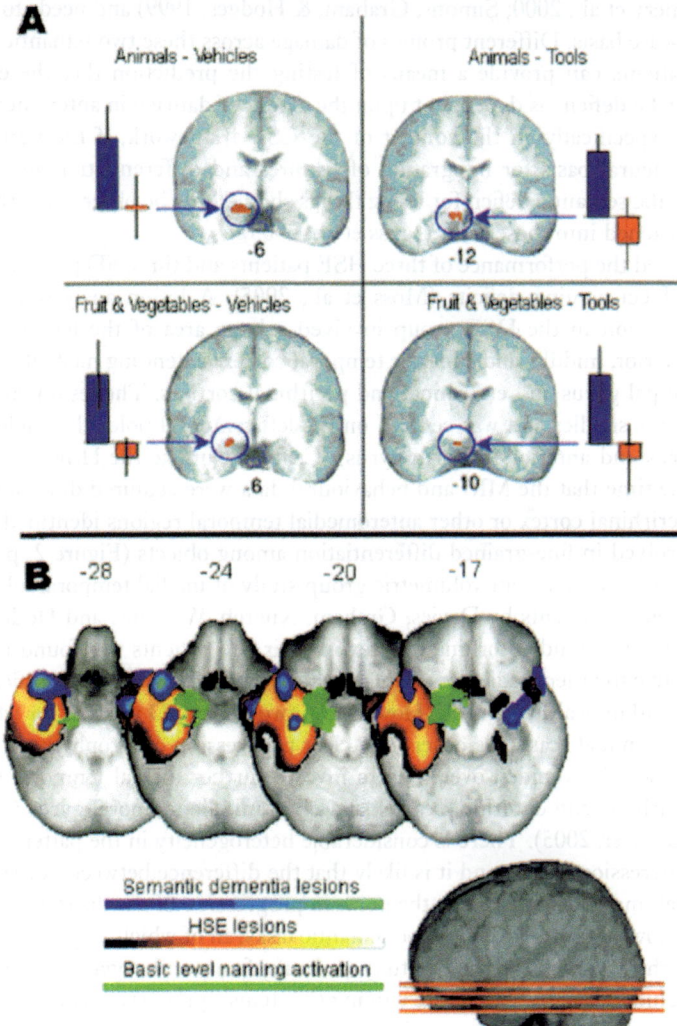

Figure 2. (A) Comparisons of basic-level naming contrasts are shown superimposed on a T1 anatomical image transformed into the standard stereotactic space of Talairach and Tournoux (1988). Animals versus vehicles resulted in a cluster in the parahippocampal area BA 28, also extending anteriorly to BA 34. Animals versus tools produced a cluster in BA 28. Fruit and vegetables versus vehicles resulted in a cluster in BA 34, and finally fruit and vegetables versus tools resulted in a cluster in BA 28. Talairach y coordinates are shown underneath each anatomical slice where left = left. Slices were selected on the basis of maximum cluster extent in this plane. Plots of signal change (with standard error bars) for the relevant contrasts are shown next to each slice (living things categories in blue; artefact categories in red). (B) Common lesions for the HSE and semantic dementia groups plus overlap of the basic-level naming activations for the living minus nonliving categories (shown in solid green) superimposed on the mean patient T1 scan. The lesions were detected by comparing each patient to a control group and conducting conjunction analyses in the two groups (conjunction T scores are shown in the colour bars). The activations overlap shows those areas where two or more of the four comparisons shows an overlap (animals vs. vehicles, animals vs. tools, fruit and vegetables vs. vehicles, and fruit and vegetables vs. tools) in the basic-level naming task, superimposed on the mean patient brain.

2004; Mummery et al., 2000; Simons, Graham, & Hodges, 1999) and needs to be evaluated on a case-by-case basis. Different profiles of damage across these two semantically impaired groups of patients can provide a means of testing the prediction that the emergence of category-specific deficits is dependent upon the extent of damage in anteromedial temporal cortex. More specifically, in the context of the CSA framework, if the perirhinal cortex provides the neural basis for integration of features and differentiation among objects, a category-specific semantic deficit for living things should only be observed to the extent that damage encroached into these areas (Moss et al., 2005).

We compared the performance of three HSE patients and three SD patients, all of whom had severe object-naming deficits (Moss et al., 2005). A lesion analysis confirmed that the common lesion in the HSE group involved a large area of the left temporal cortex (including inferior, middle, and superior temporal cortex) extending medially to encompass parahippocampal gyrus and entorhinal and perirhinal cortices. The lesion common to the SD patients was smaller and was centred on the left temporal pole (also including middle temporal gyrus and anterior fusiform gyrus). Critically, unlike the HSE patients, the SD patients at the time that the MRI and behavioural data were acquired did not have damage within the perirhinal cortex or other anteromedial temporal regions identified in the fMRI studies as involved in fine-grained differentiation among objects (Figure 2, panel B). This finding contrasts with a recent volumetric group study of medial temporal lobe volumes in SD and Alzheimer patients by Davies, Graham, Xuereb, Williams, and Hodges (2004), in which the perirhinal (and entorhinal) cortex in their SD patients was found to be severely damaged relative to other medial temporal structures and with volumes significantly smaller than those found in healthy controls. However, in two of our SD patients for which we were able to acquire annual scans, we observed a clear progression of atrophy both posteriorally and medially, which extended over time to involve further medial temporal lobe regions, including perirhinal and entorhinal cortex as well as middle temporal gyrus (Bright, Moss, Stamatakis, & Tyler, 2005). There is considerable heterogeneity in the pattern and progress of disease progression in SD, and it is likely that the difference between our study and that of Davies et al. may relate in part to the point in progression of the disorder.

Figure 3 shows performance on three semantic tasks, all of which employed the same four categories as those used in the fMRI study (animals, fruits and vegetables, tools, vehicles). The tasks included basic-level picture naming (72 items), property verification from verbal input (80 items), and category fluency. There is a clear difference in the pattern of responses made by the HSE and SD patients. On naming and property verification all HSE patients performed significantly worse ($p < .01$) on living things than on nonliving things. In contrast, SD patients did not show this pattern; they were equally poor on living and nonliving things. On category fluency, the HSE patients also showed a disproportionate advantage for nonliving things, generating almost twice as many exemplars than for living things. In contrast, there was no evidence for a selective deficit for living things in the SD patients, despite an equivalent overall semantic impairment in both groups (in fact, these patients showed a numerical advantage for living things). The fact that the HSE patients have a disproportionate deficit for living things in tasks requiring fine-grained differentiation and have anteromedial damage is consistent with the fMRI data showing significantly greater activation in the same region during the fine-grained discrimination of living things relative to artefacts.

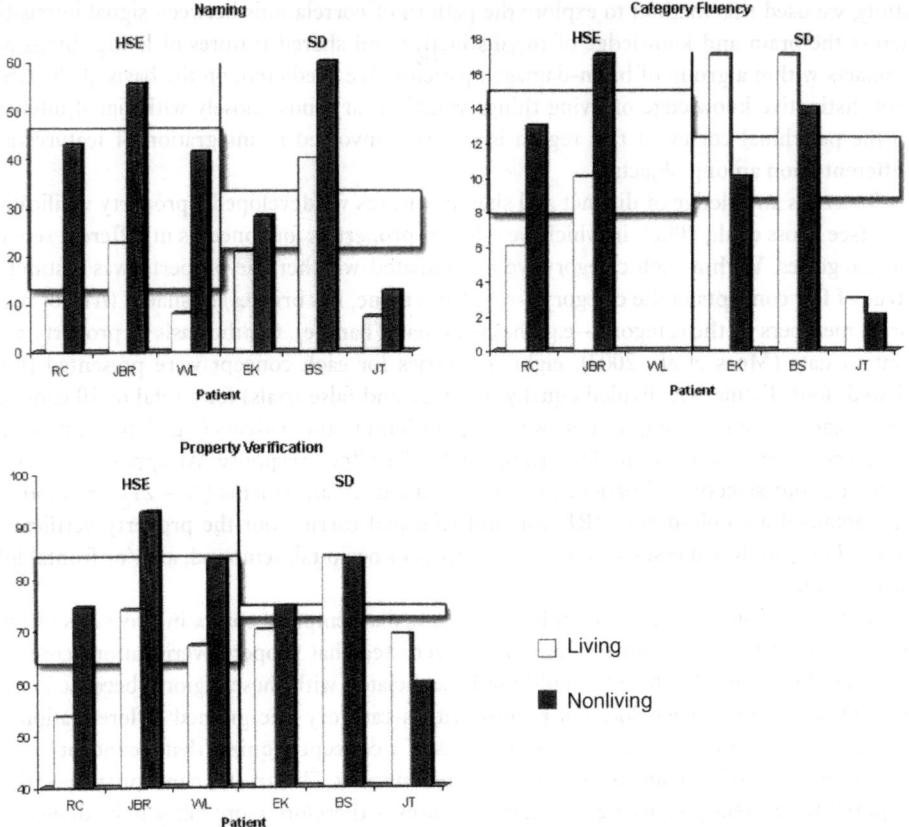

Figure 3. Evidence from clinical cases. Graphs show the overall accuracy of HSE and SD patients in three semantic tasks. In the naming task, patients named pictures at a basic level (e.g., lion, hammer). Scores shown as percentage correct. The property verification task provided a measure of conceptual knowledge from verbal input (e.g., "do cats have fur?", "Do forks have prongs?"), equally divided between true and false trials. Scores shown as percentage correct. For category fluency, each patient was asked to list as many animals, fruit, tools, and vehicles as they could, with 1 minute allowed for each category (data unavailable for patient WL). Scores for category fluency are actual number of correct responses. The horizontal lines indicate the average scores for living and nonliving categories in the two groups.

We have built on these findings in a recent study in which we employ a lesion–behaviour correlational approach in order to map the sensitivity of brain tissue integrity to perform-ance among different conditions of a semantic task. Specifically, we related lesion extent as determined by structural MRI data to variations in patients' performance. This method uses two continuous measures—signal intensity values for each voxel in the brain (a measure of tissue integrity) and continuous scores on a behavioural test, thus avoiding binarizing either the structural data as damaged or healthy, or the behavioural results as impaired or preserved (for a detailed description of the methodology, see Tyler, Marslen-Wilson, & Stamatakis, 2005). An advantage of this approach is that no a priori assumptions are made about either patient aetiology (e.g., HSE, stroke) or the functions of specific brain regions. In the current study, we used this method to explore the pattern of correlations between signal intensities across the brain and knowledge of the distinctive and shared features of living things and artefacts within a group of brain-damaged patients. We predicted, on the basis of the CSA, that distinctive knowledge of living things would covary most closely with signal intensity in the perirhinal cortex, if this region is critically involved in integration of features and differentiation among objects.

To assess knowledge of distinct and shared features we developed a property verification task (see Moss et al., 2002), in which we selected properties for concepts in different semantic categories. Within each category we manipulated whether the property was distinctive (true of few concepts in the category—e.g., has a mane, has bristles) or shared (true of many or all members of the category—e.g., has eyes, has a handle). On the basis of property generation data (Moss et al., 2002), eight properties for each concept were presented (four, shared, four distinctive, divided equally into true and false trials) for a total of 10 concepts per category. Each of the questions was read out loud to the patients (e.g., "does a cow have whiskers?"), who were required to produce a "yes" or "no" response. We applied no specific criteria in our selection of patients, but simply included all patients ($N = 21$) for whom we had obtained a whole-brain MRI scan and who had carried out the property verification study. These included cases with varying degrees of occipital, temporal, and/or frontal lobe involvement.

In line with our claims on the role of anteromedial temporal cortex in processes of fine-grained differentiation amongst objects, we predicted that property verification scores for the shared features of concepts would not be associated with these regions, because knowledge of shared properties does not require within-category fine-grained differentiation. In contrast, verification of the distinctive features of a concept requires that a concept be distinguished from other concepts within the same category. The greater conceptual specificity required for correctly verifying distinctive features is therefore more likely to be dependent on processes of fine-grained discrimination. Moreover, the greater within-category similarity (i.e., a lower proportion of distinctive to shared properties) among animals than tools is likely to render these concepts particularly susceptible to damage in anteromedial regions.

For the analysis we correlated shared and distinctive property verification scores (percentage correct) for animals and tools in each voxel across all the patient scans (with global mean signal for each scan included as a confounding covariate). The critical comparison is between the pattern of brain–behaviour correlations for the distinctive properties of animals and tools. In the context of the CSA framework, we would predict that, because animals with their high degree of within-category similarity place greater demands on processes of fine-grained discrimination, property verification for their distinctive features will be particularly sensitive to the integrity of the anteromedial temporal cortex. In contrast, tools, which are less similar to each other (and therefore easier to differentiate), would be less associated with damage to anteromedial temporal regions.

The results showed that the distinctive properties of animals correlated significantly (taking an uncorrected statistic of $p = .001$) with signal intensity across a large region of anterior temporal cortex. This extended from left anterior fusiform gyrus (BA 20), middle temporal gyrus (BA 21), and temporal pole (BA 38), medially to include parahippocampal and perirhinal cortices (BA 36, 35). No other significant effects were observed, either for the shared properties of animals or the distinctive/shared properties of tools. To explore the pattern of activation below this statistical cut-off, we reduced the threshold ($p = .01$). This produced signal–behaviour correlations for distinctive properties of animals across a larger extent of left anterior inferior temporal cortex, including more anterior temporal pole, perirhinal cortex, parahippocampal gyrus (BA 36), and the hippocampus.

Figure 4 (top panel) shows areas of behaviour–signal covariation in the four conditions at the lower threshold. A more extensive region of left anterior temporal cortex was significantly correlated with the performance on distinctive properties in both animals and tools, relative to shared properties. Critically, however, for tools the regions associated with knowledge of distinctive properties did not correlate with signal intensity in anteromedial temporal cortex, as predicted on the basis that these regions are engaged when more

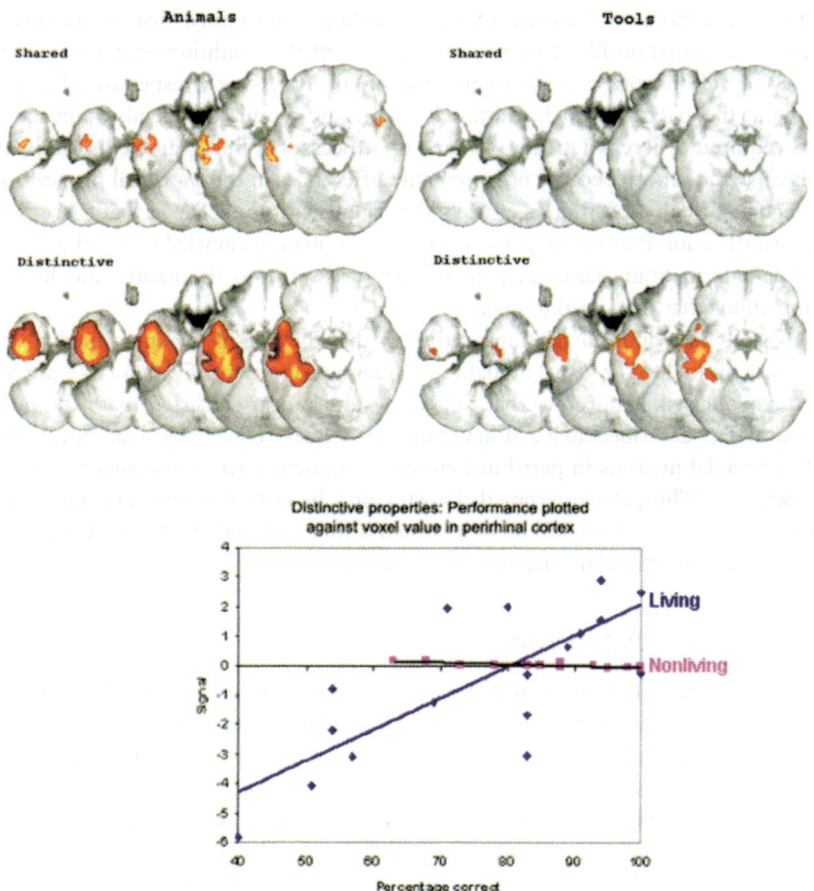

Figure 4. Areas that correlated with knowledge for the shared and distinctive properties of living things in a group of 21 brain-damaged patients thresholded at .01 uncorrected for multiple comparisons. The plot shows voxel values from every patient scan in the perirhinal cortex (at −20 −4 −29, after correction for global mean differences and scaling) plotted against property verification scores for the distinctive properties of living and nonliving things (each modelled separately).

fine-grained discriminations are required. Instead, significant correlations were associated with fusiform gyrus (BA 20/37) and middle temporal gyrus (BA 21), in areas partly overlapping with those found for the verification of shared properties of animals.

In Figure 4 (bottom panel) performance on distinctive properties is shown for each patient plotted against voxel values in the perirhinal cortex (BA 35, peak signal: -20 -4 -29 in Montreal Neurological Institute [MNI] space). As can be seen, there was greater variability in signal and a strong, positive correlation with the distinctive properties of animals (higher signal intensity = better performance). In contrast, there was relatively little variability in signal and no evidence for a positive correlation with the distinctive properties of tools.

These findings are consistent with the idea that the left anteromedial temporal cortex (and the perirhinal cortex in particular) provides the neural basis for the integration of complex conjunctions of features, thereby enabling fine-grained discriminations among concepts. The spatial profile of correlations in each of the conditions suggest that much of the same neural network is involved in conceptual processing irrespective of category or domain, but the extent to which additional regions are recruited depends upon the level of analysis required. There is a greater demand on processes of fine-grained discrimination for animals relative to tools, because of systematic differences in the statistical properties of the two categories. Such a position is entirely consistent with the CSA in that there is no true neural specialization for one category or domain of semantic knowledge, but that disproportionate deficits in brain-damaged patients may arise due to the nature and level of the discriminations that they need to make.

The signal–behaviour correlations also show that the role of the anteromedial temporal cortex is not limited to the visual modality. Although property verification tasks can be based on pictures, our task is purely verbal, yet the results appear consistent with the fMRI and behavioural data described above. A key claim of the hierarchical object-processing account is that polymodal neurons in perirhinal cortex integrate information across a range of sensory modalities. Thus, the anteromedial cortex may be critical to semantic processing and not just to visual object recognition, as it supports the combination of many types of feature into meaningful multimodal conceptual representations.

Summary and conclusions

The results of these studies provide important insights into the way in which information about objects may be processed within the brain. In combining theoretical cognitive accounts of the representation and processing of objects with models of the neural infrastructure that supports these processes, we have extended the CSA, incorporating differences in regional recruitment during object processing as a function of the nature of the discriminations required. We claim that a disproportionate deficit for the identification of living things over artefacts is critically related to the level of damage in anteromedial temporal cortex, but only when fine-grained representations are required. The likelihood or degree of cortical involvement in conceptual processing will be shaped by the nature of the category from which an object is drawn and the level of representation required. Consistent with a process-driven involvement of anteromedial temporal cortex in object processing, objects do not invariably recruit this region: In tasks that do not require fine-grained discrimination, there are no differences in anteromedial activation between living and nonliving things and no category-specific deficits in patients with anteromedial temporal lobe damage.

The fMRI data are consistent with a unified conceptual system, because anteromedial temporal cortex is differentially activated according to the statistical properties of a concept rather than by object category or domain per se. Thus, both living things and artefacts may be associated with anteromedial temporal cortex during basic-level naming, but the higher degree of within-category similarity for living things will place greater demands on the processes of fine-grained discrimination associated with this region.

This notion of "graded" or "disproportionate" effects also applies to the deficits observed in patients with semantic deficits: We rarely, if ever, see an all-or-nothing deficit for a specific category or domain—the pattern is one of impairment across semantic categories, with milder deficits for some categories over others. For example, in the Moss et al. (2005) study, the HSE and SD patients, on average, were able to identify fewer than half the pictures of common artefacts (see Figure 3, left panel), even though performance in the HSE patients was disproportionately poorer for naming living things. Thus, we argue against true specificity both in terms of the category-specific deficits observed in semantically impaired patients and in terms of the neural correlates of object processing in the healthy brain. In both cases (patient performance and fMRI findings), category-specific effects arise due to the differences in content and structure of concepts in those categories.

In our view, neuroimaging studies of healthy individuals undertaken in parallel with behavioural investigations of patients with clear, anatomically defined lesions can provide a powerful basis for building upon claims made in the animal literature. This approach has enabled us to extend our cognitive theory concerning the structure of semantic memory, by incorporating evidence from hierarchical models of object processing in nonhuman primates (Bussey et al., this issue) and testing its implications for conceptual processing in the human brain.

REFERENCES

Aggleton, J. P., & Shaw, C. (1996). Amnesia and recognition memory: A reanalysis of psychometric data. *Neuropsychologia, 34,* 51–62.

Alvarez, P., & Squire, L. R. (1994). Memory consolidation and the medial temporal lobe: A simple network model. *Procedings of the National Academy of Sciences (USA), 91,* 7041–7045.

Barense, M. D., Bussey, T. J., Lee, A. C. H., Rogers, T. T., Hodges, J. R., Saksida, L. M., et al. (2005). Feature ambiguity influences performance on novel object discriminations in patients with damage to perirhinal cortex. *Journal of Cognitive Neuroscience* (supplement), D117.

Bright, P., Moss, H. E., Stamatakis, E. A., & Tyler, L. K. (in press). The time course of semantic dementia: MRI and behavioural findings. *Journal of Cognitive Neuroscience* (supplement), B237.

Buckley, M. J. (this issue). The role of the perirhinal cortex and tippocampus in learning, memory, and perception. *Quarterly Journal of Experimental Psychology, 58B,* 246–268.

Buckley, M. J., Booth, M. C. A., Rolls, E. T., & Gaffan, D. (2001). Selective perceptual impairments after perirhinal cortex ablation. *Journal of Neuroscience, 21* 9824–9836.

Buckley, M. J., & Gaffan, D. (1998) Perirhinal cortex ablation impairs visual object identification. *Journal of Neuroscience, 18,* 2268–2275.

Bussey, T. J., Muir, J. L., & Aggleton, J. P. (1999). Functionally dissociating aspects of event memory: The effects of combined perirhinal and postrhinal cortex lesions on object and place memory in the rat. *Journal of Neuroscience, 19,* 495–502.

Bussey, T. J., & Saksida, L. M. (2002). The organization of visual object representations: A connectionist model of effects of lesions in perirhinal cortex. *European Journal of Neuroscience, 15*, 355–364.

Bussey, T. J., Saksida, L. M., & Murray, E. A. (2002). Perirhinal cortex resolves feature ambiguity in complex visual discriminations. *European Journal of Neuroscience, 15*, 365–374.

Bussey, T. J., Saksida, L. M., & Murray, E. A. (this issue). The perceptual-mnemonic/feature conjunction model of perirhinal cortex function. *Quarterly Journal of Experimental Psychology, 58B*, 269–282.

Caramazza, A., & Shelton, J. R. (1998). Domain-specific knowledge systems in the brain: The animate–inanimate distinction. *Journal of Cognitive Neuroscience, 10*, 1–34.

Damasio, A. R. (1989). The brain binds entities and events by multiregional activation from convergence zones. *Neural Computation, 1*, 123–132.

Davies, R. R., Graham, K. S., Xuereb, J. H., Williams, G. B., & Hodges, J. R. (2004). The human perirhinal cortex and semantic memory. *European Journal of Neuroscience, 20*, 2441–2446.

De Renzi, E., & Lucchelli, F. (1994). Are semantic systems separately represented in the brain? The case of living category impairment. *Cortex, 30*, 3–25.

Durrant-Peatfield, M ., Tyler, L. K., Moss, H. E., & Levy, J. (1997). The distinctiveness of form and function in category structure: A connectionist model. In M.G. Shafto & P. Langley (Eds.), *Proceedings of the Nineteenth Annual Conference of the Cognitive Science Society* (pp. 193–198). Mahwah, NJ: Lawrence Erlbaum, Associates, Inc.

Eacott, M. J., Gaffan, D., & Murray, E. A. (1994). Preserved recognition memory for small sets, and impaired stimulus identification for large sets, following rhinal cortex ablations in monkeys. *European Journal of Neuroscience, 6*, 1466–1478.

Gaffan, D., & Heywood, C. A. (1993). A spurious category-specific visual agnosia for living things in normal human and nonhuman primates. *Journal of Cognitive Neuroscience, 5*, 118–128.

Gauthier, I., Curran, T., Curby, K. M., & Collins, D. (2003). Perceptual interference evidence for a non-modular account of face processing. *Nature Neuroscience, 6*, 428–432.

Gauthier, I., Skudlarski, P., Gore, J. C., & Anderson, A. W. (2000). Expertise for cars and birds recruits brain areas involved in face recognition. *Nature Neuroscience, 3*, 191–197.

Gauthier, I., Tarr, M. J., Anderson A. W., Skudlarski, P., & Gore, J. C. (1999). Activation of the middle fusiform 'face area' increases with expertise recognizing novel objects. *Nature Neuroscience, 2*, 568–573.

Gilbert, P. E., & Kesner, R. P. (2003). Recognition memory for complex visual discriminations is affected by stimulus interference in rodents with perirhinal cortex damage. *Learning and Memory, 10*, 525–530.

Grabowski, T. J., Damasio, H., Tranel, D., Ponto, L. L., Hichwa, R. D., & Damasio, A. R. (2001). A role for left temporal pole in the retrieval of words for unique entities. *Human Brain Mapping, 13*, 199–212.

Greer, M., van Casteren, M., McClellan, S., Moss, H. E., Rodd, J., Rogers, T. et al. (2001). The emergence of semantic categories from distributed featural representations. In J. D. Moore & K. Stenning (Eds.), *Proceedings of the 23rd Annual Conference of the Cognitive Science Society* (pp. 358–363). London: Lawrence Erlbaum Associates, Inc.

Hillis, A. E., & Caramazza, A. (1991). Category-specific naming and comprehension impairment: A double dissociation. *Brain & Language, 114*, 2081–2094.

Hampton, R. R. (this issue). Monkey perirhinal cortex is critical for visual memory, but not for visual perception: Reexamination of the behavioural evidence from monkeys. *Quarterly Journal of Experimental Psychology, 58B*, 283–299.

Holdstock, J. S. (this issue). The role of the human medial temporal labe in recognition and object discrimination. *Quarterly Journal of Experimental Psychology, 58B*, 326–339.

Holdstock, J. S., Gutnikov, S. A., Gaffan, D., & Mayes, A. R. (2000). Perceptual and mnemonic matching-to-sample in humans: Contributions of hippocampus, perirhinal and other medial temporal lobe cortices. *Cortex, 36*, 301–322.

Humphreys, G. W., Riddoch, M. J., & Quinlan, P. (1988). Cascade processes in picture identification. *Cognitive Neuropsychology, 5*, 67–103.

Kanwisher, N., McDermott, J., & Chun, M. (1997). The fusiform face area: A module in human extrastriate cortex specialized for the perception of faces. *Journal of Neuroscience, 17*, 4302–4311.

Kanwisher, N., Stanley, D., & Harris, A. (1999). The fusiform face area is selective for faces not animals. *Neuroreport, 10*, 183–187 (1999).

Lambon Ralph, M. A., Patterson, K., Garrard, P., & Hodges, J. R. (2003). Semantic dementia with category specificity: A comparative case-series study. *Cognitive Neuropsychology, 20*, 307–326.

Lee, A. C. H., Barense, M. D., & Graham, K. S. (this issue). The contribution of the human medial temporal lobe to perception: Bridging the gap between animal and human studies. *Quarterly Journal of Experimental Psychology, 58B*, 300–325.

Lerner, Y., Hendler, T., Ben-Bashat, D., Harel, M., & Malach, R. (2001). A hierarchical axis of object processing stages in the human visual cortex. *Cerebral Cortex, 11*, 287–297.

Levy, D. A., Bayley, P. J., & Squire, L. R. (2004). The anatomy of semantic knowledge: Medial vs. lateral temporal lobe. *Proceedings of the National Academy of Sciences, (USA), 101*, 6710–6715.

Moss, H. E., Rodd, J. M., Stamatakis, E. A., Bright, P., & Tyler, L. K. (2005). Anteromedial temporal cortex supports fine-grained differentiation among objects. *Cerebral Cortex, 15*, 616–627.

Moss, H. E., Tyler, L. K., & Devlin, J. (2002). The emergence of category specific deficits in a distributed semantic system. In E. Forde & G. W. Humphreys (Eds.), *Category-specificity in brain and mind* (pp. 115–148). Hove, UK: Psychology Press.

Moss, H. E., Tyler, L. K., Durrant-Peatfield, M., & Bunn, E. (1998). Two eyes of a see-through: Impaired and intact semantic knowledge in a case of selective deficit for living things. *Neurocase, 4*, 291–310.

Moss, H. E., Tyler, L. K., & Jennings, F. (1997). When leopards lose their spots: Knowledge of visual properties in category-specific deficits for living things. *Cognitive Neuropsychology, 14*, 511–547.

Mummery, C. J., Patterson, K., Price, C. J., Ashburner, J., Frackowiak, R. S. J., & Hodges, J. R. (2000). A voxel based morphometry study of semantic dementia: The relationship between temporal lobe atrophy and semantic memory. *Annals of Neurology, 47*, 36–45.

Murray, E. A., & Bussey, T. J. (1999). Perceptual-mnemonic functions of the perirhinal cortex. *Trends in Cognitive Sciences, 3*, 142–151.

Murray, E. A., & Richmond, B. J. (2001). The role of perirhinal cortex in object perception, memory and associations. *Current Opinion in Neurobiology, 11*, 188–193.

Nadel, L., & Moscovitch, M. (1997). Memory consolidation, retrograde amnesia and the hippocampal complex. *Current Opinion in Neurobiology, 7*, 217–227.

Randall, B., Moss, H. E., Rodd, J., Greer, M., & Tyler, L. K. (2004). Distinctiveness and correlation in conceptual structure: Behavioural and computational studies. *Journal of Experimental Psychology: Language, Memory and Cognition, 30*, 393–406.

Reed, J. M., & Squire, L. R. (1997). Impaired recognition memory in patients with lesions limited to the hippocampal formation. *Behavioral Neuroscience, 111*, 667–675.

Sacchett, C., & Humphreys, G. W. (1992). Calling a squirrel a squirrel but a canoe a wigwam: A category-specific deficit for artifactual objects and body parts. *Cognitive Neuropsychology, 9*, 73–86.

Sato, N., & Nakamura, K. (2003). Visual response properties of neurons in parahippocampal cortex of monkeys. *Journal of Neurophysiology, 90*, 876–886.

Shelton, J. R., & Caramazza, A. (1999). Deficits in lexical and semantic processing: Implications for models of normal language. *Psychonomic Bulletin & Review, 6*, 5–27.

Simmons, K., & Barsalou, L. W. (2003). The similarity-in-topography principle: Reconciling theories of conceptual deficits. *Cognitive Neuropsychology, 3–6*, 451–486.

Simons, J. S., Graham, K. S., & Hodges, J. R. (1999). What does semantic dementia reveal about the functional role of the perirhinal cortex? *Trends in Cognitive Sciences, 3*, 248–249.

Talairach, J., & Tournoux, P. (1988). *A co-planar stereotactic atlas of the human brain.* Stuttgart: Thieme Verlag.

Tarr, M. J., & Gauthier, I., (2000). FFA: A flexible fusiform area for subordinate-level visual processing automatized by expertise. *Nature Neuroscience, 3*, 764–769.

Tranel, D., Damasio, H., & Damasio, A. (1997). A neural basis for the retrieval of conceptual knowledge. *Neuropsychologia, 35*, 1319–1327.

Tyler, L. K., Marslen-Wilson, W. D., & Stamatakis, E. A. (2005). Dissociating neurocognitive component processes: Voxel based correlational methodology. *Neuropsychologia, 43*, 771–778.

Tyler, L. K., & Moss, H. E. (2001). Towards a distributed account of conceptual knowledge. *Trends in Cognitive Science, 5*, 244–252.

Tyler, L. K., Moss, H. E., Durrant-Peatfield, M., & Levy, J. (2000). Conceptual structure and the structure of categories: A distributed account of category-specific deficits. *Brain & Language, 75*, 195–231.

Tyler, L. K., Stamatakis, E. A., Bright, P., Acres, K., Abdallah, S., Rodd, J. M., et al. (2004). Processing objects at different levels of specificity. *Journal of Cognitive Neuroscience, 16*, 351–362.

Warrington, E. K., & Shallice, T. (1984). Category specific semantic impairments. *Brain, 107*, 829–854.

Zola-Morgan, S., Squire, L. R., Amaral, D. G. & Suzuki, W. A. (1989). Lesions of perirhinal and parahippocampal cortex that spare the amygdala and hippocampal formation produce severe memory impairment. *Journal of Neuroscience, 9*, 4355–4370.

THE QUARTERLY JOURNAL OF EXPERIMENTAL PSYCHOLOGY
2005, 58B (3/4), 378–396

Perirhinal cortex and its neighbours in the medial temporal lobe: Contributions to memory and perception

Elisabeth A. Murray

National Institute of Mental Health, Bethesda, MD, USA

Kim S. Graham

MRC Cognition and Brain Sciences Unit, Cambridge, UK

David Gaffan

Oxford University, UK

As promised in the Introduction, this Special Issue presents several recurring themes concerning the perirhinal cortex and its neighbours within the medial temporal lobe (MTL). First, although orthodoxy insists that the diverse constituents of the MTL operate as a single functional entity, several papers presented here challenge that idea, although some defend it. Second, although many experts hold that the MTL subserves memory but not perception, several papers presented here point to a role for certain MTL structures in both. Third, although some researchers have invoked "species differences" to account for discrepant findings, several papers presented here document a striking convergence of findings in humans, nonhuman primates, and rodents. We close this Special Issue by high-lighting these recurring themes, acknowledging discrepant findings and pointing to future research that might resolve some current controversies.

Does the MTL contain multiple functional subdivisions?

All the articles in this Special Issue discuss, directly or indirectly, the idea of functional specialization for the processing of object and spatial information by different parts of the MTL (D. Gaffan, 1994a). Converging evidence from many experimental techniques indicates

Correspondence should be addressed to Dr Elisabeth A. Murray, Laboratory of Neuropsychology, National Institute of Mental Health, 49 Convent Drive, Building 49, Room 1B80, Bethesda, MD 20892-4415, USA. Email: murraye@mail.nih.gov

DOI:10.1080/02724990544000077

dissociations of function between different MTL structures, with perirhinal cortex and hippocampus being important for object and spatial processing, respectively. The perirhinal cortex has been proposed to mediate *object identification*: the knowledge that a particular object or class of objects is one and the same across the different instances in which it is experienced (Buckley & Gaffan, 1998c; Murray, Bussey, Hampton, & Saksida, 2000). Consistent with this idea, lesion studies in rats and monkeys suggest that perirhinal cortex is especially dedicated to object processing. Eacott and E. Gaffan (this issue) review lesion studies in rats using paradigms involving either visual discrimination for food reward or measures of spontaneous recognition memory. The two paradigms yield similar patterns of results. In one of their studies, rats were trained to make visual discriminations in an automated apparatus. One stimulus was a "constant negative"—that is, the stimulus was the same across trials, and it was never rewarded. The other stimulus differed from the constant negative either in its object properties (but not location on a video monitor screen) or in its location on the screen (but not object properties). Approach towards the variable stimulus was the correct (rewarded) response. Tests of object processing (e.g., trials in which the correct stimulus varied from the negative only in object properties) were found to depend on the perirhinal cortex but not the postrhinal cortex. Conversely, tests of egocentric locations of objects (e.g., trials in which the correct stimulus varied from the negative only in location on the screen) depended on the postrhinal cortex but not the perirhinal cortex (E.A. Gaffan, Healey, & Eacott, 2004). Thus, perirhinal cortex appears specialized for object processing whereas postrhinal cortex appears specialized for some aspect of egocentric spatial processing. Because in this task the stimuli available for choice are always present, these data also suggest that the perirhinal and postrhinal cortex are responsible for representing some aspects of object and egocentric spatial information, respectively. We return to this topic later in Section 2.

Articles by Bussey, Saksida, and Murray (this issue) and by Buckley (this issue) review findings from lesion studies in monkeys that employ tests placing large demands on object processing. As was the case for the data discussed by Eacott and E. Gaffan (2005 this issue), these authors, too, find that perirhinal cortex is essential for learning about the properties of objects—that is, for object identification. In earlier work, Bussey and his colleagues (Bussey & Saksida, 2002; Murray & Bussey, 1999) offered the specific proposal that the role of perirhinal cortex in object processing might be explained by a consideration of the hierarchical organization of receptive fields in the ventral visual stream, or object-analyser pathway. Anatomical and physiological evidence indicates that neurons in caudal visual fields of the ventral visual stream represent simple visual features, whereas neurons in more rostral fields represent conjunctions of features. If so, then monkeys with lesions of perirhinal cortex, which is located at the rostral extremity of the ventral visual stream, should have difficulty representing conjunctions of visual features. To test this idea, Bussey et al. (this issue) designed visual discrimination problems that contained varying degrees of feature ambiguity, a property of visual discrimination problems that emerges when a given feature (or set of features) appears in both rewarded (S+) and unrewarded (S−) objects. For example, in one experiment, objects were constructed by combining complex greyscale pictures to yield compound stimuli AB+, CD+, AD−, BC−, where each capital letter represents a picture, and the + and − represent correct (rewarded) and incorrect (unrewarded), respectively, in the context of a discrimination problem. The underlying logic is that visual discriminations possessing feature ambiguity cannot be solved on the basis of any single feature alone, because any particular feature may appear in either the S+ or the S−. Instead, their solution depends on the ability to represent the conjunctions of features. Thus, the prediction is that monkeys with perirhinal cortex lesions would be impaired in acquiring visual discriminations under conditions of feature ambiguity but not under conditions of no (or low) feature ambiguity. Bussey et al. (this issue) review evidence that monkeys with perirhinal cortex lesions are indeed selectively impaired in the acquisition and performance of visual discriminations possessing high levels of feature ambiguity. Furthermore, this result is also obtained when feature ambiguity is achieved using a different method. Not only was feature ambiguity manipulated by constructing objects, as already described, but it was also manipulated by blending images using commercially available software. Morphed (or blended) stimuli share more features, and therefore possess greater feature ambiguity, than the original images from which they were derived. The foregoing data are consistent with the idea that perirhinal cortex is specialized for processing object information. As for other regions within the MTL, preliminary data suggest that monkeys with selective hippocampal lesions perform as well as controls (Saksida, Bussey, Buckmaster, & Murray, 2005) or even better (Saksida, Bussey, Buckmaster, & Murray, 2003) under conditions of high feature ambiguity on these and related object discrimination tasks.

The findings reviewed by Buckley (this issue) also provide compelling evidence for a role for perirhinal cortex in object identification. One line of evidence involves oddity tasks in which monkeys are rewarded for selecting and touching the odd one of several images on a touch-sensitive monitor screen (Buckley, Booth, Rolls, & Gaffan, 2001). There were two main types of oddity judgement: discrimination of the one object that differed from the others along a single dimension (e.g., colour, size) and discrimination of one different object from among several different views of the same object. Monkeys with perirhinal cortex lesions, like controls, were able to select the odd object on the basis of colour alone, shape alone, or size alone, even when the task was made perceptually difficult. These authors suggest that the monkeys can make the colour, size, or shape oddity judgements on the basis of simple features, without taking into account the entire object. By contrast, when monkeys were required to make oddity judgements by selecting the different item from among a set of identical objects or faces seen from different views, monkeys with perirhinal cortex lesions were impaired relative to controls.

Lee, Barense, and Graham (this issue) tested patients with static lesions of the MTL using some of the same tests of object processing as those used by Buckley et al. (2001) in monkeys. In addition, they extended their investigation to include a spatial version of oddity, in which subjects were required to choose the one different scene from among three different views of the same scene. Patients with damage restricted to the hippocampal formation were found to be impaired on tests of scene oddity, but not on tests of face or object oddity. By contrast, patients with MTL lesions that included the hippocampus as well as the neighbouring entorhinal and perirhinal cortex were found to be impaired on all three versions of the oddity task—scene, faces, and objects. As predicted, neither group was impaired on colour oddity, which had been included as a control task. These data converge with the findings from nonhuman primates in supporting a role for perirhinal cortex in object processing. In addition, however, they highlight a selective role for the hippocampus in spatial processing and point to functional dissociations within the MTL.

Physiological methods yield results consistent with the findings from lesion studies. Rolls et al. (Hölscher, Rolls, & Xiang, 2003; Rolls, Franco, & Stringer, this issue), using unit-recording techniques in monkeys, show that the activity of perirhinal cortex neurons in monkeys reflects long-term familiarity of visual images. Although there are now several mnemonic correlates that have been reported in the activity of perirhinal cortex neurons, long-term familiarity, unlike some other correlates, is reflected by an increase in the overall level of neuronal firing to familiar relative to novel pictures. As familiarity builds across repeated exposures, both within and across days, neurons in perirhinal cortex tend to become more and more active during the period of the stimulus presentation. How this new physiological correlate relates to other types of neuronal correlates described for the perirhinal cortex is at present unknown, but it nevertheless provides strong evidence for perirhinal involvement in object processing.

Aggleton and Brown (this issue) describe results from another physiological measure—namely, the activation of the immediate early gene *c-fos*, following presentation to rats of novel and familiar objects. Expression of *c-fos* is thought to provide an index of neuronal activity for the roughly 30-minute time period of a test session; although the temporal resolution of the technique is poor, the anatomical resolution is good. In this case, there is evidence not only in favour of perirhinal cortex involvement in the processing of objects, but also against the involvement of the hippocampus in the same processes. These authors review compelling evidence for a double-dissociation of function of the perirhinal cortex and the hippocampus. The experimental design involved rats passively viewing "objects" (two-dimensional images) presented on a monitor screen. Due to crossed projections in the visual system, novel images could be presented to one hemisphere while at the same time familiar images were presented to the other hemisphere. The critical comparison was the amount of *c-fos* activation in a given structure in the hemisphere that had viewed novel objects relative to the same region in the hemisphere that had viewed familiar objects. Exposure to novel relative to familiar images resulted in activation of perirhinal cortex and the adjacent visual cortex area TE but not the hippocampus. By contrast, when the same passive viewing design was used but a spatial manipulation was introduced, the converse pattern of results was obtained. In the spatial task, rats viewed familiar objects in familiar spatial arrangements (one hemisphere) or familiar objects in novel spatial arrangements (other hemisphere). Under these conditions significant *c-fos* changes were found in several parts of the hippocampus and in the postrhinal cortex but not in the perirhinal cortex or TE. This striking pattern of results confirms that the perirhinal cortex and hippocampus make selective contributions to memory, with the former engaged in object processing and the latter engaged in spatial processing.

Bright, Moss, Stamatakis, and Tyler (this issue) discuss processing of object information in humans. These authors review several lines of evidence obtained from either fMRI studies of healthy human subjects or studies of patients with damage to the ventromedial temporal cortex following herpes simplex encephalitis (HSE) infection. In some experiments, subjects were required to silently name objects at either a category level (e.g., living or manmade) or an item level (e.g., lion, table). Importantly, the actual stimulus material presented to the subjects was identical in the two conditions, and this was true for both the fMRI studies and the neuropsychological studies. The fMRI studies revealed that the item naming condition led to activation of more rostrally located regions of anteromedial temporal cortex than did the category naming condition. Furthermore, the more rostral regions—those involved exclusively in the item level naming condition—overlapped significantly with the regions of brain damage in HSE patients, who were found to be impaired in item level but not category level naming. These data are consistent with the idea that anteromedial temporal cortex provides the neural basis for representing the complex conjunctions of features, thereby enabling representation, discrimination, and identification of objects.

Two contributions (Henson, this issue, Holdstock, this issue) offer some evidence at odds with the idea that there are functional subdivisions within the MTL. Henson reviews evidence from event-related fMRI studies in humans with an eye towards any specializations that might be evident for item recognition. If some structures in the MTL are more important for object processing than others, then one might expect this to be reflected in fMRI studies of item recognition memory. Indeed, in monkeys, the "match-suppression effect"— thought to be a neural correlate of stimulus recognition—has been found in perirhinal cortex, entorhinal cortex, and area TE of inferior temporal cortex (Brown, Wilson, & Riches, 1987; Miller & Desimone, 1994) but not in hippocampus. Henson considers several factors, including site of activations (e.g., hippocampus vs. extrahippocampal MTL; left vs. right hemisphere), type of stimulus material, whether the activations were related to successful encoding or to successful recognition of items, and task instructions (e.g., forced choice vs. yes/no), among other things. Although MTL structures showed memory-related differences in haemodynamic activity, overall, there was no clear evidence in support of functional dissociations within the MTL. Specifically, no clear patterns of activation emerged that might suggest a special role in recognition memory for hippocampus versus surrounding medial temporal cortex, or of anterior versus posterior MTL, or of left versus right hemisphere. Henson did, however, note some trends in the data. One of these was for a potentially greater role for hippocampus and parahippocampal cortex in encoding and retrieving source information and associations between distinct items, and for a role for anterior MTL, in the region of perirhinal cortex, for encoding item information. Another trend was for greater activation in the MTL with scenes relative to words, which might reflect the greater novelty, complexity, or spatial content of scenes. Finally, a third trend was for a decreased activation of medial temporal cortex for old relative to new items, a finding that is consistent with the aforementioned unit-recording studies in monkeys. On the basis of these trends, he suggests that future fMRI studies should employ a range of stimulus materials, including nonverbal materials and materials varying in degree of familiarity, and the use of different types of source information; in addition, where possible, the MTL should be scanned at a greater spatial resolution. In summary, although trends in the pattern of findings from human fMRI hint at a division of labour within the MTL, at present there is no compelling evidence from human fMRI for such a division, at least not for stimulus recognition.

Holdstock reviews findings from a series of neuropsychological studies carried out in two patients with damage thought to be limited to the hippocampus and in a group of amnesic patients of mixed etiology. In support of the idea of functional divisions in the MTL, she reports that the patients with hippocampal damage have recognition scores in the normal range, but recall scores that are significantly impaired relative to controls. This same pattern of spared recognition but impaired recall has been reported by Vargha Khadem et al. (Vargha-Khadem, Gadian, & Mishkin, 2001) in humans with early hippocampal damage. By contrast, associative tests of recognition (i.e., tests of stimulus pairs) yield a slightly different pattern. Here, although within–modality pairs (e.g., word pairs, face pairs) seem to be spared after hippocampal damage, across–modality pairs (e.g., object–location pairs, face–voice pairs) are not. Hence, only some aspects of stimulus recognition may be spared after hippocampal damage.

Does the MTL contribute to perception?

Many of the contributions to this Special Issue emphasize a role for perirhinal cortex in visual perception as well as memory. In particular, the reviews by Eacott and E. Gaffan (this issue), Bussey et al. (this issue), and Buckley (this issue) describe studies in which rats or monkeys with selective removals of the perirhinal cortex were given tasks specifically designed to tax visual perception. The findings from all three sets of studies led the authors to much the same conclusion—namely, that perirhinal cortex is important for representing the conjunctions of features that comprise an object. Taken together, the findings strongly suggest that the perirhinal cortex is important not only for stimulus memory, for which the evidence is indisputable, but also for representing the complex conjunctions of features that allow us to discriminate even highly similar objects (e.g., faces) from one another. On this view the perirhinal cortex is held to participate in both perception and memory; it is important for *perception* in that it houses the machinery enabling the representation of complex conjunctions of features; it subserves *memory* in that it serves as a site of storage of stimulus representations in both the short and the long term.

Until recently, a central tenet of cognitive neuroscience was that the MTL mediates long-term information storage by a process of consolidation. On this view, the MTL operates in a time-limited manner such that the final repository of the long-term memories lies outside the MTL, in the neocortical processing areas that were initially involved in the processing. A corollary of this time-limited role for MTL in memory is that the MTL is not important for perceptual processing, but only for memory. For the visual system, in particular, this has led to the specific proposal that the perirhinal cortex is not involved in perception (Buffalo, Reber, & Squire, 1998; Sakai & Miyashita, 1993). At least four articles in this Special Issue argue for a role for perirhinal cortex in visual perception, as well as in memory. Some of this evidence will be reviewed below, together with the evidence and arguments against this view.

In the studies reported by Eacott and E. Gaffan (this issue), discussed earlier, it was found that rats with perirhinal cortex lesions were impaired despite the fact that the correct image and the comparison image (the "constant negative") were both present at the time of choice. These authors therefore argued that the perirhinal cortex is responsible for the "representation" of object information. Such a role would necessarily mean that perirhinal cortex is contributing to perception. In a similar vein, Buckley (this issue) noted that the oddity tasks used in monkeys possess, on each trial, all the information required to make the oddity judgement. There is no requirement to hold information over a delay period and, hence, no overt memory component. Thus, the impairment after perirhinal cortex lesions on the oddity tasks strongly suggests a visual perceptual impairment rather than a memory impairment.

As Hampton (this issue) notes, measures of perception are often confounded with a requirement for new learning. He therefore argues that evaluation of the hypothesis that perirhinal cortex makes a critical contribution to perception should make use of tests that measure perceptual abilities independent of learning and memory. In one study, Buckley and Gaffan (1998a) used digitized images of objects in a series of object discrimination problems presented to monkeys on a touch screen. Six different views of each object, derived from digital photographs of the objects taken from six different vantage points, were prepared. Monkeys were trained on a set of visual discrimination problems using three of the six views of objects. After learning, the monkeys were evaluated for transfer to the same discrimination problems using the new views of the same objects. Buckley and Gaffan found that monkeys with perirhinal cortex lesions were impaired relative to controls in reattaining criterion on the discrimination problems with new views. As Hampton notes, the size of the deficit in monkeys with perirhinal cortex lesions, estimated as a ratio of the number of errors to criterion made by operated monkeys relative to controls, appears to be the same for transfer as it was in initial learning. Hampton therefore suggests that deficits observed in monkeys with perirhinal cortex lesions in transferring to new views of objects could arise from a learning impairment. Buckley (this issue) counters by showing that on the first trial with each object after transfer, controls scored the same as they had previously whereas monkeys with perirhinal cortex lesions made many more errors. These new analyses therefore argue against Hampton's idea that the deficit in this transfer task after perirhinal cortex damage is due to an impairment in learning. Hampton argues that the same possibility needs to be addressed for the many other instances where measures of perception and learning are confounded. For example, in both the oddity design (Buckley et al., 2001) and the discrimination learning design (Bussey, Saksida, & Murray, 2002), deficits after perirhinal cortex ablation could in principle be due to learning rather than to a deficit in perceptual processing per se. While this is a valid point, it remains difficult to see why, in the oddity paradigm for example, an impairment in memory for the learned oddity rule should manifest itself with complex stimuli and not with simple stimuli of equal perceptual difficulty, except as a result of a perceptual impairment.

The studies reviewed by Lee et al. (this issue) also speak to this issue. To assess the role of MTL in perception, this group used a version of discrimination learning that provided the subjects with the correct exemplar (original S+) on each trial. Perceptual difficulty was manipulated using the morphing procedure, a method also used by Bussey et al. (this issue) as discussed in the previous section. After only three discrimination learning trials the subjects were required to choose the one of two morphed images that most closely resembled the sample (S+). Under these conditions, Lee et al. (2005) found that patients with damage limited to the hippocampus were impaired in discriminating scenes, whereas patients with MTL lesions were impaired in discriminating faces and, to a lesser extent, objects, in addition to scenes. Because all three stimuli (the sample S+ and two choice stimuli) were present on the monitor screen at the time of choice, and the patients were nevertheless impaired, these data strongly suggest a role for perirhinal cortex in perception. They also reveal the surprising finding that a similar deficit in the simultaneous condition holds for hippocampal contributions to spatial processing. Thus both the "morph" and "oddity" paradigms, in addition to providing support for the idea of functional subdivisions within the MTL, suggest a role for the perirhinal cortex and hippocampus in object and spatial perception, respectively.

An additional criticism of the perceptual account of perirhinal cortex function has been articulated by Squire and his colleagues. For example, Buffalo et al. (1998) argued that because patients with MTL damage that includes perirhinal cortex are unimpaired on delayed nonmatching-to-sample (DNMS) with short delays between sample presentation and choice test, perception must be intact. Bussey et al. (this issue) counter by noting that before one can conclude that perception is intact, one must tax perception in some way. Indeed, Eacott et al. (Eacott, Gaffan, & Murray, 1994) reported that monkeys with perirhinal and entorhinal cortex ablations perform well on delayed matching-to-sample (DMS) at short delays. It was only when perception was taxed by making the stimuli more difficult to discriminate that performance was brought down from ceiling. Under these conditions, the same operated monkeys were impaired on simultaneous and 0-s delay versions of matching-to-sample.

A more compelling argument against a role for perirhinal cortex in perception comes from the use of probe tests. After monkeys have learned discrimination problems, their perception can be taxed by giving probe tests in which the stimuli presented for choice are shrunken, enlarged, or degraded, among other manipulations. Because the performance is evaluated in single probes, this is thought to provide an estimate of perceptual abilities independent of learning. Hampton (this issue) notes that when monkeys are given perceptually demanding tasks using such probe tests, and perception is challenged to a degree that reduces performance of intact monkeys, monkeys with perirhinal cortex lesions can still perform as well as intact monkeys (Hampton & Murray, 2002). Bussey et al. (this issue) respond by explaining that their perceptual-mnemonic/feature-conjunction model predicts impairments only when perception is taxed in a specific way—namely, in conditions of high feature ambiguity. Because Hampton and Murray (2002) did not manipulate feature ambiguity, no impairment would be predicted. On this view, the lack of impairment after perirhinal cortex lesions cited by Hampton is not in conflict with the idea that perirhinal cortex represents conjunctions of features.

Two published studies directly contradict some of the data presented in the Special Issue. Stark and Squire (2000) used the oddity tasks developed by Buckley et al. (2001) in humans. They found that patients with MTL lesions that included perirhinal cortex performed just as well as did controls. These findings are in apparent conflict with the findings from monkeys with perirhinal cortex lesions trained on the same tasks (Buckley et al., 2001) and disagree with the results reported by Lee et al. (this issue). While there is no obvious way in which these results can be reconciled, Lee et al. note two significant differences between the studies: (a) differences in the number of trials given to the subjects (greater in Lee et al. than in Stark and Squire) and (b) notably different scores of the control groups in the two studies.

Recently, Levy et al. (Levy, Shrager, & Squire, 2005) reported results from humans with MTL lesions that had been tested using the same feature ambiguity manipulations as those used by Bussey et al. (Bussey, Saksida, & Murray, 2003) in monkeys and similar to those used by Lee et al. (2005) in humans. As was the case for the oddity tasks, this group found that patients with MTL lesions that included perirhinal cortex were unimpaired in tests of object processing; the tests included same–different judgements, simultaneous matching-to-sample, and discrimination learning under conditions of high feature ambiguity. The authors conclude, therefore, that the perirhinal cortex is not essential for perception. Given that the "simultaneous" version of discrimination used by Lee et al. (2005) converges on simultaneous matching-to-sample, the lack of concordance between the two investigations is surprising. Unfortunately, in the study by Levy et al., unlike the investigation of Lee et al., the scores obtained by the control subjects, even in the most difficult conditions, were very close to ceiling. A more compelling negative result would come from a study in which perception was taxed in such a way as to bring down control scores from ceiling.

How does perception interact with memory?

A separate issue is whether a perceptual impairment might account in part or whole for the deficits observed after MTL damage that are traditionally interpreted as mnemonic. Although this was proposed for perirhinal cortex in the context of DNMS (Murray & Bussey, 1999), it might now also be asked in terms of the hippocampus and spatial memory. As we have seen, some of the evidence most often cited in favour of the mnemonic account of perirhinal cortex function are the findings of delay-dependent deficits in recognition memory after damage that includes perirhinal cortex (e.g., Eacott et al., 1994; Meunier, Bachevalier, Mishkin, & Murray, 1993, in monkeys; Buffalo et al., 1998; Holdstock, Gutnikov, Gaffan, & Mayes, 2000, in humans). As argued by Hampton (this issue), this finding suggests a role for perirhinal cortex in memory, independent of perception. Interestingly, a reanalysis of data from studies of recognition memory in monkeys carried out by Ringo (1991) argues against the idea of a delay-dependent recognition memory deficit. Using published data from a large number of studies that were carried out in several different laboratories, Ringo converted scores originally provided as percent correct (obtained from monkeys trained on the DNMS task) to d-prime, thereby allowing direct comparison of recognition loss across experiments and across delay conditions. There were two main findings. First, the overall magnitude of the impairment differed according to the type of lesion. Second, he found that the magnitude of the recognition impairment remained stable across all delays investigated. Thus, rather than revealing a delay-dependent memory deficit, Ringo's reanalysis indicated that lesions within the MTL led to a stable memory loss that was apparent even at short delays. Although the studies included in the reanalysis examined the effects of TE lesions, fornix transection, amygdala lesions, hippocampal lesions, and combined amygdalo-hippocampal lesions, they did not address the effects of lesions limited to perirhinal cortex, as these studies had not yet been carried out at the time Ringo's study was published. More recently, however, exactly the same pattern of findings has been reported after reanalysis of studies involving lesions restricted to perirhinal cortex and hippocampus (Baxter & Murray, 2001). These data thus raise doubts about the idea that the MTL structures contribute to memory, independent of perception. Although they support the notion that perirhinal cortex plays a role in perception, whether a perceptual account can replace a mnemonic account remains to be seen.

What are the boundaries among the concepts of object, context, and place?

In discussing the neural basis for object, context, and spatial processing, the question arises as to the boundaries for these constructs. There is overwhelming evidence that the brain treats things with regularly co-occurring elements as a special class: objects. As we have seen, whether examined with lesions, unit recording methods, fMRI, or other methods such as c-fos, it is clear that object processing involves the inferior temporal cortex, including the perirhinal cortex. For example, in monkeys, neurons in inferior temporal cortex are active in relation to object features (Desimone & Ungerleider, 1989; Logothetis & Scheinberg, 1996; Tanaka, 1997). Furthermore, perirhinal cortex shows neuronal correlates of object recognition (Brown et al., 1987), object–object associations (Baker, Behrmann, & Olson, 2002; Sakai & Miyashita, 1991), and long-term familiarity for objects (Holscher et al., 2003). Lesion studies in nonhuman primates show that perirhinal cortex is necessary for these same types of object memory: recognition memory (Buffalo, Ramus, Squire, & Zola, 2000; Meunier et al., 1993), object–object association memory (Buckley & Gaffan, 1998b; Murray, Gaffan, & Mishkin, 1993), and memory for retention of object discriminations (Thornton, Rothblat, & Murray, 1997) and object–object associations (Higuchi & Miyashita, 1996) over several weeks.

Although "objects", "contexts", and "places" may have regularly co-occurring elements, and there may in fact be several dimensions that distinguish between stimuli that we typically refer to as "objects" and "contexts", it may nevertheless be useful to consider work that has explored the boundaries between objects, contexts, and places. In studies of recognition memory in rats, Cassaday and Rawlins (1997) explored the boundaries of objects and their locations. Rats were trained to approach and to enter the one of two goal boxes that differed from the sample (DNMS). Goal boxes could be removed and their positions in the apparatus interchanged, so contents of the goal boxes rather than the location of the goal box within the test room provided the critical information for making a correct response. These authors found that rats with hippocampal lesions performed as well as controls when tested with small goal boxes containing discrete objects but were impaired relative to controls when tested with larger goal boxes that contained an array of objects (Cassaday & Rawlins, 1997). These data align with those discussed earlier in suggesting that when information about discrete objects can guide responses, as in the case of the small goal boxes, structures outside the hippocampus can represent and store that information. In addition, however, the results also suggest a role for the hippocampus in processing spatial relations among objects or spatial contextual processing, as presumably was required with the large goal boxes.

The data reviewed by Eacott and E. Gaffan (this issue) provide new insights into the way the brain processes what might be conveniently termed object, contextual, and spatial information. In their studies, "objects" are small, three-dimensional items, "context" was defined as the type of surface (smooth vs. wire mesh) covering the floor, and "place" is defined as one of two possible locations within a testing arena. Manipulations included putting objects upon other objects, objects on different contexts, and objects in different places within the arena. In a series of cleverly designed experiments, it was found that object–object associations depended on the perirhinal cortex, object–context associations depended on the postrhinal cortex, and associations of all three kinds of information—objects, contexts, and place—required the fornix. This striking dissociation implies that the role of postrhinal cortex is in processing objects in relation to local cues, perhaps registering the appearance of items in a particular view. Note that this is consistent with the finding discussed earlier, that postrhinal cortex was important for coding the egocentric locations of objects. The fornix, by contrast, would be involved in the integration of object information across multiple views to form a larger scale scene memory (D. Gaffan, 1994b).

Interestingly, this account is consistent with several other findings. For example, recent studies have revealed that contrary to earlier findings based on aspiration lesions of the hippocampus and subjacent parahippocampal cortex, selective hippocampal lesions in monkeys fail to yield deficits in several tests of spatial memory, including spatial DNMS

(Murray & Mishkin, 1998), spatial reversal learning (Murray, Baxter, & Gaffan, 1998a), and object–place association (Malkova & Mishkin, 2003). Because all these studies involve objects appearing on test trays, these data are consistent with the idea that the hippocampus is not necessary for representation and storage of information that might be viewed in a single scene. An even more striking set of findings has emerged for one of these tasks. In the object–place association task, trials are composed of two phases: acquisition and test. In the acquisition phase of each trial, monkeys are presented with two different objects overlying two of the three food wells on a test tray. A few seconds later, in the test phase, the monkeys are confronted with one of these same objects plus its duplicate, but now, only one of the objects occupies the same location as it had earlier. The monkey can obtain a food reward by displacing the object that occupies the same location as it had in acquisition. Malkova and Mishkin (2003) found that selective lesions of the hippocampus fail to affect performance on this task. By contrast, lesions of the parahippocampal cortex yield a significant impairment, one just as severe as that observed after aspiration removal of the hippocampal formation and subjacent parahippocampal cortex (Parkinson, Murray, & Mishkin, 1988). The findings reviewed by Eacott and E. Gaffan (this issue) suggest a new way to account for these data. The location of objects on a test tray, which can be seen in a single view, is perhaps stored and represented by the parahippocampal cortex rather than the hippocampus. Tasks that require memory for movements within a scene, or for locations on a scale that cannot be represented in a single view, require the hippocampus proper. Findings derived from matching-to-location tasks are consistent with this account. Monkeys with selective excitotoxic hippocampal lesions are impaired in matching-to-location in conditions in which they were moving about on a tether in a large room (Hampton, Hampstead, & Murray, 2004), but not on a formally analogous delayed nonmatching-to-location task administered with items on a test tray (Murray & Mishkin, 1998).

Consistent with the foregoing account, the posterior ventral visual cortex has been implicated in representing the spatial layout of scenes and in topographic maps (Aguirre, Detre, Alsop, & D'Esposito, 1996). In one study (Epstein & Kanwisher, 1998), subjects passively viewed pictures of objects, faces, or scenes, and a parahippocampal place area (PPA) was identified as the region of the brain in which scenes produced greater activation relative to either objects or faces. Clearly, this study would not have involved representation of a large-scale map, and perhaps that accounts for the lack of hippocampal involvement in these studies. When subjects are required to encode and represent a larger scale spatial map, as occurs when subjects are asked either to move through an environment or to integrate partially overlapping scenes (local maps) with one another in a virtual reality environment, hippocampal involvement is typically found (Maguire, Frackowiak, & Frith, 1996).

Although not all experimental findings can be reconciled with this account (Beason-Held, Rosene, Killiany, & Moss, 1999; spatial span task), the set of findings reviewed by Eacott and E. Gaffan (this issue) provides a principled basis for further fruitful exploration of the neural basis for object, contextual, and spatial processing.

What types of memory are subserved by the MTL?

The contributions in this Special Issue highlight several types of memory thought to be subserved by structures within the MTL. As indicated in the Introduction to this Special Issue

(Graham & Gaffan, this issue), there is general agreement that the MTL is critical for the acquisition of event memories and factual knowledge, which, at least in humans, are thought to reflect episodic and semantic memories, respectively. The prevailing view (e.g., Squire, Stark, & Clark, 2004), predicts no specialization for the processing of these types of memory in the MTL. For example, Manns et al. (Manns, Hopkins, & Squire, 2003) studied a group of patients with damage thought to be largely restricted to the hippocampus. They found that the patients were impaired in acquisition of semantic memory. Likewise, Stark and Squire (Stark & Squire, 2003), studying a patient group with similarly restricted hippocampal damage, found that the subjects were impaired on tests of recognition memory. Taken together with evidence that patients with larger lesions within the MTL are likewise impaired in these types of memory, these findings argue against the idea that different structures within the MTL make selective contributions to stimulus recognition and semantic memory. Other dichotomies, for example familiarity versus recollection, likewise do not seem to map onto distinct structures within the MTL (Wixted & Squire, 2004).

Many studies in this Special Issue, however, challenge this unitary view. Studies in animals provide some of the best evidence. As discussed in earlier sections, studies of perirhinal cortex neurons in monkeys have implicated several types of neuronal activity in stimulus memory. For example, perirhinal cortex neuronal activity can reflect item recognition, as a response decrement (Brown et al., 1987; Miller & Desimone, 1994; Sobotka & Ringo, 1993) or enhancement (Li, Miller, & Desimone, 1993), item recency (Fahy, Riches, & Brown, 1993), object–object associative memory (Miyashita, 1988; Erickson & Desimone, 1999), long-term item familiarity (Holscher et al., 2003), and an item's temporal proximity to reward (Liu & Richmond, 2000). Similar findings have been reported in rats (Zhu, Brown, & Aggleton, 1995). The studies of long-term stimulus–stimulus associations and of long-term familiarity, in particular, suggest that perirhinal cortex serves as a site of permanent storage of stimulus representations. Lesion studies in nonhuman primates are consistent with the idea. For example, monkeys with rhinal cortex lesions were impaired in retention of either visual stimulus–stimulus associations (Murray et al., 1993) or tactual–visual associations (Goulet & Murray, 2001) that had been learned over the course of several months before surgery. Furthermore, in an elegant disconnection design, Higuchi and Miyashita (1996) found that lesions of perirhinal cortex in one hemisphere disrupted the pair-coding signals that before surgery had been recorded in area TE, the region just laterally adjacent to perirhinal cortex. Thus, perirhinal cortex appears to work together with area TE in representing and storing stimulus representations.

Because the perirhinal cortex is the first region within the ventral visual stream to receive sensory inputs from multiple sensory modalities (see Suzuki, 1996, for review), it is in a pivotal position to link sensory representations across sensory modalities. In addition, the studies of long-term stimulus memory (e.g., stimulus–stimulus associations) reviewed above suggest that perirhinal cortex is critical for learning about and storing the co-occurrence of stimulus features across presentations. For these and other reasons it has been suggested that perirhinal cortex operates as the kernel of a system for semantic memory (Eacott et al., 1994; Murray, 1996; Murray, Malkova, & Goulet, 1998b). Bright et al. (this issue), in reviewing evidence from humans, suggests that anteromedial temporal cortex provides the neural basis for making fine-grained discriminations among objects. As discussed earlier, these authors find fMRI evidence for a role for anteromedial temporal cortex in identification and (silent) naming of items as an object level. Complementing these findings are the deficits in object-level naming found in patients with damage to the anteromedial temporal cortex and the involvement of perirhinal cortex in patients with semantic dementia (Davies, Graham, Xuereb, Williams, & Hodges, 2004). Thus, damage to perirhinal cortex may underlie the emergence of category-specific semantic deficits in brain-damaged patients. Consistent with a role for anteromedial temporal cortex including perirhinal cortex in semantic processing, Vargha-Khadem et al. (2001) have found that humans who sustained early damage to the hippocampus are severely impaired in episodic memory but relatively spared in semantic memory. In some of these patients, stimulus recognition is as good as that in controls. This echoes the findings reviewed by Holdstock (this issue), who reported that the patient YR is within 0.5 standard deviations of controls on tests of recognition, but 3.5 standard deviations below the control average on tests of recall. Recent findings from fMRI also point to a role for inferotemporal cortex and hippocampus in item familiarity and recollection, respectively (Davachi, Mitchell, & Wagner, 2003; Ranganath et al., 2004).

What are fruitful directions for future research?

"Ours"—the traditional answer—has its appeal, but here we offer a less restrictive reply. If we intend to understand the organization of memory, and the way in which the various MTL structures contribute to it, we need to know whether the many different types of impairment that follow perirhinal cortex ablation derive from a single functional deficit. Could apparent difficulties with perception, recognition memory, long-term familiarity, and stimulus–stimulus association all result from a single cognitive disability? A major step toward answering that question might involve resolving whether genuine perceptual impairments account for apparent memory impairments. This parsimonious solution to a number of current problems has been suggested (Buckley, this issue; Murray & Bussey, 1999), but exactly how various impairments could arise from the disruption of a single cognitive process, even one as central as perception, remains an important topic for future research. Deficits in stimulus recognition and other aspects of long-term memory could easily arise from a degraded ability to represent complex conjunctions of features. According to this model, visual representations are degraded in the absence of perirhinal cortex and are incomplete from the time of encoding. If so, then deficits in both "memory" and "perception" would be predicted, and any deficits in "memory" could result from the incompleteness of the object representations that were established during the encoding process. The deficits in crossmodal memory, however, require a more sophisticated account because of their all-or-none nature, which contrasts with the graded deficits on recognition and other long-term memory tasks (Murray et al., 1998a). Perhaps the perirhinal cortex serves as an obligatory anatomical route—an anatomical bottleneck—through which non-visual inputs, such as tactual or flavour signals, need to gain access to visual representations. Support for this conjecture would constitute an important step in reconciling apparently discrepant findings. However, as D. Gaffan (2001, 2002) has argued, it is difficult to see the perceptual functions of cortex independently from its functions in memory, since all cortical areas demonstrate plasticity. It may be, therefore, that progress in this general area of the relation between perception and memory in cortical function will be best served by acknowledging that perception and memory are psychological rather than physiological terms, rather than by attempting to explain memory impairments as consequences of perceptual impairments.

Another impediment to progress in this area is the lack of a direct link between the putative neural correlates of memory and its behavioural expression. For example, in the case of stimulus recognition, the "match suppression" effect has been proposed to mediate stimulus recognition, yet there is no evidence tying disruption of this process to any disruption of recognition. Indeed, there is some evidence of the independence of this type of neural coding and recognition memory (Sobotka & Ringo, 1996). A second avenue for future research, then, involves studies that address whether such a link exists and explore the relationship between perception and memory in perirhinal cortex.

More in-depth study of the interactions among the disparate components of the MTL could also provide much-needed insight into these neural systems. In schematic diagrams of the MTL the hippocampus is often placed at the top of a pyramid, with the base of the pyramid composed of the perirhinal and parahippocampal/postrhinal cortex and the middle level of the pyramid occupied by the entorhinal cortex. This ideation of the MTL implies that the cortical fields near the hippocampus only function through interaction with the hippocampus proper. The double dissociations reviewed by Aggleton and Brown (this issue) and by Eacott and E. Gaffan (this issue), together with other reviews and findings (D. Gaffan, 1994a; Murray & Wise, 2004; Saksida et al., 2003), show that the structures comprising the MTL can operate relatively independently. Indeed, the question might be asked: "Under what conditions do the perirhinal cortex or the parahippocampal/postrhinal cortex need to interact with the hippocampus?" There have been few direct investigations of this question, but they are worth noting. D. Gaffan and Parker (1996) found that monkeys with perirhinal cortex ablation in one hemisphere and fornix transection in the other hemisphere were impaired in learning object-in-place scenes. This result indicates that the perirhinal cortex must functionally interact with the fornix in mediating this kind of learning. Consistent with this idea, Murray et al. (1998a) found that bilaterally symmetrical lesions of either the rhinal (i.e., perirhinal plus entorhinal) cortex or the hippocampus led to an impairment in learning the spatial version of the scenes task. In other tasks, such as DNMS, perirhinal cortex needs to interact with prefrontal cortex (Parker & Gaffan, 1998). A third avenue for future research, then, might be to assess the role of hippocampal interactions with perirhinal and parahippocampal/postrhinal cortex in tasks that vary object, contextual, and spatial processing demands.

Anatomical studies also reveal insights into neural networks. Studies in rats (Burwell, 2000) indicate that perirhinal cortex projects preferentially to the lateral entorhinal cortex whereas postrhinal cortex projects preferentially to medial entorhinal cortex. In monkeys (Suzuki & Amaral, 1994) the perirhinal cortex is preferentially related to the anterolateral portion of entorhinal cortex and parahippocampal cortex to the caudal and medial portion of the entorhinal cortex. This division of labour is further carried into the hippocampal formation and subicular complex, highlighting potential processing streams that might be devoted to different functions. In rats, for example, the dorsal portion of the hippocampus, which receives heavy inputs from medial entorhinal cortex, plays an essential role in efficient performance on tests of spatial navigation (Bannerman et al., 1999; Kjelstrup et al., 2002). The ventral portion of the hippocampus, which receives heavy inputs from lateral entorhinal cortex, apparently does not. Furthermore, in this Special Issue, Aggleton and Brown (this issue) review evidence that circuits underlying spatial working memory (water maze with submerged platform) differ from those underlying spatial reference memory (water maze with visible landmark). Both kinds of task require the extrahippocampal regions, specifically the perirhinal, postrhinal, and entorhinal cortex, but the working-memory task also requires dentate gyrus and the reference-memory task depends on the subiculum. A fourth avenue for future research, then, involves a continual re-evaluation of current doctrine in the context of finer anatomical distinctions than most research has used to date.

Improved testing methods might also yield important advances. The idea that parahippocampal/postrhinal cortex represents objects together with local cues is consistent with the idea that the hippocampus is necessary for allocentric as opposed to egocentric spatial processing. Interestingly, egocentric spatial tasks in monkeys typically involve left or right responses on a test tray (e.g., the spatial delayed-response task), but this factor (e.g., egocentric versus allocentric spatial framework) is often confounded with representation of a scene (context). It would be instructive to design tasks in which egocentric versus allocentric task solutions were pitted against contextual versus spatial task solutions as discussed above in Section 4. This information might shed new light on the conditions in which the hippocampus is necessary for spatial processing. A fifth avenue for future research, then, involves the continued effort to develop better experimental controls and manipulations.

This Special Issue highlights yet another fruitful goal for the long term: that we should continue to attempt to undertake cross-disciplinary research utilizing—as much as possible—similar paradigms and approaches. When such an attempt is made (Bright et al., this issue; Buckley; this issue; Bussey et al., this issue; Eacott & E. Gaffan, this issue; Lee et al., this issue), much progress can be made in understanding the circumstances that lead to discrepancies across species (e.g., nonhuman primates vs. humans) and methodologies (e.g., patient lesion work vs. functional neuroimaging). A sixth avenue for future research, then, involves manipulations that take into account the factors highlighted here as influencing MTL function (e.g., objects vs. scenes; minimum vs. maximum feature ambiguity; egocentric vs. allocentric spatial processing, etc.).

Finally, the newly emerging view of the MTL as a collection of variously specialized cortical areas, each involved both in perception and in memory, should have important implications for our approach to the study of other cortical areas, outside the MTL. For example, it will be necessary to investigate the possible contribution of prefrontal cortex to a wider range of processes in perception and in memory acquisition than would have been envisaged in an era when perception and memory were thought to be rigidly demarcated cortical functions.

In summary, we imagine a future in which behavioural tasks become increasingly suited to the underlying cognitive processes, and studies address the link between neural activity and behaviour, to explore the relationship between perception and memory and to assess the role of hippocampal interactions with extrahippocampal temporal cortex and with prefrontal cortex, all in the context of finer anatomical distinctions and improved experimental control.

REFERENCES

Aggleton, J. P., & Brown, M. W. (this issue). Contrasting hippocampal and perirhinal cortex function using immediate early gene imaging. *Quarterly Journal of Experimental Psychology, 58B*, 218–233.

Aguirre, G. K., Detre, J. A., Alsop, D. C., & D'Esposito, M. (1996). The parahippocampus subserves topographical learning in man. *Cerebral Cortex, 6*, 823–829.

Baker, C. I., Behrmann, M., & Olson, C. R. (2002). Impact of learning on representation of parts and wholes in monkey inferotemporal cortex. *Nature Neuroscience, 5,* 1210–1216.

Bannerman, D. M., Yee, B. K., Good, M. A., Heupel, M. J., Iversen, S. D., & Rawlins, J. N. (1999). Double dissociation of function within the hippocampus: a comparison of dorsal, ventral, and complete hippocampal cytotoxic lesions. *Behavioral Neuroscience, 113,* 1170–1188.

Baxter, M. G., & Murray, E. A. (2001). Opposite relationship of hippocampal and rhinal cortex damage to delayed nonmatching-to-sample deficits in monkeys. *Hippocampus, 11,* 61–71.

Beason-Held, L. L., Rosene, D. L., Killiany, R. J., & Moss, M. B. (1999). Hippocampal formation lesions produce memory impairment in the rhesus monkey. *Hippocampus, 9,* 562–574.

Bright, P., Moss, H. E., Stamatakis, E. A., & Tyler, L. K. (this issue). The anatomy of object processing: The role of anteromedial temporal cortex. *Quarterly Journal of Experimental Psychology, 58B,* 361–377.

Brown, M. W., Wilson, F. A. W., & Riches, I. P. (1987). Neuronal evidence that inferomedial temporal cortex is more important than hippocampus in certain processes underlying recognition memory. *Brain Research, 409,* 158–162.

Buckley, M. J. (this issue). The role of the perirhinal cortex and hippocampus in learning, memory, and perception. *Quarterly Journal of Experimental Psychology, 58B,* 246–268.

Buckley, M. J., Booth, M. C., Rolls, E. T., & Gaffan, D. (2001). Selective perceptual impairments after perirhinal cortex ablation. *Journal of Neuroscience, 21,* 9824–9836.

Buckley, M. J., & Gaffan, D. (1998a). Learning and transfer of object–reward associations and the role of the perirhinal cortex. *Behavioral Neuroscience, 112,* 15–23.

Buckley, M. J., & Gaffan, D. (1998b). Perirhinal cortex ablation impairs configural learning and paired-associate learning equally. *Neuropsychologia, 36,* 535–546.

Buckley, M. J., & Gaffan, D. (1998c). Perirhinal cortex ablation impairs visual object identification. *Journal of Neuroscience, 18,* 2268–2275.

Buffalo, E. A., Ramus, S. J., Squire, L. R., & Zola, S. M. (2000). Perception and recognition memory in monkeys following lesions of area TE and perirhinal cortex. *Learning & Memory, 7,* 375–382.

Buffalo, E. A., Reber, P. J., & Squire, L. R. (1998). The human perirhinal cortex and recognition memory. *Hippocampus, 8,* 330–339.

Burwell, R. D. (2000). The parahippocampal region: Corticocortical connectivity. In H. E. Scharfman, M. P. Witter, & R. Schwarcz (Eds.), *The parahippocampal region. Implications for neurological and psychiatric diseases* (pp. 25–42). New York: New York Academy of Sciences.

Bussey, T. J., & Saksida, L. M. (2002). The organization of visual object representations: A connectionist model of effects of lesions in perirhinal cortex. *European Journal of Neuroscience, 15,* 355–364.

Bussey, T. J., Saksida, L. M., & Murray, E. A. (2002). Perirhinal cortex resolves feature ambiguity in complex visual discriminations. *European Journal of Neuroscience, 15,* 365–374.

Bussey, T. J., Saksida, L. M., & Murray, E. A. (2003). Impairments in visual discrimination after perirhinal cortex lesions: Testing 'declarative' vs. 'perceptual-mnemonic' views of perirhinal cortex function. *European Journal of Neuroscience, 17,* 649–660.

Bussey, T. J., Saksida, L. M., & Murray, E. A. (this issue). The perceptual-mnemonic/feature conjunction model of perirhinal cortex function. *Quarterly Journal of Experimental Psychology, 58B,* 269–282.

Cassaday, H. J., & Rawlins, J. N. (1997). The hippocampus, objects, and their contexts. *Behavioral Neuroscience, 111,* 1228–1244.

Davachi, L., Mitchell, J. P., & Wagner, A. D. (2003). Multiple routes to memory: Distinct medial temporal lobe processes build item and source memories. *Proceedings of the National Academy of Sciences, USA, 100,* 2157–2162.

Davies, R. R., Graham, K. S., Xuereb, J. H., Williams, G. B., & Hodges, J. R. (2004). The human perirhinal cortex and semantic memory. *European Journal of Neuroscience, 20,* 2441–2446.

Desimone, R., & Ungerleider, L. G. (1989). Neural mechanisms of visual processing in monkeys. In F. Boller & J. Graham (Eds.), *Handbook of neuropsychology* (Vol. 2, pp. 267–299). New York: Elsevier Science.

Eacott, M. J., & Gaffan, E. A. (this issue). The roles of the perirhinal cortex, postrhinal cortex, and the fornix in memory for objects, contexts, and events in the rat. *Quarterly Journal of Experimental Psychology, 58B,* 202–217.

Eacott, M. J., Gaffan, D., & Murray, E. A. (1994). Preserved recognition memory for small sets, and impaired stimulus identification for large sets, following rhinal cortex ablations in monkeys. *European Journal of Neuroscience, 6,* 1466–1478.

Epstein, R., & Kanwisher, N. (1998). A cortical representation of the local visual environment. *Nature*, *392*, 598–601.

Erickson, C. A., & Desimone, R. (1999). Responses of macaque perirhinal neurons during and after visual stimulus association learning. *Journal of Neuroscience*, *19*, 10404–10416.

Fahy, F. L., Riches, I. P., & Brown, M. W. (1993). Neuronal activity related to visual recognition memory: Long-term memory and the encoding of recency and familiarity information in the primate anterior and medial inferior temporal and rhinal cortex. *Experimental Brain Research*, *96*, 457–472.

Gaffan, D. (1994a). Dissociated effects of perirhinal cortex ablation, fornix transection and amygdalectomy: Evidence for multiple memory systems in the primate temporal lobe. *Experimental Brain Research*, *99*, 411–422.

Gaffan, D. (1994b). Scene-specific memory for objects: A model of episodic memory impairment in monkeys with fornix transection. *Journal of Cognitive Neuroscience*, *6*, 305–320.

Gaffan, D. (2001). What is a memory system? Horel's critique revisited. *Behavioural Brain Research*, *127*, 5–11.

Gaffan, D. (2002). Against memory systems. *Philosophical Transactions of the Royal Society of London*, *357*, 1111–1121.

Gaffan, D., & Parker, A. (1996). Interaction of perirhinal cortex with the fornix-fimbria: Memory for objects and "object in place" memory. *Journal of Neuroscience*, *16*, 5864–5869.

Gaffan, E. A., Healey, A. N., & Eacott, M. J. (2004). Objects and positions in visual scenes: Effects of perirhinal and postrhinal cortex lesions in the rat. *Behavioral Neuroscience*, *118*, 992–1010.

Goulet, S., & Murray, E. A. (2001). Neural substrates of crossmodal association memory in monkeys: The amygdala versus the anterior rhinal cortex. *Behavioral Neuroscience*, *115*, 271–284.

Graham, K. S., & Gaffan, D. (this issue). The role of the medial temporal lobe in memory and perception: Evidence from rats, nonhuman primates and humans. *Quarterly Journal of Experimental Psychology*, *58B*, 193–201.

Hampton, R. R. (this issue). Monkey perirhinal cortex is critical for visual memory, but not for visual perception: Reexamination of the behavioural evidence from monkeys. *Quarterly Journal of Experimental Psychology*, *58B*, 283–299.

Hampton, R. R., Hampstead, B. M., & Murray, E. A. (2004). Selective hippocampal damage in rhesus monkeys impairs spatial memory in an open-field test. *Hippocampus*, *14*, 808–818.

Hampton, R. R., & Murray, E. A. (2002). Learning of discriminations is impaired, but generalization to altered views is intact, in monkeys (Macaca mulatta) with perirhinal cortex removal. *Behavioral Neuroscience*, *116*, 363–377.

Henson, R. (this issue). A mini-review of fMRI studies of human medial temporal lobe activity associated with recognition memory. *Quarterly Journal of Experimental Psychology*, *58B*, 340–360.

Higuchi, S., & Miyashita, Y. (1996). Formation of mnemonic neuronal responses to visual paired associates in inferotemporal cortex is impaired by perirhinal and entorhinal cortex lesions. *Proceedings of the National Academy of Sciences, USA*, *93*, 739–743.

Holdstock, J. S. (this issue). The role of the human medial temporal lobe in object recognition and object discrimination. *Quarterly Journal of Experimental Psychology*, *58B*, 326–339.

Holdstock, J. S., Gutnikov, S. A., Gaffan, D., & Mayes, A. R. (2000). Perceptual and mnemonic matching-to-sample in humans: Contributions of the hippocampus, perirhinal and other medial temporal lobe cortices. *Cortex*, *36*, 301–322.

Hölscher, C., Rolls, E. T., & Xiang, J. (2003). Perirhinal cortex neuronal activity related to long-term familiarity memory in the macaque. *European Journal of Neuroscience*, *18*, 2037–2046.

Kjelstrup, K. G., Tuvnes, F. A., Steffenach, H. A., Murison, R., Moser, E. I., & Moser, M. B. (2002). Reduced fear expression after lesions of the ventral hippocampus. *Proceedings of the National Academy of Sciences, USA*, *99*, 10825–10830.

Lee, A. C. H., Barense, M. D., & Graham, K. S. (this issue). The contribution of the human medial temporal lobe to perception: Bridging the gap between animal and human studies. *Quarterly Journal of Experimental Psychology*, *58B*, 300–325.

Lee, A. C., Bussey, T. J., Murray, E. A., Saksida, L. M., Epstein, R. A., Kapur, N., et al. (2005). Perceptual deficits in amnesia: Challenging the medial temporal lobe 'mnemonic' view. *Neuropsychologia*, *43*, 1–11.

Levy, D. A., Shrager, Y., & Squire, L. R. (2005). Intact visual discrimination of complex and feature-ambiguous stimuli in the absence of perirhinal cortex. *Learning & Memory*, *12*, 61–66.

Li, L., Miller, E. K., & Desimone, R. (1993). The representation of stimulus familiarity in anterior inferior temporal cortex. *Journal of Neurophysiology*, *69*, 1918–1929.

Liu, Z., & Richmond, B. J. (2000). Response differences in monkey TE and perirhinal cortex: Stimulus association related to reward schedules. *Journal of Neurophysiology*, *83*, 1677–1692.

Logothetis, N. K., & Scheinberg, D. L. (1996). Visual object recognition. *Annual Review of Neuroscience*, *19*, 577–621.

Maguire, E. A., Frackowiak, R. S. J., & Frith, C. D. (1996). Learning to find your way: A role for the human hippocampal formation. *Proceedings of the Royal Society of London: Biological Sciences*, *263*, 1745–1750.

Malkova, L., & Mishkin, M. (2003). One-trial memory for object–place associations after separate lesions of hippocampus and posterior parahippocampal region in the monkey. *Journal of Neuroscience*, *23*, 1956–1965.

Manns, J. R., Hopkins, R. O., & Squire, L. R. (2003). Semantic memory and the human hippocampus. *Neuron*, *38*, 127–133.

Meunier, M., Bachevalier, J., Mishkin, M., & Murray, E. A. (1993). Effects on visual recognition of combined and separate ablations of the entorhinal and perirhinal cortex in rhesus monkeys. *Journal of Neuroscience*, *13*, 5418–5432.

Miller, E. K., & Desimone, R. (1994). Parallel neuronal mechanisms for short-term memory. *Science*, *263*, 520–522.

Miyashita, Y. (1988). Neuronal correlate of visual associative long-term memory in the primate temporal cortex. *Nature*, *335*, 817–820.

Murray, E. A. (1996). What have ablation studies told us about the neural substrates of stimulus memory? *Seminars in the Neurosciences*, *5*, 10–20.

Murray, E. A., Baxter, M. G., & Gaffan, D. (1998a). Monkeys with rhinal cortex damage or neurotoxic hippocampal lesions are impaired on spatial scene learning and object reversals. *Behavioral Neuroscience*, *112*, 1291–1303.

Murray, E. A., & Bussey, T. J. (1999). Perceptual-mnemonic functions of the perirhinal cortex. *Trends in Cognitive Science*, *3*, 142–151.

Murray, E. A., Bussey, T. J., Hampton, R. R., & Saksida, L. M. (2000). The parahippocampal region and object identification. In H. E. Scharfman, M. P. Witter, & R. Schwarcz (Eds.), *The Parahippocampal region. Implications for neurological and psychiatric diseases* (pp. 166–174). New York: New York Academy of Sciences.

Murray, E. A., Gaffan, D., & Mishkin, M. (1993). Neural substrates of visual stimulus–stimulus association in rhesus monkeys. *Journal of Neuroscience*, *13*, 4549–4561.

Murray, E. A., Malkova, L., & Goulet, S. (1998b). Crossmodal associations, intramodal associations, and object identification. In A. D. Milner (Ed.), *Comparative neuropsychology* (pp. 51–69). Oxford, UK: Oxford University Press.

Murray, E. A., & Mishkin, M. (1998). Object recognition and location memory in monkeys with excitotoxic lesions of the amygdala and hippocampus. *Journal of Neuroscience*, *18*, 6568–6582.

Murray, E. A., & Wise, S. P. (2004). What, if anything, is the medial temporal lobe, and how can the amygdala be part of it if there is no such thing? *Neurobiology of Learning & Memory*, *82*, 178–198.

Parker, A., & Gaffan, D. (1998). Interaction of frontal and perirhinal cortices in visual object recognition memory in monkeys. *European Journal of Neuroscience*, *10*, 3044–3057.

Parkinson, J. K., Murray, E. A., & Mishkin, M. (1988). A selective mnemonic role for the hippocampus in monkeys: Memory for the location of objects. *Journal of Neuroscience*, *8*, 4159–4167.

Ranganath, C., Yonelinas, A. P., Cohen, M. X., Dy, C. J., Tom, S. M., & D'Esposito, M. (2004). Dissociable correlates of recollection and familiarity within the medial temporal lobes. *Neuropsychologia*, *42*, 2–13.

Ringo, J. L. (1991). Memory decays at the same rate in macaques with and without brain lesions when expressed in d' or arcsine terms. *Behavioural Brain Research*, *42*, 123–134.

Rolls, E. T., Franco, L., & Stringer, S. M. (this issue). The perirhinal cortex and long-term familiarity memory. *Quarterly Journal of Experimental Psychology*, *58B*, 234–245.

Sakai, K., & Miyashita, Y. (1991). Neural organization for the long-term memory of paired associates. *Nature*, *354*, 152–155.

Sakai, K., & Miyashita, Y. (1993). Memory and imagery in the temporal lobe. *Current Opinion in Neurobiology*, *3*, 166–170.

Saksida, L. M., Bussey, T. J., Buckmaster, C. A., & Murray, E. A. (2003). Perirhinal cortex lesions can impair, and hippocampal lesions can facilitate, transverse patterning in rhesus monkeys. Program No. 939.12. *2003 abstract viewer/itinerary planner*. Washington, DC: Society for Neuroscience.

Saksida, L. M., Bussey, T. J., Buckmaster, C. A., & Murray, E. A. (2005). *The role of the hippocampus in feature ambiguous discriminations.* Unpublished raw data.

Sobotka, S., & Ringo, J. L. (1993). Investigation of long-term recognition and association memory in unit responses from inferotemporal cortex. *Experimental Brain Research*, *96*, 28–38.

Sobotka, S., & Ringo, J. L. (1996). Mnemonic responses of single units recorded from monkey inferotemporal cortex, accessed via transcommissural vs. direct pathways. *Journal of Neuroscience*, *16*, 4222–4230.

Squire, L. R., Stark, C. E., & Clark, R. E. (2004). The medial temporal lobe. *Annual Reviews of Neuroscience*, *27*, 279–306.

Stark, C. E. L., & Squire, L. R. (2000). Intact visual perceptual discrimination in humans in the absence of perirhinal cortex. *Learning & Memory*, *7*, 273–278.

Stark, C. E., & Squire, L. R. (2003). Hippocampal damage equally impairs memory for single items and memory for conjunctions. *Hippocampus*, *13*, 281–292.

Suzuki, W. A. (1996). Neuroanatomy of the monkey entorhinal, perirhinal and parahippocampal cortices: Organization of cortical inputs and interconnections with amygdala and striatum. *Seminars in the Neurosciences*, *8*, 3–12.

Suzuki, W. A., & Amaral, D. G. (1994). Topographic organization of the reciprocal connections between monkey entorhinal cortex and the perirhinal and parahippocampal cortices. *Journal of Neuroscience*, *14*, 1856–1877.

Tanaka, K. (1997). Mechanisms of visual object recognition: Monkey and human studies. *Current Opinion in Neurobiology*, *7*, 523–529.

Thornton, J. A., Rothblat, L. A., & Murray, E. A. (1997). Rhinal cortex removal produces amnesia for preoperatively learned discrimination problems but fails to disrupt postoperative acquisition and retention in rhesus monkeys. *Journal of Neuroscience*, *17*, 8536–8549.

Vargha-Khadem, F., Gadian, D. G., & Mishkin, M. (2001). Dissociations in cognitive memory: The syndrome of developmental amnesia. *Philosophical Transactions of the Royal Society of London. Series B: Biological Sciences*, *356*, 1435–1440.

Wixted, J. T., & Squire, L. R. (2004). Recall and recognition are equally impaired in patients with selective hippocampal damage. *Cognitive Affective & Behavioral Neuroscience*, *4*, 58–66.

Zhu, X. O., Brown, M. W., & Aggleton, J. P. (1995). Neuronal signaling of information important to visual recognition memory in rat rhinal and neighboring cortices. *European Journal of Neuroscience*, *7*, 753–765.

THE QUARTERLY JOURNAL OF EXPERIMENTAL PSYCHOLOGY
2005, 58B (3/4), 397–400

Subject index